AF597959

METHODS IN MOLECULAR BIOLOGY™

Series Editor
John M. Walker
School of Life Sciences
University of Hertfordshire
Hatfield, Hertfordshire, AL10 9AB, UK

For further volumes:
http://www.springer.com/series/7651

Tissue-Protective Cytokines

Methods and Protocols

Edited by

Pietro Ghezzi

Brighton and Sussex Medical School, Falmer, UK

Anthony Cerami

Leiden University Medical Center, Leiden, The Netherlands

Editors
Pietro Ghezzi
Brighton and Sussex Medical School
Falmer, UK

Anthony Cerami
Leiden University Medical Center
Leiden, The Netherlands

ISSN 1064-3745 ISSN 1940-6029 (electronic)
ISBN 978-1-62703-307-7 ISBN 978-1-62703-308-4 (eBook)
DOI 10.1007/978-1-62703-308-4
Springer New York Heidelberg Dordrecht London

Library of Congress Control Number: 2013932704

Printed on acid-free paper

Humana Press is a brand of Springer
Springer is part of Springer Science+Business Media (www.springer.com)

Preface

In 2004, Pietro Ghezzi asked me to write a short introductory chapter for a book on TNF he was putting together for this very same series. The focus of what I wrote then was on TNF as an inflammatory agent. In fact, our finding, in the 1980s, that cachexia associated with inflammatory and infectious diseases was due to TNF opened the way to the development of anti-TNF drugs that are now standard treatment for chronic inflammatory disease.

Now, I am asked again to write about the continuation of this saga. In 1998, with Mike Brines and Carla Hand, I started investigating why patients who receive EPO feel well soon after the first treatment. That led to the discovery, published in 2000, that systemically administered EPO is neuroprotective in animal models of ischemic and traumatic brain injury. That paper opened the field of the neuroprotective action of EPO. The protective action of EPO was soon demonstrated in other tissues, hence the use of the term "tissue-protective cytokine" (1). As in the case of TNF, we had to fight the commonplace that EPO has only erythropoietic actions, that its receptor is present only in erythroid progenitor cells, and that EPO is produced only by the kidney (2).

As in the case of TNF, originally identified for its antitumor activities, we had to work against the common belief that EPO is solely an erythropoietic cytokine acting solely on erythroid progenitors. Several investigators also documented the expression of EPO in the central nervous system and other tissues, against the common belief that only the kidney and the foetal liver produce EPO.

From the perspective of pharmacological use, the erythropoietic action of EPO, by increasing the haematocrit and activating platelets, has some undesired side effects as a tissue-protectant, and this led to the development of novel non-erythropoietic EPO-derived tissue-protective molecules some of which are described here.

I believe that tissue-protection will be a new field of interest of cytokine biology, both in discovering novel actions of known cytokines and in developing new drugs. In this context, this book is a valuable collection of methodological papers that describe in detail the key models that have been used to characterize the tissue-protective actions of EPO and derivatives and will, hopefully, be of use in the discovery of new tissue-protective molecules.

Leiden, The Netherlands *Anthony Cerami*

References

1. Cerami A (2011) The value of failure: the discovery of TNF and its natural inhibitor erythropoietin. J Intern Med 269(1):8–15
2. Ghezzi P, Bernaudin M, Bianchi R, Blomgren K, Brines M, Campana W, Cavaletti G, Cerami A, Chopp M, Coleman T, Digicaylioglu M, Ehrenreich H, Erbayraktar S, Erbayraktar Z, Gassmann M, Genc S, Gokmen N, Grasso G, Juul S, Lipton SA, Hand CC, Latini R, Lauria G, Leist M, Newton SS, Petit E, Probert L, Sfacteria A, Siren AL, Talan M, Thiemermann C, Westenbrink D, Yaqoob M, Zhu C (2010) Erythropoietin: not just about erythropoiesis. Lancet 375(9732):2142

Contents

Contributors

ALBERTO AURICCHIO • *Telethon Institute of Genetics and Medicine (TIGEM), Naples, Italy; Medical Genetics, Department of Medical Translational Science, Federico II University, Naples, Italy*

MYRIAM BERNAUDIN • *CERVOxy team "Hypoxia, cerebrovascular and tumoral pathophysiologies", UMR 6301-ISTCT, CNRS, CEA, Université de Caen Basse-Normandie, CYCERON, Caen, France*

ANNE-LISE BIENVENU • *Malaria Research Unit, SMITH ICBMS UMR CNRS, UCBL, INSA Lyon, Lyon, France*

ANNA YU. BOGDANOVA • *Institute of Veterinary Physiology, Vetsuisse Faculty and Zurich Center for Integrative Human Physiology (ZIHP), University of Zurich, Zurich, Switzerland*

MICHAEL BRINES • *Araim Pharmaceuticals, Inc., Ossining, NY, USA*

PORRETTA-SERAPIGLIA CARLA • *Neuromuscular Diseases Unit, IRCCS Foundation, "Carlo Besta" Neurological Institute, Milan, Italy*

ANTHONY CERAMI • *Araim Pharmaceuticals, Ossining, NY, USA; Leiden University Medical Center, Leiden, The Netherlands*

ILARIA CERVELLINI • *Brighton & Sussex Medical School, Falmer, UK*

PASQUALINA COLELLA • *Telethon Institute of Genetics and Medicine (TIGEM), Naples, Italy*

DARRELL CONKLIN • *Department of Computer Science and Artificial Intelligence, University of the Basque Country UPV/EHU, San Sebastián, Spain; IKERBASQUE, Basque Foundation for Science, Bilbao, Spain*

JOVANY CRUZ • *Baylor College of Medicine, Houston, TX, USA*

ALBERT DAHAN • *Department of Anesthesiology, Leiden University Medical Center, Leiden, The Netherlands*

MURAT DIGICAYLIOGLU • *Departments of Neurosurgery and Physiology, University of Texas Health Science Center, San Antonio, TX, USA*

ANDREW DILLEY • *Brighton and Sussex Medical School, Falmer, UK*

CATHARINE H. DUMAN • *Department of Psychiatry, Yale University School of Medicine, New Haven, CT, USA*

SERHAT ERBAYRAKTAR • *Department of Neurosurgery, School of Medicine, Dokuz Eylul University, İnciralti, İzmir, Turkey*

ZÜBEYDE ERBAYRAKTAR • *Department of Medical Biochemistry, School of Medicine, Dokuz Eylul University, İnciralti, İzmir, Turkey*

SAMSON KUMAR GADDAM • *Baylor College of Medicine, Houston, TX, USA*

EITHAN GALUN • *Goldyne Savad Inst. of Gene Therapy, Hadassah Hebrew University Hospital, Jerusalem, Israel*

MAX GASSMANN • *Institute of Veterinary Physiology, Vetsuisse Faculty and Zurich Center for Integrative Human Physiology (ZIHP), University of Zurich, Zurich, Switzerland*

PIETRO GHEZZI • *Brighton & Sussex Medical School, Falmer, UK*
LAURIA GIUSEPPE • *Neuromuscular Diseases Unit, IRCCS Foundation, "Carlo Besta" Neurological Institute, Milan, Italy*
NECATI GOKMEN • *Department of Anesthesia and Reanimation, School of Medicine, Dokuz Eylul University, İnciralti, İzmir, Turkey*
SANDRA E. JUUL • *Division of Neonatology, Department of Pediatrics, University of Washington, Seattle, WA, USA*
ROBERTO LATINI • *Department of Cardiovascular Research, Istituto Mario Negri, Milan, Italy*
ANNELISE LETOURNEUR • *CERVOxy team "Hypoxia, cerebrovascular and tumoral pathophysiologies", UMR 6301-ISTCT, CNRS, CEA, Université de Caen Basse-Normandie, CYCERON, Caen, France*
MANUELA MENGOZZI • *Brighton & Sussex Medical School, Falmer, UK*
SAMUEL S. NEWTON • *Department of Psychiatry, Yale University School of Medicine, New Haven, CT, USA*
MARIEKE NIESTERS • *Department of Anesthesiology, Leiden University Medical Center, Leiden, The Netherlands*
OMOLARA O. OGUNSHOLA • *Institute of Veterinary Physiology, Vetsuisse Faculty and Zurich Center for Integrative Human Physiology (ZIHP), University of Zurich, Zurich, Switzerland*
EDWIGE PETIT • *CERVOxy team "Hypoxia, cerebrovascular and tumoral pathophysiologies", UMR 6301-ISTCT, CNRS, CEA, Université de Caen Basse-Normandie, CYCERON, Caen, France*
STEPHANE PICOT • *Malaria Research Unit, SMITH ICBMS UMR CNRS, UCBL, INSA Lyon, Lyon, France*
LOMBARDI RAFFAELLA • *Neuromuscular Diseases Unit, IRCCS Foundation, "Carlo Besta" Neurological Institute, Milan, Italy*
DOMENICO RIBATTI • *Department of Basic Medical Sciences, Neurosciences, and Sensory Organs, University of Bari Medical School, Bari, Italy*
BIANCHI ROBERTO • *Neuromuscular Diseases Unit, IRCCS Foundation, "Carlo Besta" Neurological Institute, Milan, Italy*
CLAUDIA ROBERTSON • *Department of Neurosurgery, Baylor College of Medicine, Houston, TX, USA*
STEFAN ROSE-JOHN • *Institut für Biochemie, Christian-Albrechts-Universität zu Kiel, Kiel, Germany*
SIMON ROUSSEL • *CERVOxy team "Hypoxia, cerebrovascular and tumoral pathophysiologies", UMR 6301-ISTCT, CNRS, CEA, Université de Caen Basse-Normandie, CYCERON, Caen, France*
TOMMY SEABORN • *Faculty of Medicine, Department of Pediatrics, Centre de Recherche de l'Hôpital St-François d'Assise (CR-SFA), Centre Hospitalier Universitaire de Québec (CHUQ), Laval University, Québec, QC, Canada*
JORGE SOLIZ • *Faculty of Medicine, Department of Pediatrics, Centre de Recherche de l'Hôpital St-François d'Assise (CR-SFA), Centre Hospitalier Universitaire de Québec (CHUQ), Laval University, Québec, QC, Canada*
MAARTEN SWARTJES • *Department of Anesthesiology, Leiden University Medical Center, Leiden, The Netherlands*

MARK I. TALAN • *Laboratory of Cardiovascular Sciences, National Institute on Aging, NIH, Baltimore, MD, USA*

OMAR TOUZANI • *CERVOxy team "Hypoxia, cerebrovascular and tumoral pathophysiologies", UMR 6301-ISTCT, CNRS, CEA, Université de Caen Basse-Normandie, CYCERON, Caen, France*

CHRISTOPHER M. TRAUDT • *Division of Neonatology, Department of Pediatrics, University of Washington, Seattle, WA, USA*

OSMAN YILMAZ • *Animal Research Center, School of Medicine, Dokuz Eylul University, İnciralti, İzmir, Turkey*

Chapter 1

Erythropoietin and Engineered Innate Repair Activators

Michael Brines and Anthony Cerami

Abstract

Erythropoietin (EPO) is a pleiotropic type I cytokine that has been identified as a major endogenous tissue protective molecule. In response to injury, EPO and a distinct receptor are expressed with a characteristic temporal and spatial expression pattern. Together, these serve to limit injury and to initiate repair. Administration of EPO in the setting of injury has been shown to be beneficial in a multitude of preclinical models. However, translation into the clinic has been hampered by EPO's adverse effects, including promotion of thrombosis. Recently, engineered molecules based on EPO's structure–activity relationships have been developed that are devoid of hematopoietic effects. These compounds are promising candidates for treatment of a wide variety of acute and chronic diseases.

Key words Innate immune response, Inflammation, Tissue damage, Tissue protection, Healing, Drug design, Type 1 cytokine

1 Erythropoietin

Erythropoietin (EPO) is a 165-amino-acid protein (MW ~ 30.4 kDa) and a member of the type I cytokine superfamily (reviewed in ref. 1). The molecule contains four alpha helices that self-associate to form a relatively compact, globular shape. For many years it was thought that the sole function of EPO is to maintain an adequate number of erythrocytes within the circulation. This physiological process operates as a classical hormonal negative feedback system that is activated by relative hypoxia that is detected by the kidney. If red cell number significantly declines, oxygenation also falls, and EPO is produced by medullary tubular interstitial cells. The plasma concentrations of EPO required to maintain normal red cell numbers is in the picomolar range. When secreted into the circulation, EPO targets colony forming units erythroid within the bone marrow to mature into erythrocytes. The principal effect of EPO in its erythropoietic role is to function as a molecular switch that turns off programmed cell death in developing erythrocyte precursors. Since erythrocytes are constantly lost due to

Pietro Ghezzi and Anthony Cerami (eds.), *Tissue-Protective Cytokines: Methods and Protocols*, Methods in Molecular Biology, vol. 982, DOI 10.1007/978-1-62703-308-4_1, © Springer Science+Business Media, LLC 2013

senescence, new red cells must be continuously produced in order to maintain an optimum erythrocyte mass, but these express the EPO receptor only transiently. Therefore, EPO concentrations within the circulation are not episodic, but rather are sustained. The plasma half life of EPO made by the kidney is correspondingly prolonged (~4–6 h), due to the presence of four oligosaccharide chains capped by terminal sialic acids.

2 Tissue Protection

In the 1990s it was discovered that tissues other than the kidney (as well as the liver and spleen in some species and in neonates) can synthesize EPO and further, that non-hematopoietic tissues (e.g., the brain (2) and kidney (3)) can express a receptor for EPO. Results of the earliest work indicated that hypoxia was a potent stimulus of EPO production in the brain, e.g., by astrocytes (4), and further, endogenous EPO or recombinant EPO administered locally was neuroprotective (5). However, the EPO molecule produced for neuroprotection by astrocytes was found to be smaller than renal EPO, having less fully sialyted oligosaccharide chains (hypoglycosylated EPO; hypoEPO) (4). This structural feature results in a much shorter serum half-life of hypoEPO than for renal EPO. The hypoEPO variant may have evolved to limit cross-talk between the tissue protective system and the hematopoietic system.

Work performed in the nervous system showed that unlike hematopoiesis, which required constant stimulation by EPO to be effective, neuroprotection by EPO required only a brief exposure. Specifically, using an in vitro system of neuronal injury, it was shown that if injured cells which were programmed for apoptosis were exposed to renal EPO, programmed cell death was effectively prevented. The required duration of exposure to EPO, however, was surprisingly very short: only 5 min produced a degree of neuroprotection equivalent to a continued presence. Further, this neuroprotection required upon gene expression, as interference with RNA synthesis abolished the neuroprotective effects observed (5). Similar to the mechanism of action of EPO in the hematopoietic system, hypoEPO operates a molecular switch, turning on long-lasting biological actions.

At the time of the initial discoveries of EPO's protective role in tissue injury, it was thought that such a large protein could not cross the blood brain barrier. But, in fact, peripherally administered EPO was highly effective in protecting against not only hypoxic injury, but also other forms of damage, including blunt trauma, excessive stimulation by excitotoxins, and immune-mediated disease (6). Notably, however, doses substantially higher than those needed for erythropoiesis were consistently required for tissue protection.

The diametrically opposed pharmacodynamics of renal EPO-hematopoiesis compared to hypoEPO-tissue protection suggested that systemic use of hypoEPO could provide tissue protection, but would not trigger effective erythropoiesis. To test this hypothesis, renal EPO was fully desialyted (asialoEPO), which was shown using a variety of tissue injury models to be a potent protective agent but totally devoid of hematopoietic effects (7).

The structure of the EPO receptor that controls the survival of erythrocyte precursors was defined in the 1990s. Similar to the structure of other members of the type I cytokine family, the EPO receptor was found to consist of subunits, in this case a dimer of two identical subunits, $(EPOR)_2$ (1). The EPO molecule has two regions localized on opposite sides of the molecule (one of high affinity and the other lower) that recognize docking sites within each of the subunit receptor proteins. When EPO binds to this receptor, occupancy of these binding sites leads to a conformational change that triggers phosphorylation of Janus kinase 2 (Jak-2), which then subsequently phosphorylates multiple other molecules downstream and activates a constellation of biological hematopoietic responses (1).

Although it was originally believed that erythrocyte precursors were the only cells that expressed the hematopoietic homodimer, accumulated evidence revealed a variety of other cell types also express this receptor. In a general way, other cells that express $(EPOR)_2$ are also involved in maintaining adequate circulating numbers of red cells. Specifically, these are cell types that are involved in limiting blood loss during hemorrhage by promoting thrombosis (platelets and endothelial cells) or maintaining adequate circulating pressure and blood volume (via vascular smooth muscle) (8). In the subacute time frame, EPOR is also expressed by hepatocytes that secrete hepcidin, the major modulator of iron absorption (9). In the setting of red cell losses and attendant relative hypoxia, EPO serum levels become elevated from the normal range of 1–20 mIU/ml to as high as 200–300 mIU/ml.

3 Alternative EPO Signaling

The endogenous tissue protective system is one aspect of an evolutionarily ancient innate immune response to injury or infection (reviewed in ref. 10). The "logic" of this system is simple: danger from the invasion of pathogens and parasites or damage caused by diverse injuries is sensed by surveillance cells in the vicinity. A variety of molecular signatures, some representative of the foreign organism, others endogenous molecules expressed and released from damaged cells, quickly activates a proinflammatory reaction.

This response is driven by TNFα which amplifies the response by stimulating other inflammatory cytokines and attracting immune competent cells into the vicinity of the injury. However, this highly amplified positive control system could easily spin out of control unless very tightly regulated. EPO made in and around the locale of injury has been found to play this major role by limiting the production of inflammatory cytokines as well as protect cells that have been exposed to molecular signals that would otherwise activate self-destructive properties, e.g., apoptosis. The temporal-spatial aspects of this endogenous protective system is distinctive. TNFα (and probably other proinflammatory cytokines) stimulate the expression EPO and of a receptor for EPO in the immediate region surrounding the lesion. However, TNFα and EPO are mutually suppressive of each other's synthesis and biological activities. Therefore, cells closest to a lesion that express EPOR do not highly express hypoEPO. At the periphery of the lesion (where TNFα concentrations are lower) hypoEPO is synthesized, but with a significant time delay. This mismatch between abundant receptor and limited ligand suggests that exogenous receptor activators could significantly attenuate damage over that mediated by the endogenous system.

The first non-hematopoietic activity for EPO was observed for endothelial cells. Here it was observed that EPO exposure in vitro stimulated endothelial cells to undergo mitosis and migrate—the first steps in neoangiogenesis (11). Interestingly, the concentration of EPO that elicited this behavior was observed to be substantially above that required for erythropoiesis (~2 nM versus 100 pM). Subsequent work showed that pheochromocytoma (neuronal-like) cells also expressed a receptor that bound EPO (12). Biochemical analysis showed, however, that this receptor appeared to be different, having a molecular weight and was associated with different proteins than was the homodimeric receptor $(EPOR)_2$ that mediated erythropoiesis. Further, the affinity of this receptor for EPO was lower (~16 nM) than for $(EPOR)_2$ (~100–200 pM). Subsequent in vivo and in vitro work revealed that the affinity for the EPO receptor that mediated tissue protection was uniformly lower than for erythropoiesis. These observations suggested that the EPO receptor that provides tissue protection is an alternate form of EPOR.

The existence of an additional EPO receptor raised the possibility that receptor-specific ligands might be developed, as has been accomplished in many other biological systems characterized by receptor isoforms. AsialoEPO was clearly not such a ligand. Although asialoEPO had been functionally shown to preferentially signal the tissue protective system, this molecule none-the-less binds to $(EPOR)_2$. In fact, with the loss of the negative charged terminal sialic acids, asialoEPO actually has an increased affinity for the hematopoietic receptor, and therefore a higher specific activity than EPO itself (13): asialoEPO does not trigger erythropoiesis in vivo because its serum half-life is only several minutes. In contrast, asialoEPO is more active in vitro than EPO as an erythropoietic agent. Therefore, asialoEPO is not a specific ligand for the tissue protective receptor.

In the late 1990s, there was intense interest in the structure of the hematopoietic receptor and the corresponding structure–activity relationships of EPO. Elegant mutagenesis studies coupled with detailed NMR and X-ray receptor structure studies defined which regions of EPO bound to the receptor and further, which amino acids were key for binding (14). It was determined that similar to the growth hormone receptor homodimer that had been elucidated previously, $(EPOR)_2$ formed a dimeric structure which subsequently bound different parts of EPO sequentially to each subunit (8). With the knowledge of specific essential regions of EPO needed to bind to the assembled homodimer, specific changes to these regions could be accomplished (e.g., by chemical or molecular biological methods) that would abolish binding to $(EPOR)_2$. Such engineered molecules might still bind to the hypothesized tissue protective receptor isoform. A number of modified EPO molecules were constructed which as expected, did not bind to the hematopoietic receptor and therefore were not erythropoietic (15). Excitingly, many of these molecules retained tissue protective activity similar to EPO in a wide variety of animal models, thereby confirming the existence of an alternative receptor system (15).

4 The Injury-Induced Cytoprotective Receptor

The identity of this alternate EPO receptor was discovered to consist of a heteromeric structure (16). As background, EPO is a member of the type I cytokine family that includes a host of other cytokines, including GM-CSF, IL-3, IL-5, prolactin, growth hormone, ciliary neurotrophic factor, and leukemia inhibitory factor, among others. In contrast to the homodimeric hematopoietic receptor for EPO, other members of this family typically form heteromers. Additionally, receptor subunits that assemble to form an active receptor can share subunits. For example, GM-CSF, IL-3, and IL-5 all share a receptor subunit, termed the beta common receptor (βCR)—the specificity of the receptor is conferred by an additional unique alpha receptor subunit (17). It was particularly notable that the βCR had been reported to associate with EPOR in vitro (18). Based on this knowledge, one possible structure for the tissue protective receptor could be EPOR complexed with βCR.

Search for the alternative receptor system was initiated by constructing an affinity column from carbamylated EPO (CEPO), a non-erythropoietic, tissue protective molecule in which lysines had been chemically converted to homocitrulline (15). Membrane

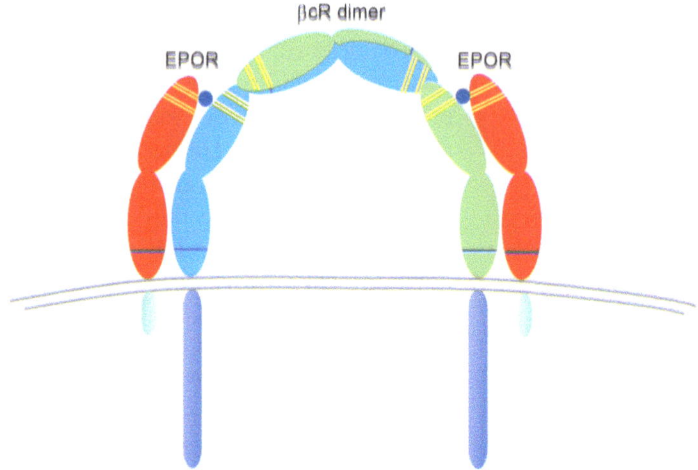

Fig. 1 The innate repair receptor. Available evidence indicates that the receptor mediating tissue protection and repair is assembled from subunits consisting of EPOR and βCR (for review, see ref. 19)

fractions prepared from tissues exhibiting tissue protection (e.g., brain, kidney, liver) and were subsequently passed over the column (16). After extensive washing, the retained proteins were removed from the column, run on a gel and probed with anti-EPOR or anti-βCR antibodies. The results clearly showed that both βCR and EPOR subunits were among the proteins retained on the CEPO column. Confirmation that the beta βCR was indeed a component of the tissue protective receptor was confirmed by studying mice in which the βCR was knocked-out. Although these mice are phenotypically normal while young, as predicted their tissues do not respond to either EPO or non-erythropoietic tissue protective molecules following injury (16). Based on the biology of the related GM-CSF receptor, the proposed structure of the tissue protective receptor is illustrated in Fig. 1 (reviewed in ref. 19). The signaling pathways activated by EPO in tissue protection include the same JAK2-signal transducer and activator of transcription (STAT) utilized in erythropoiesis. However, multiple alternative pathways have also been identified, including the protein kinase B-Akt system, as well as mitogen-activated kinases (reviewed in ref. 19).

5 Exogenous EPO for Tissue Protection?

Although many acute experimental models show impressive protective effects following administration of exogenous EPO, there are major potential roadblocks for clinical utility. Clearly, the use of systemically administered EPO for tissue protection requires a

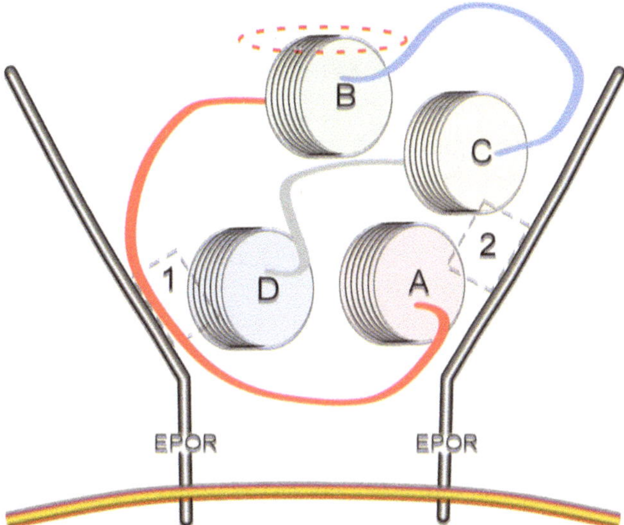

Fig. 2 The tissue protective region of EPO resides on the outer portion of helix B. Helix B is not involved in binding to the hematopoietic receptor dimer. *Boxed sites 1 and 2* correspond to the high and low affinity binding sites of the receptor for EPO respectively. The *circled region* of helix B delimits the tissue protective region from which peptide analogues have been designed (reproduced from ref. 21)

concentration range such that EPO occupies and activates both receptor types. As previously mentioned, the hematopoietic receptor is not restricted only to red cell development, as the endothelium is also activated into a prothrombotic state, and highly reactive, young platelets are produced by the bone marrow. Therefore, several potential side effects of the use of renal EPO for tissue protection are excessive erythropoiesis and thromboses. These adverse effects have been clearly observed in large clinical trials using EPO. As one example, a multicenter trial examining the use of EPO in critical medical and surgical cases showed a significant tissue protective effect in trauma patients, but at the expense of an increased incidence of thromboses (20). These adverse effects would be especially problematic in any chronic use of EPO as a treatment of tissue injury.

A solution to this therapeutic conundrum was developed by successfully altering the EPO molecule so as not to bind to the hematopoietic receptor in a way that retained its full tissue protective activities (15). The existence of selective protective molecules also suggested that a specific region of EPO was responsible for tissue protection. Of note, when EPO binds to the hematopoietic receptor, a large portion of the molecule is either buried or involved in binding to the receptor. The remaining portion of the molecule (helix B) faces away from the interior of the (Fig. 2; ref. 21). A biological significance for helix B can be further appreciated by

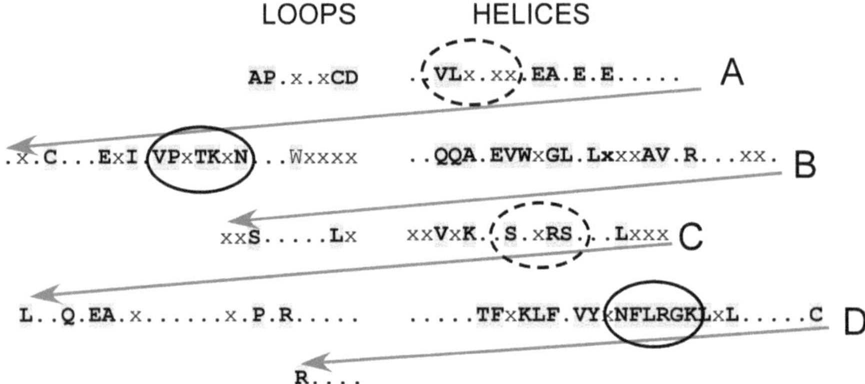

Fig. 3 Helix B of the EPO molecule is highly conserved over 360 million years of natural selection. Homology between human and amphibian (*Xenopus laevis*) EPOs: helices are aligned to the *right* and *arrows* indicate continuation of the protein sequence into the loops adjacent to the helices (*left*). *Shaded uppercase bold letters* correspond to identical amino acids, whereas *x* represents strong conservative substitutions. Binding sites to the hematopoietic receptor (*solid ellipse* high affinity; *dashed ellipse* low affinity) as well as helix B are highly conserved (N-terminal is at the *upper left* and C-terminal at the *lower left*)

comparing the sequence homology of evolutionarily distant vertebrates. For example, an EPO molecule has been recently identified from *Xenopus laevis*, the African clawed frog, that shared a common ancestor with humans approximately 360 million years ago (22). Examining for sequence similarities, a number of regions of the EPO molecule are highly conserved (Fig. 3). Most of these correspond to the regions of EPO that bind to the dimeric hematopoietic receptor. However, helix B, which does not bind to the hematopoietic receptor, is also highly conserved, consistent with an important biological function requiring conservation of structure to maintain function. We hypothesized that this region bound to the tissue protective receptor.

To test the hypothesis that the tissue protective activity of EPO resided within helix B, a 25 amino acid peptide constituting the sequence of helix B was synthesized (21). As expected, this peptide was not erythropoietic. However, it was fully tissue protective in a wide variety of in vitro and in vivo animal models (21). Additional consideration of the molecular structure of EPO led to a simplification of the structure needed for tissue protection. Namely, as previously noted EPO is a globular molecule—the helices interact to provide structure for the molecule. Therefore, a specific portion of helix B always faces out into the aqueous medium (Fig. 2). Utilizing crystallographic data, the specific amino acids that are on the aqueous face of helix B were identified and a small 11 AA peptide was constructed, termed Helix B Surface Peptide (HBSP). This peptide was fully tissue protective in a wide variety of models of tissue injury (21), confirming the importance of helix B in the endogenous response to injury.

6 Regeneration and Repair

One of the earliest studies that identified a non-erythropoietic activity for EPO showed that EPO was trophic for cholinergic neurons in vitro and in vivo (23). Shortly thereafter, a widespread presence of EPO and EPO receptor was reported for the brain (2) and neuroprotective effects of EPO discovered (5). Much of the subsequent study has evaluated the importance of EPO in the protection of tissue from damage/insults. However, this focus is too restricted, as much work has also shown that EPO and non-erythropoietic tissue protective derivatives play a critical role in repair of damaged tissues. For example, tissue protective derivatives of EPO have been shown to play a significant role in the improvement of function when administered following experimental brain injury, even when given with a significant temporal delay (24). A major role has also been documented for EPO-induced endothelial cell growth of new blood vessels into damaged tissues, allowing for an accelerated healing phase, e.g., within infarcted myocardium (25), and for wounds in general (26). Endothelial nitric oxide synthase plays a critical role in the healing response and the βCR-EPOR has been shown to be the receptor mediating this effect (27). Recently, vascular endothelial growth factor receptor 2 (VEGFR2) has also been shown to be a component of the receptor complex as well (28). These observations illustrate the multiple ways the βCR can associate with different receptor subunits to provide specific physiological functions (reviewed in ref. 19).

7 Looking Forward

Organisms respond to tissue injury by activating an inflammatory cascade that typically results in amplified damage that extends far beyond the locus of the initial injury. Therapeutics that target specific components of the inflammatory cascade, e.g., anti-TNFα therapy, have been spectacularly successful in a number of chronic inflammatory diseases. However, as successful as these approaches have been, they target only a component of the damage cascade and further, do not specifically activate endogenous repair mechanisms. Work performed by many investigators over the last 20 years have uncovered the critical role the EPO plays in counteracting injury at a level higher than individual proinflammatory cytokines, as well as effects that occur at multiple lower levels that involve attenuating the damaging effects of otherwise self-amplifying proinflammatory cytokines, reversing programmed cell death, and stimulating repair mechanisms (Fig. 4). In this sense, endogenous hypoglycosylated-EPO activates a molecular switch that shuts off inflammation, as well as activates healing and regenerative processes.

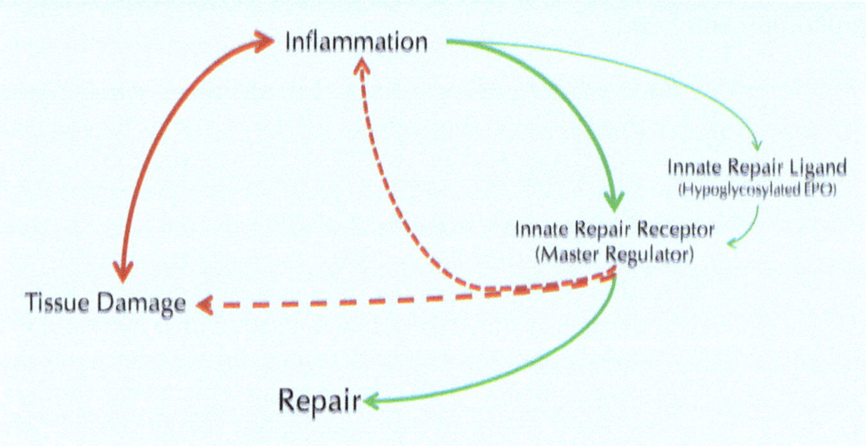

Fig. 4 The Innate Repair System. The endogenous tissue protective system is activated by inflammation caused by tissue damage. In turn, inflammation induces an innate repair receptor immediately following injury, as well as a delayed, less vigorous expression of hypoEPO. Activation of the innate repair receptor by an innate repair activator damps further tissue damage and inflammation, as well as initiates multiple molecular pathways that promote repair

In essence, this molecular switch functions as an innate repair receptor (and its ligands-innate repair activators) that have been shown to be composed of the βCR and EPOR, as well as other receptor subunits relevant for specific tissues, e.g., VEGFR2 in endothelial cells. Because the endogenous EPO system is activated with a temporal delay and further, is also attenuated by proinflammatory cytokines, the molecular switch that controls inflammation and healing is an attractive target for therapeutics. However, the presence of an additional receptor for EPO mediating hematopoiesis possessing an affinity significantly higher than that for tissue protection and regeneration, makes renal EPO itself a poor innate repair activator. Several different approaches have produced engineered molecules that do not interact with the hematopoietic-thrombotic system and therefore activate only tissue protection and regeneration. This new class of therapeutics—innate repair activators (IRAs)—targets the fundamental processes of tissue injury at a level that controls both the self-damaging as well as regenerative components. IRAs therefore offer much promise for beneficial therapeutic effects in a wide range of diseases and injuries.

Disclosure of Interest

The authors are officers of Araim Pharmaceuticals and currently hold stocks/shares in the company.

References

1. Jelkmann W (2007) Erythropoietin after a century of research: younger than ever. Eur J Haematol 78:183–205
2. Marti HH, Wenger RH, Rivas LA et al (1996) Erythropoietin gene expression in human, monkey and murine brain. Eur J Neurosci 8:666–676
3. Westenfelder C, Biddle DL, Baranowski RL (1999) Human, rat, and mouse kidney cells express functional erythropoietin receptors. Kidney Int 55:808–820
4. Masuda S, Okano M, Yamagishi K, Nagao M, Ueda M, Sasaki R (1994) A novel site of erythropoietin production. Oxygen-dependent production in cultured rat astrocytes. J Biol Chem 269:19488–19493
5. Morishita E, Masuda S, Nagao M, Yasuda Y, Sasaki R (1997) Erythropoietin receptor is expressed in rat hippocampal and cerebral cortical neurons, and erythropoietin prevents in vitro glutamate-induced neuronal death. Neuroscience 76:105–116
6. Brines ML, Ghezzi P, Keenan S et al (2000) Erythropoietin crosses the blood–brain barrier to protect against experimental brain injury. Proc Natl Acad Sci U S A 97:10526–10531
7. Erbayraktar S, Grasso G, Sfacteria A et al (2003) Asialoerythropoietin is a nonerythropoietic cytokine with broad neuroprotective activity in vivo. Proc Natl Acad Sci U S A 100:6741–6746
8. Fisher JW (2003) Erythropoietin: physiology and pharmacology update. Exp Biol Med (Maywood) 228:1–14
9. Pinto JP, Ribeiro S, Pontes H et al (2008) Erythropoietin mediates hepcidin expression in hepatocytes through EPOR signaling and regulation of C/EBPalpha. Blood 111:5727–5733
10. Brines M, Cerami A (2008) Erythropoietin-mediated tissue protection: reducing collateral damage from the primary injury response. J Intern Med 264:405–432
11. Anagnostou A, Lee ES, Kessimian N, Levinson R, Steiner M (1990) Erythropoietin has a mitogenic and positive chemotactic effect on endothelial cells. Proc Natl Acad Sci U S A 87:5978–5982
12. Masuda S, Nagao M, Takahata K et al (1993) Functional erythropoietin receptor of the cells with neural characteristics. Comparison with receptor properties of erythroid cells. J Biol Chem 268:11208–11216
13. Imai N, Higuchi M, Kawamura A et al (1990) Physicochemical and biological characterization of asialoerythropoietin. Suppressive effects of sialic acid in the expression of biological activity of human erythropoietin in vitro. Eur J Biochem 194:457–462
14. Cheetham JC, Smith DM, Aoki KH et al (1998) NMR structure of human erythropoietin and a comparison with its receptor bound conformation. Nat Struct Biol 5:861–866
15. Leist M, Ghezzi P, Grasso G et al (2004) Derivatives of erythropoietin that are tissue protective but not erythropoietic. Science 305:239–242
16. Brines M, Grasso G, Fiordaliso F et al (2004) Erythropoietin mediates tissue protection through an erythropoietin and common beta-subunit heteroreceptor. Proc Natl Acad Sci U S A 101:14907–14912
17. Lopez AF, Hercus TR, Ekert P et al (2010) Molecular basis of cytokine receptor activation. IUBMB Life 62:509–518
18. Jubinsky PT, Krijanovski OI, Nathan DG, Tavernier J, Sieff CA (1997) The beta chain of the interleukin-3 receptor functionally associates with the erythropoietin receptor. Blood 90:1867–1873
19. Brines M, Cerami A (2012) The receptor that tames the innate immune response. Mol Med 18(1):486–496
20. Corwin HL, Gettinger A, Fabian TC et al (2007) Efficacy and safety of epoetin alfa in critically ill patients. N Engl J Med 357:965–976
21. Brines M, Patel NS, Villa P et al (2008) Non erythropoietic, tissue-protective peptides derived from the tertiary structure of erythropoietin. Proc Natl Acad Sci U S A 105:10925–10930
22. Nogawa-Kosaka N, Hirose T, Kosaka N et al (2010) Structural and biological properties of erythropoietin in Xenopus laevis. Exp Hematol 38:363–372
23. Konishi Y, Chui DH, Hirose H, Kunishita T, Tabira T (1993) Trophic effect of erythropoietin and other hematopoietic factors on central cholinergic neurons in vitro and in vivo. Brain Res 609:29–35
24. Mahmood A, Lu D, Qu C et al (2007) Treatment of traumatic brain injury in rats with erythropoietin and carbamylated erythropoietin. J Neurosurg 107:392–397
25. van der Meer P, Lipsic E, Henning RH et al (2005) Erythropoietin induces neovascularization and improves cardiac function in rats with heart failure after myocardial infarction. J Am Coll Cardiol 46:125–133
26. Erbayraktar Z, Erbayraktar S, Yilmaz O, Cerami A, Coleman T, Brines M (2009) Nonerythropoietic tissue protective compounds are highly effective facilitators of wound healing. Mol Med 15:235–241
27. Su KH, Shyue SK, Kou YR et al (2011) Beta common receptor integrates the erythropoietin signaling in activation of endothelial nitric oxide synthase. J Cell Physiol 226(12):3330–3339
28. Sautina L, Sautin Y, Beem E et al (2010) Induction of nitric oxide by erythropoietin is mediated by the beta common receptor and requires interaction with VEGF receptor 2. Blood 115:896–905

Chapter 2

Epo and Non-hematopoietic Cells: What Do We Know?

Omolara O. Ogunshola and Anna Yu. Bogdanova

Abstract

The hematopoietic growth factor erythropoietin (Epo) circulates in plasma and controls the oxygen carrying capacity of the blood (Fisher. Exp Biol Med (Maywood) 228:1–14, 2003). Epo is produced primarily in the adult kidney and fetal liver and was originally believed to play a role restricted to stimulation of early erythroid precursor proliferation, inhibition of apoptosis, and differentiation of the erythroid lineage. Early studies showed that mice with targeted deletion of Epo or the Epo receptor (EpoR) show impaired erythropoiesis, lack mature erythrocytes, and die in utero around embryonic day 13.5 (Wu et al. Cell 83:59–67, 1995; Lin et al. Genes Dev. 10:154–164, 1996). These animals also exhibited heart defects, abnormal vascular development as well as increased apoptosis in the brain suggesting additional functions for Epo signaling in normal development of the central nervous system and heart. Now, in addition to its well-known role in erythropoiesis, a diverse array of cells have been identified that produce Epo and/or express the Epo-R including endothelial cells, smooth muscle cells, and cells of the central nervous system (Masuda et al. J Biol Chem. 269:19488–19493, 1994; Marti et al. Eur J Neurosci. 8:666–676, 1996; Bernaudin et al. J Cereb Blood Flow Metab. 19:643–651, 1999; Li et al. Neurochem Res. 32:2132–2141, 2007). Endogenously produced Epo and/or expression of the EpoR gives rise to autocrine and paracrine signaling in different organs particularly during hypoxia, toxicity, and injury conditions. Epo has been shown to regulate a variety of cell functions such as calcium transport (Korbel et al. J Comp Physiol B. 174:121–128, 2004) neurotransmitter synthesis and cell survival (Velly et al. Pharmacol Ther. 128:445–459, 2010; Vogel et al. Blood. 102:2278–2284, 2003). Furthermore Epo has neurotrophic effects (Grimm et al. Nat Med. 8:718–724, 2002; Junk et al. Proc Natl Acad Sci U S A. 99:10659–10664, 2002), can induce an angiogenic phenotype in cultured endothelial cells, is a potent angiogenic factor in vivo (Ribatti et al. Eur J Clin Invest. 33:891–896, 2003) and might enhance ventilation in hypoxic conditions (Soliz et al. J Physiol. 568:559–571, 2005; Soliz et al. J Physiol. 583, 329–336, 2007). Thus multiple functions have been identified breathing new life and exciting possibilities into what is really an old growth factor.

This review will address the function of Epo in non-hematopoietic tissues with significant emphasis on the brain and heart.

Key words Non-hematopoietic cells, Adult kidney, Fetal liver, HIF

Pietro Ghezzi and Anthony Cerami (eds.), *Tissue-Protective Cytokines: Methods and Protocols*, Methods in Molecular Biology, vol. 982, DOI 10.1007/978-1-62703-308-4_2, © Springer Science+Business Media, LLC 2013

1 Epo Expression Is Regulated by Hypoxia-Inducible Factors

Epo expression is hypoxia inducible and regulation occurs via the hypoxia responsive element (HRE) present in the 3′ region of the gene which is bound by heterodimeric transcription factors namely hypoxia-inducible factors (HIFs). Three members of the HIF transcription factor family HIF-1, -2, and -3 have now been identified. HIF-1 was discovered in 1991 by its ability to bind and stimulate transcription of the Epo gene during hypoxia (16, 17) and for several years, was assumed to be the primary stimulus for Epo production in response to acute hypoxia. Later a second hypoxia-inducible transcription factor termed HIF-2 was discovered (18–20). Subsequent data from in vivo (21) and in vitro (22) experiments suggested that despite the fact that HIF-1 clearly binds the HRE of the Epo gene in response to hypoxia and both have the potential to bind many of the same genes, in vivo HIF-2 is the primary mediator of Epo expression in kidneys in response to hypoxia. In agreement downregulation of HIF-2 in the brain, but not HIF-1, drastically reduced hypoxia-induced Epo expression (23) and more recently Haase and colleagues (24) clearly demonstrated the primary role of HIF-2 in promoting the hypoxic renal Epo response.

The HIFs are heterodimers composed of a constitutively expressed β subunit (also known as aryl hydrocarbon receptor nuclear translocation, ARNT) and an oxygen-regulated α subunit (reviewed by ref. 25–27). Regulation of HIF activity occurs at different levels including protein stability, phosphorylation, nuclear translocation, and activity, all being influenced by alterations in oxygen levels. Under normoxic conditions the α subunit is degraded. In contrast, under hypoxic conditions the α subunit is stabilized and translocated to the nucleus where it dimerizes with ARNT and subsequently binds to hypoxic binding sites (HBS) of target genes. The HBS is a conserved consensus sequence (A/G) CGTG within the HRE present in oxygen-regulated target genes involved in cell survival, glycolysis, angiogenesis, erythropoiesis, and iron metabolism (25). Degradation of HIF-α is triggered by oxygen-dependent hydroxylation of prolyl residues located in the oxygen-dependent degradation domain by a family of prolyl hydroxylases, namely PHD1, PHD2, and PHD3. These enzymes are specific HIF prolyl hydroxylases that require Fe(II) as a cofactor as well as oxygen and 2-oxoglutarate as co-substrates (28, 29). Prolyl hydroxylation promotes the recruitment of the tumor suppressor protein von Hippel Lindau, which is part of the E3 ligase ubiquitination complex, priming HIFs for degradation in the proteosomes (reviewed by ref. 30, 31).

Other regulatory elements in the 5′ promoter of the Epo gene include a highly conserved GATA sequence as well as NFκB binding motifs (32, 33). Both these sites seem to have inhibitory effects on Epo expression. The GATA site preferentially binds the transcription factor GATA-2, which has been reported to inhibit Epo gene expression (34, 35). NFκB binding to a site adjacent to the minimal HRE of the Epo promoter also inhibits Epo expression. Although activities of GATA-2 and NFκB in HepG2 cells decrease in hypoxia compared to normoxia conditions both transcription factors were shown to be involved in the suppression of Epo gene expression by IL-1β and TNFα (35). Thus these pathways may be responsible for impaired Epo synthesis in a variety of inflammatory diseases and cancers.

2 EpoR Is Expressed in Multiple Tissues

Hypoxia and anemia are major events known to induce Epo gene expression, however it should be noted that many different injuries induce Epo expression (36, 37). Once the signals are transduced erythropoietin is released into the circulating blood flow and finally binds cells expressing the Epo receptor (EpoR).

The EpoR is a member of the type 1 superfamily of single-transmembrane cytokine receptors (38, 39). Expression of the EpoR is located in progenitor cells from hematopoietic, endothelial, skeletal muscle, and neuronal compartments (40–42). EpoR is downregulated during differentiation of erythroid cells and not expressed on skeletal muscle. Interestingly, despite being significantly downregulated in developing neuronal tissues until embryonic day 17, EpoR expression persists in select vascular and neuronal compartments. EpoR has been observed in brain during development and adulthood (37, 43–46). More recent studies have demonstrated expression of EpoR on cells from a variety of tissues including heart (47), kidney (48), pancreas (49), and uterus (50).

3 Classical Erythroid EpoR Signaling

Erythropoiesis is stimulated by generating a complex network of molecular signals involved in the control of cell proliferation, differentiation, and death. EpoR homodimers are expressed on the erythroid progenitor cell surface (51) and binding of Epo to the EpoR triggers conformational changes in the receptor extracellular domain that consequently activates JAK2 by autophosphorylation (52, 53). JAK2 activation results in the phosphorylation of tyrosine residues on the cytoplasmic region of EpoR and recruits a variety of Src homology-2 (SH2) domain-containing proteins that initiate downstream cascades via different signaling pathways including signal transducer and activator of transcription (STAT), phosphatidylinositol-3 kinase (PI3K)/Akt (also known as protein kinase B) and mitogen-activated

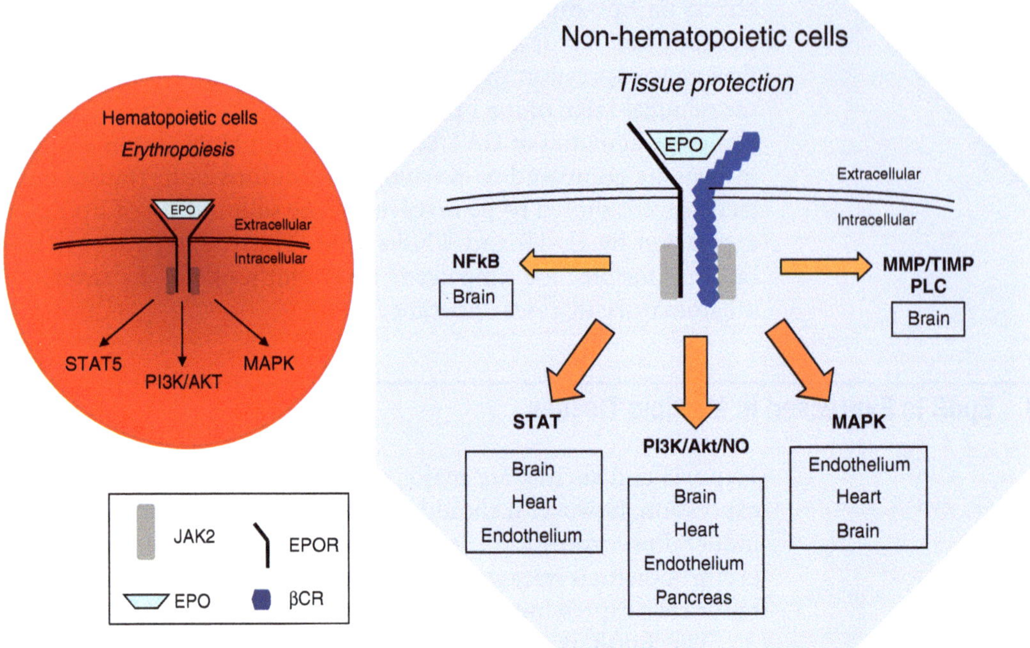

Fig. 1 Downstream pathways activated by Epo signaling in hematopoietic and non-hematopoietic cells. In non-hematopoietic cells the βCR subunit makes a functional receptor with a classic EpoR. In the absence of βCR it is postulated that the homodimer configuration will occur. Note the similarities of the downstream pathways activated by both hetero- and homodimers

protein kinase (MAPK) (54, 55). Although Epo can activate STAT1, STAT3, and STAT5a/b, JAK2/STAT5 is the classical pathway activated in erythroid cells (summarized in Fig. 1 and reviewed in ref. 56). Epo-mediated activation of this pathway leads to the upregulation of the antiapoptotic Bcl2 and Bcl-X_L gene, thereby protecting precursors from apoptosis (56, 57).

The PI3K/Akt pathway has been shown to be necessary, but not solely sufficient, for erythroid cell survival by protecting them from apoptosis (58). The PI3K/Akt cascade phosphorylates serine residue 310 of GATA-1 both in vitro and in erythroid cells thereby enhancing GATA-1 transcriptional activity (55). GATA-1 binds to a consensus GATA motif present in the *cis*-regulatory elements of most erythroid genes and is a key transcription factor for antiapoptotic Bcl-X_L and erythroid-specific gene transcription, and terminal differentiation of erythroid precursors into red blood cells (59–61). Notably PI3K can also be indirectly recruited to EpoR by other proteins such as Grb-2. PI3K-mediated Akt phosphorylation inhibits cytochrome c release from mitochondria (62) and facilitates NFκB activation by enhancing inhibitor of NF-kappaB (IκB) degradation (63). Additionally, Akt can inhibit activity of Foxo3A thereby downregulating target proteins having antiproliferative or proapoptotic functions (64, 65).

Another important Epo-mediated signaling pathway is the MAPK pathway. MAPKs are serine/threonine kinases activated by extracellular signals of which there are at least three distinct types: the classical ERK1/ERK2 kinases, the p38MAPKs (p38), and the stress-activated protein kinase/Jun kinase (SAPK/JNK) subfamily. All play important roles in Epo-induced differentiation or apoptosis (66–70).

Soon after stimulation of the receptor by its ligand, mechanisms integral to downregulation of these signaling pathways are also activated, returning signaling proteins to their basal levels. This process is crucial to prevent hyperstimulation and, consequently, the dysregulation of cellular machinery (reviewed in ref. 63). Notably, EpoR is also synthesized in a soluble form (sEpoR) that corresponds to the extracellular domain of the complete receptor as a result of alternative splicing of EpoR mRNA (71). The sEpoR is secreted into the extracellular fluid and acts as a sink, sequestering Epo and preventing its ability to activate EpoR and downstream signaling cascades (see Fig. 2, pathway 8). The presence of sEpoR has been reported in plasma and several tissues including liver, spleen, kidney, heart, brain, and bone marrow (15, 72).

4 Epo and EpoR Signaling in Non-hematopoietic Tissues

Production of Epo and expression of the EpoR has been detected in non-hematopoietic tissues and emerging evidence suggests that Epo exerts cytoprotective effects on non-erythroid cells. Notably, a tissue-specific degree of Epo regulation has been reported. Depending on the severity of hypoxia, Epo mRNA levels can increase up to 20-fold in the brain in contrast to 200-fold in the kidney (5) and remain high much longer (73). Also brain Epo, purified from primary neuronal cell cultures, was shown to have lower molecular weight and be more active than recombinant Epo and serum Epo at low concentrations (74). Importantly, tissue protection in vivo and in vitro appears to require nanomolar concentrations of Epo that are not normally reached in the circulation, in contrast to low picomolar concentrations required for erythropoiesis (75) underlining the fact that paracrine/autocrine signaling likely results in high local concentrations of Epo. The EpoR expressed by PC12 cells also had lower affinity than EpoR on erythroid cells and required different accessory proteins compared to erythrocyte precursors (4). Lower binding affinities of EpoR expressed by non-erythroid cells was also reported in vivo (76). Thus, differential activity and affinity allows specific activation of erythroid and non-hematopoietic receptors thus preventing cross-talk between the endocrine and paracrine systems of Epo.

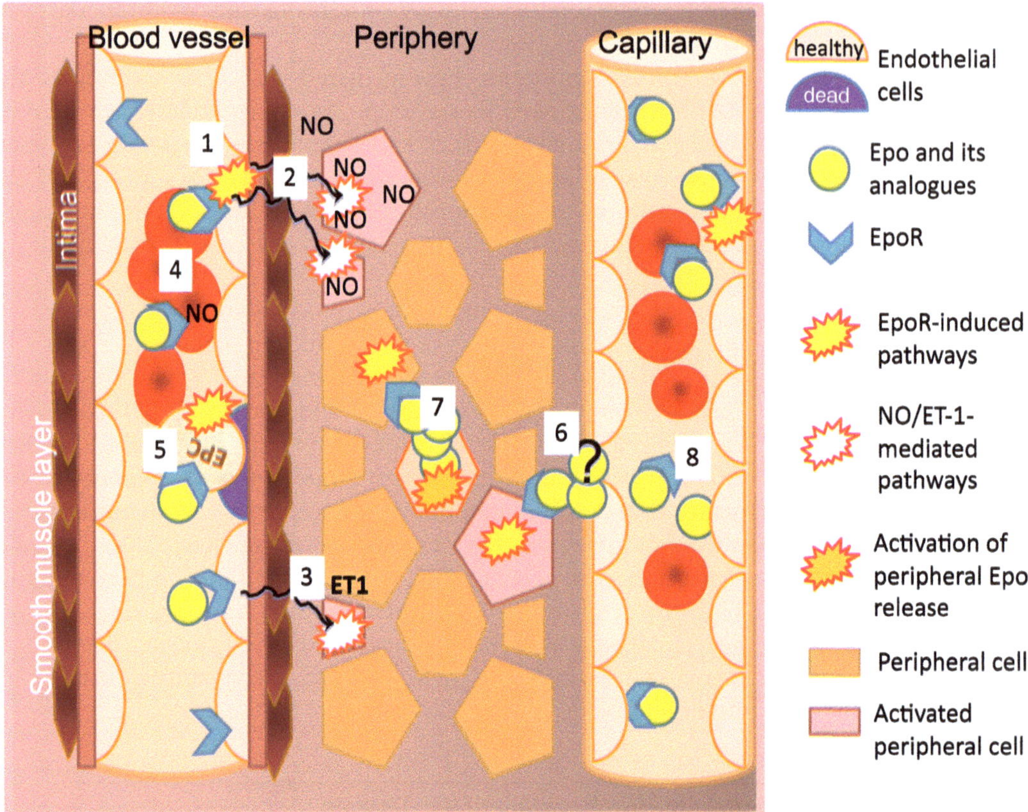

Fig. 2 Schematic representation of the multiple putative cytoprotective effects of Epo in non-erythropoietic tissues. Interaction of blood-borne Epo with heterodimeric Epo receptors on endothelial cells activates the PI3K-Akt pathway (*1*) leading to NO production by eNOS and its translocation to the periphery where it induces cytoprotective effects (*2*). Another second messenger known to be released by endothelial cells upon their stimulation with Epo is endothelin 1 (ET-1) which also elicits protective effects in peripheral cells (*3*). Further targets of circulating Epo are blood cells, including red blood cells and macrophages. Similar to endothelial cells, Epo binding to red blood cells triggers production of NO by eNOS (*4*). Endothelial precursor cells (EPCs) are very sensitive to Epo. Epo controls their number, recruitment to the site of injury, homing, and the quality of resulting mature endothelial cells (*5*). Peripheral cells were shown to respond to Epo stimulation directly. Blood vessels are largely impermeable for Epo when undamaged. However the blood-tissue barrier is less tight in capillaries and although leakage of Epo from the capillary system into the peripheral tissue has never been demonstrated convincingly, it cannot be excluded (*6*). Alternatively, peripheral cells may produce their own Epo. Indeed induction of Epo expression has been demonstrated in hypoxic brain and heart. Thus once produced the cytokine is released causing autocrine and paracrine effects (*7*). Action of Epo is transient and the cytokine is internalized and degraded upon binding to the receptor. Free Epo pools in the plasma may also be regulated by sequestration of the circulating soluble Epo receptor (*8*). For more details of these mechanisms please see main text

The functional EpoR that attenuates tissue damage is not normally, or only weakly, expressed in most tissues and is strongly induced following injury (36, 37). EpoR expression level and the number of receptors per cell is significantly lower than observed in erythropoetic precursor cells and for that reason was reported as "undetectable" in one publication (77) - an opinion not shared by the majority of researchers working in the field (78). Recent data advocates that the tissue protective non-hematopoietic receptor is distinct from the hematopoietic receptor responsible for erythropoiesis being a heterodimer consisting of the beta common receptor subunit (βCR also known as CD131) in combination with the EpoR subunit (see Fig. 1 and reviewed by ref. 75). A variety of tissues have been found to express βCR and EpoR including the central and peripheral nervous system, retina, heart, kidney, muscle, and endothelium. Notably, the important role of the βCR in Epo-mediated protection has been demonstrated in brain injury models using βCR knockout mice (79, 80) as well as in endothelium using siRNA technology (81). However the downstream signaling mechanisms activated by βCR are still to be elucidated. When EpoR is not colocalized with βCR it presumably self-associates forming the classical EpoR homodimer that also supports signaling (reviewed by ref. 75).

The importance of EpoR specifically in non-hematopoietic tissues has been recently investigated using transgenic mice with EpoR expression restricted to hematopoietic tissues and the vascular endothelium. These mice survive without any gross abnormalities but become obese and insulin resistant due to loss of Epo regulation of energy homeostasis (82). It should be noted however that because endothelial cells have the same origin as hematopoietic cells these mice still express EpoR on vascular endothelium. Recent studies using these mice in heart ischemia–reperfusion injury model (83) and traumatic brain injury model (84) identify the endothelium as a major contributor to Epo-mediated protection and supporter of significant tissue recovery from injury. More experiments are now needed in various injury paradigms to better understand the contribution of the homoreceptor, heterodimer, and the endothelium per se to tissue protection during Epo treatment.

5 Brain

5.1 Endogenous Production of Epo in CNS

Epo and EpoR have been detected during early brain development in rodent models. Both are also expressed during human fetal development starting around 7 weeks and increase from 8 to 24 weeks (43). After birth Epo was detected in human cerebral spinal fluid and found to be induced by hypoxia (5). Notably, Epo and EpoR expression persist in the human brain throughout adulthood.

Mouse models showed that knockout of either gene caused embryonic death not only due to erythropoiesis failure but also as a result of compromised brain development. In these models the neurons exhibited intrinsic defects such as slowed proliferation and increased sensitivity to hypoxic stress (85). Additionally a specific deficit in post-stroke neurogenesis by the impaired migration of

NPC to the peri-infarct cortex was also observed in adult mice stroke models. Thus a clear role for coordinated Epo signaling in early brain development is evident.

5.2 Neuroprotection by Epo In Vitro

Different neural cells express Epo and the EpoR including neurons, astrocytes, and oligodendrocytes (6, 74, 86, 87). Epo appears to be mainly produced by astrocytes (4, 88), while EpoR is expressed by neurons (43). During injury however it seems all cells are capable of upregulating the Epo signaling cascade eliciting both autocrine and paracrine effects (see Fig. 2, pathway 7).

Epo was shown to protect neurons from hypoxic and toxic insults in different cell culture and ex vivo models (see Fig. 2, pathway 6). Epo supplementation counteracted hypoxia-induced cell death in cortical and hippocampal neurons (89–91) and protected PC12 cells from serum withdrawal (92). In toxicity models Epo pretreatment protected hippocampal and cortical neurons from glutamate (93) and NMDA exposure (46), ketamin cytotoxicity (94), kainate-induced excitocytotoxicity in cultured spinal neurons (95), as well as SH-SY5Y neuroblastoma cells from staurosprine-induced cell death (96) to name but a few. Supplementation of Epo also increased neuronal survival during oxygen glucose deprivation, the in vitro model for hypoxic-ischemia (88). Epo has also been suggested to contribute to myelin recovery by enhancing generation, proliferation, and differentiation of oligodendrocytes after ischemic injury (97, 98) and inflammatory injury (99).

Generally Epo protects neuronal cells by regulating the balance between proapoptotic and antiapoptotic pathways. Similar to erythroid cells, a major mechanism occurs through JAK2/STAT activation and induction of PI3K/Akt pathways that inhibit the pro-apoptotic protein Bad and prevent release of cytochrome c and caspase activation (see Fig. 1). Akt activation also inhibits glycogen synthase kinase 3 (GSK3) (94) resulting in inhibition of the mitochondrial permeability transition pore, a major determinant of cell death, through caspase activation. However inhibition of Akt only partially prevented neuroprotection suggesting the contribution of additional signaling mechanisms (89). A unique pathway for Epo-mediated neuroprotection in the brain seems to be induction of crosstalk between JAK2 and NFκB signaling cascades (see Fig. 1). EpoR mediated activation of JAK2 led to phosphorylation of IκB, subsequent nuclear translocation of NFκB, and NFκB-dependent transcription of neuroprotective genes (88, 100). Accordingly transfection of cerebrocortical neurons with a dominant interfering form of JAK2, or an IκB super-repressor, blocked Epo-mediated prevention of neuronal apoptosis. Epo can also modulate the activity of calcium channels through phospholipase C (PLC) (101), thereby reducing the release of excitatory neurotransmitters and augmenting nitric oxide production (92, 102). Very recent data suggests that Epo-mediated neuroprotection is also associated with increased

TIMP-1 activity and decreased MMP-9 activity in vivo and in vitro, and can be reversed by inhibition of JAK2 or TIMP-1 (103).

A couple of studies have recently implicated Epo to be a mediator of the protective effects of nitric oxide (NO) in neurons. Loss of EpoR coincided with programmed cell death in neurons (104). Neuronal NO was induced during hypoxia and correlated with protection in control cells but not increased in neurons that lacked the EpoR. However when treated with a neuronal nitric oxide synthase (nNOS) inhibitor the neurons lost their ability to induce EpoR expression in hypoxia and thus were not protected (104). In line with this finding another study demonstrated that nNOS knockout mice are more susceptible to peripheral neuropathy than their wild type counterparts due to the absence of NO-mediated activation of HIF-1 and subsequent downstream neuroprotection by Epo (105). Ex vivo experiments showed that protection recovered by using low doses of NOS donors was almost completely abrogated by Epo siRNA. Thus it appears the neuroprotective effect of Epo, as well as EpoR expression on neural cells, may also be regulated by NO.

Intriguingly, what determines the specific pathways activated by Epo, or the coordination of these multiple cascades, remains till now unknown.

5.3 Neuroprotection by Epo In Vivo

Different animal models have suggested potential clinical uses of Epo to combat ischemia or trauma. Cerebroventricular infusion of Epo was shown to reduce ischemia-induced learning disabilities and rescue hippocampal CA1 neurons from lethal ischemic damage in gerbils whereas infusion of EpoR abolished neuroprotection. In various mouse and rat models of ischemia, intracerebral injection of Epo also attenuated brain damage by reducing infarct volume by up to 50% (6, 106, 107) and improved cognitive function (108–110). This was further underlined by the fact that cerebral administration of soluble EpoR reduced the protective effect of hypoxia preconditioning by up to 80% in other models (111, 112). Overall exogenous Epo administration (see Fig. 2, pathway 6) has been shown to be protective in multiple cerebral tissue injuries including neonatal ((113) and reviewed by ref. 114, 115) or adult rodent focal brain ischemia, brain trauma (116), animal models of multiple sclerosis (117, 118) as well as spinal chord injury (119, 120). Increased oligodendrogenesis and attenuated proinflammatory cell infiltration was also observed in mouse models of EAE suggesting Epo positively stimulates oligodendrogenesis and reduces the autoimmune response (117, 118). In the neonatal brain, Epo significantly reduced white matter damage during hypoxia/ischemia and increased oligodendrogenesis and maturation of oligodendrocytes despite being applied in a delayed manner (113). Notably, in models of prolonged hypoxia, Epo secretion from astrocytes was shown to play an important role in neuronal survival (4, 5) highlighting the paracrine functions of Epo (see Fig. 2, pathway 7).

Mechanistically Epo reduced infarct volume via JAK2, ERK, and PI3K/Akt pathways by elevating Bcl-xL and lowered both neuronal and inducible NOS levels in neurons (121). Upregulation of anti-apoptotic pathways was also observed in neonatal rodents submitted to focal cerebral ischemia (122). Epo-induced VEGF and BDNF have also been suggested to have an important role in angiogenesis- and neurogenesis-associated brain repair in rats treated with Epo after embolic stroke (110) similar to observations from in vitro studies (123). Epo was also shown to inhibit iNOS expression preventing the formation of excess NO and protecting facial motor neurons from death (97).

As in other neural cells Epo protects retina against cell death during injury but in contrast to other CNS regions where basal Epo is located mainly to astrocytes (4, 86), retinal neurons may express both Epo and EpoR (12). Epo prevented death of neurotrophic factor-deprived rat retinal ganglion cells (RGCs) in vitro, rescued axotomized RGCs in vivo, and prevented caspase-3 activation (124). Recently it was demonstrated that exogenous Epo significantly attenuates retinal neuronal cell death induced by glyoxal advanced glycosylation end products (AGEs) by promoting antiapoptotic and suppressing apoptotic proteins (125). Systemic administration of Epo before or immediately after retinal ischemia reduced histopathological damage and promoted functional recovery (12). When given therapeutically after light insult, Epo also mimicked the effect of hypoxic preconditioning by crossing the blood-retina barrier and preventing light-induced apoptosis via caspase-1 activation interference (11). Although transgenic overexpression of Epo with constitutively high levels of Epo in the retina protected photoreceptors against light-induced degeneration, the course or extent of retinal degeneration in genetic models was unaltered suggesting different apoptotic mechanisms exist (126).

Overall current evidence suggests that similar to erythroid cells, and as indicated by in vitro studies, phosphorylation of JAK2 is the initial step in Epo-mediated protection in the injured brain (9). Subsequently, downstream signaling modulates the transcription and activity of proteins involved in cell survival.

5.4 Neurotrophic Effects of Epo

In contrast to its neuroprotective properties, putative regeneration-enhancing effects of Epo have been less well studied. Epo was first shown to augment the activity of choline acetyltransferase in central cholinergic neurons in vitro and in vivo (127) and to enhance dopamine generation and differentiation of neuronal precursors in hypoxia. In agreement Epo was demonstrated to act directly on neural stem cells and promote the production of neuronal progenitors in forebrain (42) thus suggesting a direct contribution to neurogenesis after hypoxia. Epo-related functional recovery after spinal cord injury has also been described (119) and correlated with behavioral improvements following Epo treatment (120). During stroke models Epo also significantly improved neurogenesis and functional recovery by increasing cerebral BDNF levels (110). Epo also enhanced oligodendrogenesis (117) and recovery of neurological function after neonatal hypoxic/ischemic brain (113). In the retina, Epo promoted neurite extension from postnatal retinal ganglion cells in vitro (128), induced JAK2/STAT3 phosphorylation and activated PI3K/Akt (see Fig. 1). Inhibition of JAK2/STAT3 abolished Epo-induced growth verifying the pathway is involved in conferring regeneration-enhancing Epo functions in the retina (129).

Thus the positive effects of Epo are not limited to neuroprotection but extend to neurogenesis and differentiation. Indeed more research needs to be performed in this area.

5.5 Epo in Treatment of Brain Diseases

Studies using Epo to combat brain disease progression have been largely encouraging. In 2002 the Göttingen Epo stroke pilot study demonstrated the neuroprotective effectiveness of Epo in human stroke patients (130). Epo-treated patients showed significantly better recovery than the control group regarding the clinical outcome parameters, the evolution of infarct size, and the profile of circulating damage markers. Disappointingly, the recent German multicenter Epo Stroke Trial revealed an increased risk of serious complications such as death, intracerebral hemorrhage, brain edema, and thromboembolic events (131). This study emphasized the point that when used in combination with other drugs (in this case recombinant tissue plasminogen activator used for hemodialysis) Epo may even be detrimental for patient outcome.

Epo therapy was effective in reducing progressive atrophy and loss of gray matter in patients diagnosed with schizophrenia (132). Also in healthy volunteers Epo improved cognitive and neural processing of emotional information showing similar effects to those of serotonergic and noradrenergic antidepressant drugs (133). Together these trials suggest future clinical applications for Epo in the treatment of psychiatric disorders characterized by cognitive dysfunction. During the first phase I/IIa study of high dose Epo treatment in patients with chronic progressive multiple sclerosis significant improvement in clinical and electrophysiological motor function as well as cognitive performance was achieved (134). Epo treatment also somewhat improved outcome for patients after subarachnoid hemorrhage (135). However, in contrast, the first randomized trial of Epo in moderate traumatic brain injury patients during the resuscitative phase showed Epo did not reduce neuronal cell death compared to placebo and disappointingly injury severity was worse in the Epo group (136).

Many of the clinical studies performed show promise, however they also have a number of limitations. For example frequently the patient numbers have been small and some of the studies not blind.

Also the doses used in the different injury paradigms as well as the routes of administration vary considerably. The mechanisms that improve function, enhance regeneration and/or slow deterioration remain undetermined and similarly the reasons why some studies have been less successful or even failed is also unclear. Indeed many questions remain open and the jury is out as to whether Epo will fulfill its putative potential - based on animal studies - to be a "universal" therapy for brain diseases.

6 Heart

6.1 Endogenous Epo Acts on the Heart

Epo is important during myocardial development and knockdown of Epo or EpoR in mice results in reduction in the number of cardiomyocytes (hypoplasia) and enhanced susceptibility to left ventricular dilatation and cardiac death (137, 138). However this phenotype may be largely rescued by restoration of EpoR production in hematopoietic tissue (139). Attempts to localize the EpoR within the heart have been made by dissecting the chick embryonic heart into epicardium, myocardium, and endocardium (140). These experiments revealed that endogenous Epo is most likely produced by the epicardium whereas EpoR is present in embryonic myocardium. However, positive inotropic and lusitropic effects of Epo have been later recorded in isolated human epicardial stripes indicating that adult human and mouse epicardium responds to Epo (141). Changes in contractile force, but not in contractile rate, were reported for isolated denervated rat heart perfused with Krebs-Henseleits saline (142).

6.2 EpoR in the Heart

Epo receptors and functional responses to Epo were shown in isolated cardiomyocytes (141, 143–146) coronary endothelial cells (83, 147) and fibroblasts (148). The cardiac EpoR was shown to respond equally efficiently to Epo, carbomylated Epo (CEPO), and ARA-290 (141, 149, 150), a synthetic Epo mimetic comprised only of helix B part of the cytokine. This synthetic non-erythropoietic peptide was shown to activate the heteroreceptor, composed of an EpoR subunit and βCR, but not the classical EpoR homodimer (79). These findings suggest that the effects of Epo in the heart are most likely mediated by such a heteroreceptor. Indeed expression of βCR in the heart and the lack of Epo effect in βCR knockout myocardium was shown (79).

Whereas in hematopoietic lineage EpoR expression is induced by GATA-1, Sp1, and Wt1 transcription factors (151, 152), expression of the common EpoR subunits in the heart is under control of GATA-4 and Sp1 transcription factors (145). The role of Wt1 expressed only in epicardium in regulation of EpoR expression remains to be clarified (153). Induction of EpoR expression has been observed in the failing ischemic heart and is most likely linked to the stabilization of HIF that is downregulated in aging tissues.

In agreement heat-induced stabilization of HIF1α in the heart is also associated with an increase of EpoR in the heart (151). Thus down-regulation of various transcription factors may reduce the efficiency of myocardial Epo treatment. Changes in EpoR expression during myocardial development and as a function of age remain to be investigated. Regulators of expression of βCR in the heart have also not been studied.

6.3 Where Does Epo Act and What Are Its Targets?

The source of Epo for receptor activation in the myocardium remains unknown. Plasma-borne Epo most likely does not reach cardiomyocytes (147). Thus, the cytokine should be generated by one or more cell types within the myocardium and then be released for autocrine/paracrine receptor activation similar to that in the brain (Fig. 2, pathway 7). In zebrafish, heart and liver were shown to be the major Epo-producing organs (154). Although myocardial Epo expression may be induced by hypoxic exposure (155) the origin of endogenous Epo secreting cells in the mammalian heart is unknown.

Localization of Epo action depends on the route of its administration/secretion. When applied intravenously Epo interacts primarily with EpoR of endothelial cells of coronary vessels (Fig. 2, pathway 1) (83, 147). Thereby, cardioprotection of the plasma-borne Epo is mediated by factors secreted from the endothelium upon activation of endothelial EpoR (Fig. 2, pathways 2 and 3). Amongst these factors are endothelin-1 and NO (156). When applied directly to isolated cardiomyocytes, Epo was shown to promote mitogenesis of neonatal cardiomyocytes, affect Ca^{2+} handling in isolated cells causing an increase in the amplitude and reduction in duration of calcium transients, and protecting them from oxidative stress and doxorubicin-induced apoptosis (Fig. 2, pathways 6 and 7) (141, 157–159).

An exhaustive overview of the molecular mechanisms of cardioprotective effects of erythropoietin can be found in recent reviews (160–162). As mentioned above, the cardiac-specific receptor is most likely a heterodimer. The downstream elements of signaling cascades induced by activation of such a heteroreceptor remain largely unknown. Also current data on the molecular mechanisms of the cardioprotective action of Epo comes from observations of the downstream effects of Epo in the heart. This is characteristic of most of the studies performed to date in which observations fit into the pre-existing model of homodimer function in erythroid precursor cells (see Fig. 1). To what extent activation pathways for the homo- and heterodimer are similar remains unknown.

6.3.1 Acute Responses: PI3-Akt-eNOS Signaling

Several studies indicated that the action of Epo in the heart is associated with activation of PI3K-Akt pathway with subsequent up-regulation of NO production (83, 160, 163, 164). Endothelial NO synthase (eNOS) is localized in the caviolae of cardiomyocytes and is known to regulate the activity of L-type calcium channels by phosphorylation and S-nitrosylation. Upon eNOS activation and NO binding to soluble guanylyl cyclase, PKG-induced phosphorylation of contractile protein machinery is induced (165). These effects of Epo were confirmed for isolated cells as well as in vivo in hearts after intravenous Epo administration. In the latter case Akt and eNOS phosphorylation is restricted to the endothelial cells of coronary vessels (147). In cardiomyocytes the direct cytoprotective effect of Epo is mediated by its regulatory action on calcium handling and stabilization of the mitochondria. Epo induces activation of eNOS in cavioli by its phosphorylation at Ser 1177 by Akt. The generated NO then modulates activity of L-type Ca^{2+} channels via cGMP-sensitive phosphorylation and S-nitrosylation. Along with the Ca^{2+} release from the sarcoplasmic reticulum and SERCA2A the calcium pump is activated in response to stimulation of iNOS by Epo (166, 167). The exact molecular mechanisms of the action of Epo on calcium dynamics in the heart tissue are still unknown, however in myocardial stripes and in isolated cells (not on the vessels) they were tracked down to the PI3K-sensitive activation of PKCε (141). Stabilization of mitochondrial function in ischemic/injured myocardium by Epo is mediated by the activation of the mitochondrial KATP channels by Epo (166, 167). Furthermore, uncoupling of the mitochondrial electron transduction chain is reduced due to the interaction of iNOS-derived NO with the mitochondrial cytochromes. Mitochondrial biogenesis in cardiomyocytes is promoted by Epo which in turn induces enhancement of nuclear respiratory factor-1, PGC-1α (peroxisome proliferator-activated receptor ϒ coactivator 1α), and mitochondrial transcription factor-A gene expression in wild-type but not in $eNOS^{-/-}$ or $Akt1^{-/-}$ mice (168). Thus till now, most of the cardioprotective effects of Epo interaction with its receptor in cardiomyocytes seem to be mediated via PI3K-Akt-eNOS pathway (see Fig. 1).

Systemic induction of endogenous Epo production and release is known to occur in response to hypoxic stimulation. All the above mentioned responses of heart to Epo increase the survival probability during injury.

6.3.2 Chronic Responses: Changes in Gene Expression

Long-term activation of PI3K/Akt pathways in the heart induces activation of insulin-like growth factor binding protein-5 and downregulates peroxisome proliferator activated receptor-ϒ (PPAR-ϒ) coactivator-1 shifting metabolism from oxidative to aerobic glycolytic during long-term ischemia (169). Similar reprogramming of metabolism was observed in hypoxic heart and during pathological hypertrophic remodeling (170). Glucose delivery in cardiac myocytes is up-regulated accordingly as expression of Glut4 glucose transporter is induced along with metabolic reprogramming (171). Whether long term Epo treatment causes similar effects remains unclear. Epo binding to its receptors induces phosphorylation of Akt and eNOS - its effects are seen within 5 min (147) and can be observed for the first hour and thereafter the Epo-EpoR complex is internalized and degraded (Mihov, Tavakoli, Bogdanova unpublished observations). The internalization rate constant for Epo-EpoR complex in UT-7/Epo cells is 0.06 min^{-1} (172). Upon internalization 60% of Epo gets dissociated from the classical EpoR homodimer and recycled, whereas 40% undergoes degradation (172). This observation suggests the effect of Epo is transient with the amount of surface-based receptors decreasing upon interaction with the cytokine.

6.4 Epo in Treatment of Cardiovascular Diseases

Recent trials were performed in which very high doses of Epo were administered percutaneously in patients after they were diagnosed for myocardial infarction. The expected cardioprotective effects included pro-angiogenic, anti-inflammatory, anti-apoptotic, and anti-oxidative action of Epo which have been reported in animal models of myocardial infarction (173–175). However these trials showed no beneficial effects of Epo, and in several cases an increase in mortality and morbidity was observed due to an increased risk of thrombosis (176–179).

Possible reasons for the lack of Epo effect include the inadequate route of the cytokine administration (intravenous vs. intramyocardial vs. intraperitoneal vs. subcutaneous); lack of cofactors and ligands of NO synthases (L-arginine, tetrabiopterin, oxygen) (180–182) and a limited "window of cardioprotective effect," which was claimed to be wide, but has never been properly determined in the heart. Epo-induced activation of NOSes in their uncoupled mode, due to the shortage of substrates and cofactors, turns these enzymes from cardioprotective anti-oxidative ones to cardiotoxic and pro-oxidative (181, 183, 184). Ischemia-reperfusion of coronary vessels is associated with activation of arginase-1 in the endothelium and local reduction in arginine availability (185). Oxygen deprivation inhibits eNOS and nNOS since their affinity to this substrate is rather low (186).

As the outcome of the first Epo trials appeared to be so discouraging an alternative approach has been suggested to increase the cytokine efficacy. Cardioplegic solutions widely used in cardiac surgery to cause heart arrest are now designed to induce activation of endogenous Epo production in the arrested organ (187).

7 Pancreas

Epo deficiency and higher incidence of anemia in individuals with diabetes gave the first inkling of potential beneficial effects and therapeutic applications of Epo use in the diabetes setting. Several clinical studies reported a beneficial effect of recombinant Epo on glucose metabolism in patients undergoing hemodialysis. Epo treatment of patients with end-stage renal disease corrected lipid abnormalities and increased insulin sensitivity, with the duration of the treatment positively correlating with insulin sensitivity in these patients (188–190).

To date Epo expression by pancreatic cells has not been observed. However EpoR was expressed on islets of both human and non-human primates following Epo supplementation, or after transduction with an Adenoviral vector expressing high levels of Epo, affording protection of the islets from cytokine-induced destruction (49, 191). In addition, performance assessment of transduced islets transplanted into diabetic immunodeficient mice showed that overexpression of Epo conferred a functional advantage (191) and is also associated with a decrease in body weight (192). A number of in vitro and in vivo papers have now provided evidence that Epo is beneficial for β cell survival. In NIT-1 pancreatic cells, the PI3K inhibitor LY294002 abrogated the anti-apoptotic activity of Epo, indicating that activation of Akt was required for Epo-induced inhibition of cytokine-induced apoptosis (see Fig. 1) (193). In another study upregulation of Bcl-2, and concomitant downregulation of Bax and caspase 3, has also been suggested as a mechanism through which Epo can protect neonatal islet cells. In vivo diabetic rodent models also advocate direct effects of Epo on pancreatic β cells (see Fig. 2, pathway 6) promoting anti-apoptosis, proliferation, and angiogenesis signaling through its cognate receptor and downstream effector, JAK2, thus increasing β-cell mass (194). A very recent study administering a single dose of the novel Epo receptor agonist CNTO 530 to diet-induced obese mice resulted in improved glucose tolerance and insulin sensitivity at least in part from increased uptake of glucose by skeletal and cardiac muscle (195). The molecular mechanism(s) responsible for translating Epo receptor signaling into improved glucose tolerance are yet to be revealed and much more data is required to better understand its beneficial mechanism of action in general. However it is clear that Epo-induced pathways involving JAK2, Akt phosphorylation, and altered expression of several downstream apoptosis-related proteins, such as Bcl-2 and Bax as seen in other tissues, are likely to be a recurrent theme.

8 The Endothelium

Epo was shown to act on endothelial cells in vivo and in vitro having growth and chemotactic effects (40). In fact it has been suggested that many of the observed non-erythroid cytoprotective effects of Epo are mediated by second messengers released from endothelial cells (see Fig. 2) (196). The observation that development of the conditional non-hematopoietic EpoR knockout mouse is normal further supports this view. Equally important, Epo has been shown to facilitate vascular repair and thereby to improve blood supply to injured organs by acting on endothelial progenitor cells (EPCs; Fig. 2, pathway 5) (196). CD34+/Flk-1 (also known KDR or VEGFR2) positive cells are hematopoietic progenitor cells that may differentiate into endothelial cells and contribute to neovascularization and vascular repair (197, 198). Epo promotes proliferation (40, 196), inhibits apoptosis (199), and facilitates differentiation of EPCs (200–203). Furthermore, Epo induces mobilization of EPCs into the circulation (204, 205), and their homing (155, 206, 207). Increased eNOS expression and BH4 biosynthesis has been shown in Epo-treated EPCs and vascular cells (Fig. 2; pathway 4) (205, 208). Interestingly, recent studies on hypoxic endothelial cells have shown that VEGFR2 can also become an additional component for the EpoR/βCR complex that is essential for NO production (reviewed by ref. 75). Similar to other non-hematopoietic cells PI3K/Akt signaling cascades, induction of mitogen-activated protein kinase (MEK)/extracellular signal regulated kinase (ERK) signaling pathways (83, 147) and NO production are known to mediate Epo effects in endothelial cells in animal models and humans patients (see Figs. 1 and 2, pathway 1).

Thus indeed augmented endothelial function may play a major role in Epo-mediated protection in non-hematopoietic cells and underlie a significant amount of tissue recovery from injury. Certainly more research needs to be carried out regarding this possibility and the consequences for the future use of Epo as a treatment strategy.

9 Risks Associated with Epo Therapy

Although Epo is considered a clinically safe-to-use drug (due to its long term use by anemic patients), a number of worrying risks have been associated with its more general use as a therapeutic. The frequent use of Epo mimetics in patients with chronic kidney disease (CKD) has recently declined as randomized trials demonstrated increased incidence of cardiovascular complications and mortality without a marked benefit in quality of life (reviewed by ref. 209). Safety concerns were raised during treatment of anemia in diabetic patients with CKD when they showed a twofold higher risk of stroke, an increased risk of venous thromboembolism and cancer-related deaths (210). Several studies have suggested that exposure to high doses of Epo mimetics, when needed to achieve higher hemoglobin levels, is harmful and explains this phenomenon (211, 212). Very high doses of Epo, in conjunction with hypoxia, have also been associated with a paradoxical neurotoxic effect suggesting dose–response conditions need to be optimized. In the clinics there are also considerable concerns about potential thrombotic complications. Recent trials in which very high doses of Epo were administered to patients diagnosed with myocardial function showed an increased risk of thrombosis (176–179). Thrombotic events were also increased in critically ill patients although Epo therapy significantly reduced mortality particularly in trauma patients (213), and increased risk of venous thromboembolism was also noted in cancer patients (214). Another trial provided evidence of a possible negative interaction between short-term administration of Epo and aspirin due to its ability to modulate endothelial activation and platelet reactivity, von Willebrand factor antigen levels and factor VIII activity (215, 216). Although largely shown to improve neurodevelopmental outcome for preterm infants, Epo has been associated with a significant increase in the rate of retinopathy and may increase hypertension, coagulation, and even interfere with neuronal development in neonates (reviewed by ref. 84). Finally the therapeutic use of Epo in cancer patients remains highly controversial. A number of trials have shown that Epo treatment increases the risk for progressive disease and death although this may be dependent on the type and stage of the cancer (reviewed by ref. 217, 218). Potentially Epo could have a direct growth-promoting effect on cancer cells as they have been shown to express EpoR.

Thus it is apparent that our knowledge of the Epo signaling cascade needs to be significantly improved to be able to harness the benefits of using Epo and its mimetics as treatment for injury and disease. To a great extent its beneficial effects seem to be related to timing (the so-called "therapeutic window of opportunity"), dose and type of injury. A better understanding of these parameters would bring us significantly forward in our quest.

10 Conclusions and Outlook: What Don't We Know?

A wealth of preclinical data shows that the Epo signaling cascade is an important mediator of protection and cell survival in many different non-hematopoietic tissues as part of an innate response to injury. Many similarities exist between the mechanisms underlying its hematopoietic and non-hematopoietic functions but there are also some key differences that functionally lead to distinct outcomes. Not unexpectedly it was thought that Epo, a drug considered clinically safe, would be a trump card in most injury paradigms, however to date results from patient trials have been varied and more recently tip the balance to being negative. However, the pleiotropic and potentially beneficial biological effects of Epo signaling in non-hematopoietic tissues warrants in depth investigations of new therapeutic protocols. Clearly the generation of Epo mimetics such as asialo-Epo, CEPO, and others that are non-erythropoietic derivatives (75, 79, 149) will be instrumental in providing new options for treatment.

There are perhaps many things we do not yet know that need to be considered before being able to reliably use Epo and/or its derivatives as therapeutic drugs in different disease paradigms. For example what are the relative contributions of endogenous derived Epo and EpoR compared to exogenous recombinant Epo that is administered therapeutically? Do multiple tissue-specific Epo or EpoR isoforms exist? Is the endogenous balance between pro- and anti-apoptotic elements differentially altered by exogenous derivatives and how? What are the side effects of using low or high doses of Epo in terms of signaling pathways and negative outcomes? Can the Epo/EpoR axis be targeted clinically for therapeutic intervention in a cell or tissue-specific manner? What is the therapeutic window for treatment considering the receptor may not always be active? Is the route of administration critical to outcome? Can we prime the tissue before treatment or stimulate endogenous Epo production? And so on. The list is very long because we do not yet know enough about the non-hematopoietic mechanisms of Epo/EpoR in different tissues, or the short- and/or long-term effects of modulating the system

As more research is performed and new therapeutic applications for Epo are explored, careful consideration of potential adverse effects will need to be factored into the design of prospective clinical studies. Clearly to effectively harness the promise of Epo-an old but now pleiotropic growth factor-questions such as these need to be addressed now.

References

1. Fisher JW (2003) Erythropoietin: physiology and pharmacology update. Exp Biol Med (Maywood) 228:1–14
2. Wu H, Liu X, Jaenisch R, Lodish HF (1995) Generation of committed erythroid BFU-E and CFU-E progenitors does not require erythropoietin or the erythropoietin receptor. Cell 83:59–67
3. Lin CS, Lim SK, D'Agati V, Costantini F (1996) Differential effects of an erythropoietin receptor gene disruption on primitive and definitive erythropoiesis. Genes Dev 10:154–164
4. Masuda S, Okano M, Yamagishi K, Nagao M, Ueda M, Sasaki R (1994) A novel site of erythropoietin production. Oxygen-dependent production in cultured rat astrocytes. J Biol Chem 269:19488–19493
5. Marti HH, Wenger RH, Rivas LA, Straumann U, Digicaylioglu M, Henn V, Yonekawa Y, Bauer C, Gassmann M (1996) Erythropoietin gene expression in human, monkey and murine brain. Eur J Neurosci 8:666–676
6. Bernaudin M, Marti HH, Roussel S, Divoux D, Nouvelot A, MacKenzie ET, Petit E (1999) A potential role for erythropoietin in focal permanent cerebral ischemia in mice. J Cereb Blood Flow Metab 19:643–651
7. Li Y, Lu ZY, Ogle M, Wei L (2007) Erythropoietin prevents blood brain barrier damage induced by focal cerebral ischemia in mice. Neurochem Res 32:2132–2141
8. Korbel S, Bittorf T, Siegl E, Kollner B (2004) Recombinant human erythropoietin induces proliferation and Ca(2+)-influx in specific leukocyte subpopulations of rainbow trout (Oncorhynchus mykiss) blood and head kidney cells. J Comp Physiol B 174:121–128
9. Velly L, Pellegrini L, Guillet B, Bruder N, Pisano P (2010) Erythropoietin 2nd cerebral protection after acute injuries: a double-edged sword? Pharmacol Ther 128:445–459
10. Vogel J, Kiessling I, Heinicke K, Stallmach T, Ossent P, Vogel O, Aulmann M, Frietsch T, Schmid-Schonbein H, Kuschinsky W, Gassmann M (2003) Transgenic mice overexpressing erythropoietin adapt to excessive erythrocytosis by regulating blood viscosity. Blood 102:2278–2284

11. Grimm C, Wenzel A, Groszer M, Mayser H, Seeliger M, Samardzija M, Bauer C, Gassmann M, Reme CE (2002) HIF-1-induced erythropoietin in the hypoxic retina protects against light-induced retinal degeneration. Nat Med 8:718–724
12. Junk AK, Mammis A, Savitz SI, Singh M, Roth S, Malhotra S, Rosenbaum PS, Cerami A, Brines M, Rosenbaum DM (2002) Erythropoietin administration protects retinal neurons from acute ischemia-reperfusion injury. Proc Natl Acad Sci U S A 99:10659–10664
13. Ribatti D, Vacca A, Roccaro AM, Crivellato E, Presta M (2003) Erythropoietin as an angiogenic factor. Eur J Clin Invest 33:891–896
14. Soliz J, Joseph V, Soulage C, Becskei C, Vogel J, Pequignot JM, Ogunshola O, Gassmann M (2005) Erythropoietin regulates hypoxic ventilation in mice by interacting with brainstem and carotid bodies. J Physiol 568:559–571
15. Soliz J, Gassmann M, Joseph V (2007) Soluble erythropoietin receptor is present in the mouse brain and is required for the ventilatory acclimatization to hypoxia. J Physiol 583:329–336
16. Semenza GL, Nejfelt MK, Chi SM, Antonarakis SE (1991) Hypoxia-inducible nuclear factors bind to an enhancer element located 3′ to the human erythropoietin gene. Proc Natl Acad Sci U S A 88:5680–5684
17. Semenza GL, Wang GL (1992) A nuclear factor induced by hypoxia via de novo protein synthesis binds to the human erythropoietin gene enhancer at a site required for transcriptional activation. Mol Cell Biol 12:5447–5454
18. Ema M, Taya S, Yokotani N, Sogawa K, Matsuda Y, Fujii-Kuriyama Y (1997) A novel bHLH-PAS factor with close sequence similarity to hypoxia-inducible factor 1alpha regulates the VEGF expression and is potentially involved in lung and vascular development. Proc Natl Acad Sci U S A 94:4273–4278
19. Flamme I, Frohlich T, von Reutern M, Kappel A, Damert A, Risau W (1997) HRF, a putative basic helix-loop-helix-PAS-domain transcription factor is closely related to hypoxia-inducible factor-1 alpha and developmentally expressed in blood vessels. Mech Dev 63:51–60
20. Tian H, McKnight SL, Russell DW (1997) Endothelial PAS domain protein 1 (EPAS1), a transcription factor selectively expressed in endothelial cells. Genes Dev 11:72–82
21. Morita M, Ohneda O, Yamashita T, Takahashi S, Suzuki N, Nakajima O, Kawauchi S, Ema M, Shibahara S, Udono T, Tomita K, Tamai M, Sogawa K, Yamamoto M, Fujii-Kuriyama Y (2003) HLF/HIF-2alpha is a key factor in retinopathy of prematurity in association with erythropoietin. EMBO J 22:1134–1146
22. Warnecke C, Zaborowska Z, Kurreck J, Erdmann VA, Frei U, Wiesener M, Eckardt KU (2004) Differentiating the functional role of hypoxia-inducible factor (HIF)-1alpha and HIF-2alpha (EPAS-1) by the use of RNA interference: erythropoietin is a HIF-2alpha target gene in Hep3B and Kelly cells. FASEB J 18:1462–1464
23. Chavez JC, Baranova O, Lin J, Pichiule P (2006) The transcriptional activator hypoxia inducible factor 2 (HIF-2/EPAS-1) regulates the oxygen-dependent expression of erythropoietin in cortical astrocytes. J Neurosci 26:9471–9481
24. Kapitsinou PP, Liu Q, Unger TL, Rha J, Davidoff O, Keith B, Epstein JA, Moores SL, Erickson-Miller CL, Haase VH (2010) Hepatic HIF-2 regulates erythropoietic responses to hypoxia in renal anemia. Blood 116:3039–3048
25. Loboda A, Jozkowicz A, Dulak J (2010) HIF-1 and HIF-2 transcription factors–similar but not identical. Mol Cells 29:435–442
26. Hopfl G, Ogunshola O, Gassmann M (2003) Hypoxia and high altitude. The molecular response. Adv Exp Med Biol 543:89–115
27. Fandrey J, Gassmann M (2009) Oxygen sensing and the activation of the hypoxia inducible factor 1 (HIF-1)–invited article. Adv Exp Med Biol 648:197–206
28. Stiehl DP, Wirthner R, Koditz J, Spielmann P, Camenisch G, Wenger RH (2006) Increased prolyl 4-hydroxylase domain proteins compensate for decreased oxygen levels. Evidence for an autoregulatory oxygen-sensing system. J Biol Chem 281:23482–23491
29. Jaakkola P, Mole DR, Tian YM, Wilson MI, Gielbert J, Gaskell SJ, Kriegsheim A, Hebestreit HF, Mukherji M, Schofield CJ, Maxwell PH, Pugh CW, Ratcliffe PJ (2001) Targeting of HIF-alpha to the von Hippel-Lindau ubiquitylation complex by O2-regulated prolyl hydroxylation. Science 292:468–472
30. Webb JD, Coleman ML, Pugh CW (2009) Hypoxia, hypoxia-inducible factors (HIF), HIF hydroxylases and oxygen sensing. Cell Mol Life Sci 66:3539–3554
31. Haase VH (2009) The VHL tumor suppressor: master regulator of HIF. Curr Pharm Des 15:3895–3903
32. Lee-Huang S, Lin JJ, Kung HF, Huang PL, Lee L (1993) The human erythropoietin-encoding

gene contains a CAAT box, TATA boxes and other transcriptional regulatory elements in its 5′ flanking region. Gene 128:227–236

33. Blanchard KL, Acquaviva AM, Galson DL, Bunn HF (1992) Hypoxic induction of the human erythropoietin gene: cooperation between the promoter and enhancer, each of which contains steroid receptor response elements. Mol Cell Biol 12:5373–5385
34. Imagawa S, Yamamoto M, Miura Y (1997) Negative regulation of the erythropoietin gene expression by the GATA transcription factors. Blood 89:1430–1439
35. La Ferla K, Reimann C, Jelkmann W, Hellwig-Burgel T (2002) Inhibition of erythropoietin gene expression signaling involves the transcription factors GATA-2 and NF-kappaB. FASEB J 16:1811–1813
36. Grasso G, Sfacteria A, Cerami A, Brines M (2004) Erythropoietin as a tissue-protective cytokine in brain injury: what do we know and where do we go? Neuroscientist 10:93–98
37. Marti HH (2004) Erythropoietin and the hypoxic brain. J Exp Biol 207:3233–3242
38. D'Andrea AD, Lodish HF, Wong GG (1989) Expression cloning of the murine erythropoietin receptor. Cell 57:277–285
39. Jones SS, D'Andrea AD, Haines LL, Wong GG (1990) Human erythropoietin receptor: cloning, expression, and biologic characterization. Blood 76:31–35
40. Anagnostou A, Liu Z, Steiner M, Chin K, Lee ES, Kessimian N, Noguchi CT (1994) Erythropoietin receptor mRNA expression in human endothelial cells. Proc Natl Acad Sci U S A 91:3974–3978
41. Ogilvie M, Yu X, Nicolas-Metral V, Pulido SM, Liu C, Ruegg UT, Noguchi CT (2000) Erythropoietin stimulates proliferation and interferes with differentiation of myoblasts. J Biol Chem 275:39754–39761
42. Shingo T, Sorokan ST, Shimazaki T, Weiss S (2001) Erythropoietin regulates the in vitro and in vivo production of neuronal progenitors by mammalian forebrain neural stem cells. J Neurosci 21:9733–9743
43. Juul SE, Anderson DK, Li Y, Christensen RD (1998) Erythropoietin and erythropoietin receptor in the developing human central nervous system. Pediatr Res 43:40–49
44. Bernaudin M, Bellail A, Marti HH, Yvon A, Vivien D, Duchatelle I, Mackenzie ET, Petit E (2000) Neurons and astrocytes express EPO mRNA: oxygen-sensing mechanisms that involve the redox-state of the brain. Glia 30:271–278
45. Liu C, Shen K, Liu Z, Noguchi CT (1997) Regulated human erythropoietin receptor expression in mouse brain. J Biol Chem 272:32395–32400
46. Digicaylioglu M, Bichet S, Marti HH, Wenger RH, Rivas LA, Bauer C, Gassmann M (1995) Localization of specific erythropoietin binding sites in defined areas of the mouse brain. Proc Natl Acad Sci U S A 92:3717–3720
47. Ruifrok WP, de Boer RA, Westenbrink BD, van Veldhuisen DJ, van Gilst WH (2008) Erythropoietin in cardiac disease: new features of an old drug. Eur J Pharmacol 585:270–277
48. Brines M, Cerami A (2006) Discovering erythropoietin's extra-hematopoietic functions: biology and clinical promise. Kidney Int 70:246–250
49. Fenjves ES, Ochoa MS, Cabrera O, Mendez AJ, Kenyon NS, Inverardi L, Ricordi C (2003) Human, nonhuman primate, and rat pancreatic islets express erythropoietin receptors. Transplantation 75:1356–1360
50. Yasuda Y, Masuda S, Chikuma M, Inoue K, Nagao M, Sasaki R (1998) Estrogen-dependent production of erythropoietin in uterus and its implication in uterine angiogenesis. J Biol Chem 273:25381–25387
51. Constantinescu SN, Keren T, Socolovsky M, Nam H, Henis YI, Lodish HF (2001) Ligand-independent oligomerization of cell-surface erythropoietin receptor is mediated by the transmembrane domain. Proc Natl Acad Sci U S A 98:4379–4384
52. Witthuhn BA, Quelle FW, Silvennoinen O, Yi T, Tang B, Miura O, Ihle JN (1993) JAK2 associates with the erythropoietin receptor and is tyrosine phosphorylated and activated following stimulation with erythropoietin. Cell 74:227–236
53. Miura O, Nakamura N, Quelle FW, Witthuhn BA, Ihle JN, Aoki N (1994) Erythropoietin induces association of the JAK2 protein tyrosine kinase with the erythropoietin receptor in vivo. Blood 84:1501–1507
54. Quelle FW, Wang D, Nosaka T, Thierfelder WE, Stravopodis D, Weinstein Y, Ihle JN (1996) Erythropoietin induces activation of Stat5 through association with specific tyrosines on the receptor that are not required for a mitogenic response. Mol Cell Biol 16:1622–1631
55. Zhao W, Kitidis C, Fleming MD, Lodish HF, Ghaffari S (2006) Erythropoietin stimulates phosphorylation and activation of GATA-1 via the PI3-kinase/AKT signaling pathway. Blood 107:907–915
56. Jelkmann W, Bohlius J, Hallek M, Sytkowski AJ (2008) The erythropoietin receptor in

normal and cancer tissues. Crit Rev Oncol Hematol 67:39–61

57. Socolovsky M, Nam H, Fleming MD, Haase VH, Brugnara C, Lodish HF (2001) Ineffective erythropoiesis in Stat5a(−/−)5b(−/−) mice due to decreased survival of early erythroblasts. Blood 98:3261–3273

58. Bao H, Jacobs-Helber SM, Lawson AE, Penta K, Wickrema A, Sawyer ST (1999) Protein kinase B (c-Akt), phosphatidylinositol 3-kinase, and STAT5 are activated by erythropoietin (EPO) in HCD57 erythroid cells but are constitutively active in an EPO-independent, apoptosis-resistant subclone (HCD57-SREI cells). Blood 93:3757–3773

59. Simon MC, Pevny L, Wiles MV, Keller G, Costantini F, Orkin SH (1992) Rescue of erythroid development in gene targeted GATA-1- mouse embryonic stem cells. Nat Genet 1:92–98

60. Buck I, Morceau F, Cristofanon S, Heintz C, Chateauvieux S, Reuter S, Dicato M, Diederich M (2008) Tumor necrosis factor alpha inhibits erythroid differentiation in human erythropoietin-dependent cells involving p38 MAPK pathway, GATA-1 and FOG-1 downregulation and GATA-2 upregulation. Biochem Pharmacol 76:1229–1239

61. Ohneda K, Yamamoto M (2002) Roles of hematopoietic transcription factors GATA-1 and GATA-2 in the development of red blood cell lineage. Acta Haematol 108:237–245

62. Chong ZZ, Kang JQ, Maiese K (2003) Apaf-1, Bcl-xL, cytochrome c, and caspase-9 form the critical elements for cerebral vascular protection by erythropoietin. J Cereb Blood Flow Metab 23:320–330

63. Szenajch J, Wcislo G, Jeong JY, Szczylik C, Feldman L (2010) The role of erythropoietin and its receptor in growth, survival and therapeutic response of human tumor cells From clinic to bench—a critical review. Biochim Biophys Acta 1806:82–95

64. Mahmud DL, G-Amlak M, Deb DK, Platanias LC, Uddin S, Wickrema A (2002) Phosphorylation of forkhead transcription factors by erythropoietin and stem cell factor prevents acetylation and their interaction with coactivator p300 in erythroid progenitor cells. Oncogene 21:1556–1562

65. Chong ZZ, Maiese K (2007) Erythropoietin involves the phosphatidylinositol 3-kinase pathway, 14-3-3 protein and FOXO3a nuclear trafficking to preserve endothelial cell integrity. Br J Pharmacol 150:839–850

66. Nagata Y, Todokoro K (1999) Requirement of activation of JNK and p38 for environmental stress-induced erythroid differentiation and apoptosis and of inhibition of ERK for apoptosis. Blood 94:853–863

67. Nagata Y, Kiefer F, Watanabe T, Todokoro K (1999) Activation of hematopoietic progenitor kinase-1 by erythropoietin. Blood 93:3347–3354

68. Kolbus A, Pilat S, Husak Z, Deiner EM, Stengl G, Beug H, Baccarini M (2002) Raf-1 antagonizes erythroid differentiation by restraining caspase activation. J Exp Med 196:1347–1353

69. Uddin S, Ah-Kang J, Ulaszek J, Mahmud D, Wickrema A (2004) Differentiation stage-specific activation of p38 mitogen-activated protein kinase isoforms in primary human erythroid cells. Proc Natl Acad Sci U S A 101:147–152

70. Jacobs-Helber SM, Ryan JJ, Sawyer ST (2000) JNK and p38 are activated by erythropoietin (EPO) but are not induced in apoptosis following EPO withdrawal in EPO-dependent HCD57 cells. Blood 96: 933–940

71. Nagao M, Masuda S, Abe S, Ueda M, Sasaki R (1992) Production and ligand-binding characteristics of the soluble form of murine erythropoietin receptor. Biochem Biophys Res Commun 188:888–897

72. Fujita M, Takahashi R, Liang P, Saya H, Ashoori F, Tachi M, Kitazawa S, Maeda S (1997) Role of alternative splicing of the rat erythropoietin receptor gene in normal and erythroleukemia cells. Leukemia 11(Suppl 3):444–445

73. Chikuma M, Masuda S, Kobayashi T, Nagao M, Sasaki R (2000) Tissue-specific regulation of erythropoietin production in the murine kidney, brain, and uterus. Am J Physiol Endocrinol Metab 279:E1242–E1248

74. Masuda S, Nagao M, Takahata K, Konishi Y, Gallyas F Jr, Tabira T, Sasaki R (1993) Functional erythropoietin receptor of the cells with neural characteristics. Comparison with receptor properties of erythroid cells. J Biol Chem 268:11208–11216

75. Brines M, Cerami A (2011) The receptor that tames the innate immune response. Mol Med 18:486–496

76. Brines M, Cerami A (2005) Emerging biological roles for erythropoietin in the nervous system. Nat Rev Neurosci 6:484–494

77. Sinclair AM, Coxon A, McCaffery I, Kaufman S, Paweletz K, Liu L, Busse L, Swift S, Elliott S, Begley CG (2010) Functional erythropoietin receptor is undetectable in endothelial, cardiac, neuronal, and renal cells. Blood 115:4264–4272

78. Ghezzi P, Bernaudin M, Bianchi R, Blomgren K, Brines M, Campana W, Cavaletti G, Cerami A, Chopp M, Coleman T, Digicaylioglu M, Ehrenreich H, Erbayraktar S, Erbayraktar Z, Gassmann M, Genc S, Gokmen N, Grasso G, Juul S, Lipton SA, Hand CC, Latini R, Lauria G, Leist M, Newton SS, Petit E, Probert L, Sfacteria A, Siren AL, Talan M, Thiemermann C, Westenbrink D, Yaqoob M, Zhu C (2010) Erythropoietin: not just about erythropoiesis. Lancet 375:2142
79. Brines M, Grasso G, Fiordaliso F, Sfacteria A, Ghezzi P, Fratelli M, Latini R, Xie QW, Smart J, Su-Rick CJ, Pobre E, Diaz D, Gomez D, Hand C, Coleman T, Cerami A (2004) Erythropoietin mediates tissue protection through an erythropoietin and common beta-subunit heteroreceptor. Proc Natl Acad Sci U S A 101:14907–14912
80. Swartjes M, Morariu A, Niesters M, Brines M, Cerami A, Aarts L, Dahan A (2011) ARA290, a peptide derived from the tertiary structure of erythropoietin, produces long-term relief of neuropathic pain: an experimental study in rats and beta-common receptor knockout mice. Anesthesiology 115:1084–1092
81. Su KH, Shyue SK, Kou YR, Ching LC, Chiang AN, Yu YB, Chen CY, Pan CC, Lee TS (2011) beta Common receptor integrates the erythropoietin signaling in activation of endothelial nitric oxide synthase. J Cell Physiol 226:3330–3339
82. Teng R, Gavrilova O, Suzuki N, Chanturiya T, Schimel D, Hugendubler L, Mammen S, Yver DR, Cushman SW, Mueller E, Yamamoto M, Hsu LL, Noguchi CT (2011) Disrupted erythropoietin signalling promotes obesity and alters hypothalamus proopiomelanocortin production. Nat Commun 2:520
83. Teng R, Calvert JW, Sibmooh N, Piknova B, Suzuki N, Sun J, Martinez K, Yamamoto M, Schechter AN, Lefer DJ, Noguchi CT (2011) Acute erythropoietin cardioprotection is mediated by endothelial response. Basic Res Cardiol 106:343–354
84. Xiong Y, Mahmood A, Qu C, Kazmi H, Zhang ZG, Noguchi CT, Schallert T, Chopp M (2010) Erythropoietin improves histological and functional outcomes after traumatic brain injury in mice in the absence of the neural erythropoietin receptor. J Neurotrauma 27:205–215
85. Tsai PT, Ohab JJ, Kertesz N, Groszer M, Matter C, Gao J, Liu X, Wu H, Carmichael ST (2006) A critical role of erythropoietin receptor in neurogenesis and post-stroke recovery. J Neurosci 26:1269–1274
86. Nagai A, Nakagawa E, Choi HB, Hatori K, Kobayashi S, Kim SU (2001) Erythropoietin and erythropoietin receptors in human CNS neurons, astrocytes, microglia, and oligodendrocytes grown in culture. J Neuropathol Exp Neurol 60:386–392
87. Brines ML, Ghezzi P, Keenan S, Agnello D, de Lanerolle NC, Cerami C, Itri LM, Cerami A (2000) Erythropoietin crosses the blood–brain barrier to protect against experimental brain injury. Proc Natl Acad Sci U S A 97:10526–10531
88. Ruscher K, Freyer D, Karsch M, Isaev N, Megow D, Sawitzki B, Priller J, Dirnagl U, Meisel A (2002) Erythropoietin is a paracrine mediator of ischemic tolerance in the brain: evidence from an in vitro model. J Neurosci 22:10291–10301
89. Chong ZZ, Kang JQ, Maiese K (2002) Hematopoietic factor erythropoietin fosters neuroprotection through novel signal transduction cascades. J Cereb Blood Flow Metab 22:503–514
90. Liu R, Suzuki A, Guo Z, Mizuno Y, Urabe T (2006) Intrinsic and extrinsic erythropoietin enhances neuroprotection against ischemia and reperfusion injury in vitro. J Neurochem 96:1101–1110
91. Lewczuk P, Hasselblatt M, Kamrowski-Kruck H, Heyer A, Unzicker C, Siren AL, Ehrenreich H (2000) Survival of hippocampal neurons in culture upon hypoxia: effect of erythropoietin. Neuroreport 11:3485–3488
92. Koshimura K, Murakami Y, Sohmiya M, Tanaka J, Kato Y (1999) Effects of erythropoietin on neuronal activity. J Neurochem 72:2565–2572
93. Morishita E, Narita H, Nishida M, Kawashima N, Yamagishi K, Masuda S, Nagao M, Hatta H, Sasaki R (1996) Anti-erythropoietin receptor monoclonal antibody: epitope mapping, quantification of the soluble receptor, and detection of the solubilized transmembrane receptor and the receptor-expressing cells. Blood 88:465–471
94. Shang Y, Wu Y, Yao S, Wang X, Feng D, Yang W (2007) Protective effect of erythropoietin against ketamine-induced apoptosis in cultured rat cortical neurons: involvement of PI3K/Akt and GSK-3 beta pathway. Apoptosis 12:2187–2195
95. Yoo JY, Won YJ, Lee JH, Kim JU, Sung IY, Hwang SJ, Kim MJ, Hong HN (2009) Neuroprotective effects of erythropoietin posttreatment against kainate-induced excitotoxicity in mixed spinal cultures. J Neurosci Res 87:150–163
96. Um M, Gross AW, Lodish HF (2007) A "classical" homodimeric erythropoietin receptor is essential for the antiapoptotic effects of erythropoietin on differentiated neuroblastoma

SH-SY5Y and pheochromocytoma PC-12 cells. Cell Signal 19:634–645

97. Zhang L, Chopp M, Zhang RL, Wang L, Zhang J, Wang Y, Toh Y, Santra M, Lu M, Zhang ZG (2010) Erythropoietin amplifies stroke-induced oligodendrogenesis in the rat. PLoS One 5:e11016

98. Cho YK, Kim G, Park S, Sim JH, Won YJ, Hwang CH, Yoo JY, Hong HN (2012) Erythropoietin promotes oligodendrogenesis and myelin repair following lysolecithin-induced injury in spinal cord slice culture. Biochem Biophys Res Commun 417:753–759

99. Genc K, Genc S, Baskin H, Semin I (2006) Erythropoietin decreases cytotoxicity and nitric oxide formation induced by inflammatory stimuli in rat oligodendrocytes. Physiol Res 55:33–38

100. Digicaylioglu M, Lipton SA (2001) Erythropoietin-mediated neuroprotection involves cross-talk between Jak2 and NF-kappaB signalling cascades. Nature 412:641–647

101. Marrero MB, Venema RC, Ma H, Ling BN, Eaton DC (1998) Erythropoietin receptor-operated Ca2+ channels: activation by phospholipase C-gamma 1. Kidney Int 53:1259–1268

102. Kawakami M, Iwasaki S, Sato K, Takahashi M (2000) Erythropoietin inhibits calcium-induced neurotransmitter release from clonal neuronal cells. Biochem Biophys Res Commun 279:293–297

103. Souvenir R, Fathali N, Ostrowski RP, Lekic T, Zhang JH, Tang J (2011) Tissue inhibitor of matrix metalloproteinase-1 mediates erythropoietin-induced neuroprotection in hypoxia ischemia. Neurobiol Dis 44:28–37

104. Chen ZY, Wang L, Asavaritkrai P, Noguchi CT (2010) Up-regulation of erythropoietin receptor by nitric oxide mediates hypoxia preconditioning. J Neurosci Res 88:3180–3188

105. Keswani SC, Bosch-Marce M, Reed N, Fischer A, Semenza GL, Hoke A (2011) Nitric oxide prevents axonal degeneration by inducing HIF-1-dependent expression of erythropoietin. Proc Natl Acad Sci U S A 108:4986–4990

106. Belayev L, Saul I, Busto R, Danielyan K, Vigdorchik A, Khoutorova L, Ginsberg MD (2005) Albumin treatment reduces neurological deficit and protects blood–brain barrier integrity after acute intracortical hematoma in the rat. Stroke 36:326–331

107. Siren AL, Knerlich F, Poser W, Gleiter CH, Bruck W, Ehrenreich H (2001) Erythropoietin and erythropoietin receptor in human ischemic/hypoxic brain. Acta Neuropathol 101:271–276

108. Villa P, Bigini P, Mennini T, Agnello D, Laragione T, Cagnotto A, Viviani B, Marinovich M, Cerami A, Coleman TR, Brines M, Ghezzi P (2003) Erythropoietin selectively attenuates cytokine production and inflammation in cerebral ischemia by targeting neuronal apoptosis. J Exp Med 198:971–975

109. Ehrenreich H, Bartels C, Sargin D, Stawicki S, Krampe H (2008) Recombinant human erythropoietin in the treatment of human brain disease: focus on cognition. J Ren Nutr 18:146–153

110. Wang CH, Liang CL, Huang LT, Liu JK, Hung PH, Sun A, Hung KS (2004) Single intravenous injection of naked plasmid DNA encoding erythropoietin provides neuroprotection in hypoxia-ischemia rats. Biochem Biophys Res Commun 314:1064–1071

111. Malhotra S, Savitz SI, Ocava L, Rosenbaum DM (2006) Ischemic preconditioning is mediated by erythropoietin through PI-3 kinase signaling in an animal model of transient ischemic attack. J Neurosci Res 83:19–27

112. Prass K, Scharff A, Ruscher K, Lowl D, Muselmann C, Victorov I, Kapinya K, Dirnagl U, Meisel A (2003) Hypoxia-induced stroke tolerance in the mouse is mediated by erythropoietin. Stroke 34:1981–1986

113. Iwai M, Stetler RA, Xing J, Hu X, Gao Y, Zhang W, Chen J, Cao G (2010) Enhanced oligodendrogenesis and recovery of neurological function by erythropoietin after neonatal hypoxic/ischemic brain injury. Stroke 41:1032–1037

114. McPherson RJ, Juul SE (2010) Erythropoietin for infants with hypoxic-ischemic encephalopathy. Curr Opin Pediatr 22:139–145

115. Kumral A, Tuzun F, Oner MG, Genc S, Duman N, Ozkan H (2011) Erythropoietin in neonatal brain protection: the past, the present and the future. Brain Dev 33:632–643

116. Mammis A, McIntosh TK, Maniker AH (2009) Erythropoietin as a neuroprotective agent in traumatic brain injury review. Surg Neurol 71:527–531, discussion 531

117. Agnello D, Bigini P, Villa P, Mennini T, Cerami A, Brines ML, Ghezzi P (2002) Erythropoietin exerts an anti-inflammatory effect on the CNS in a model of experimental autoimmune encephalomyelitis. Brain Res 952:128–134

118. Zhang J, Li Y, Cui Y, Chen J, Lu M, Elias SB, Chopp M (2005) Erythropoietin treatment improves neurological functional recovery in EAE mice. Brain Res 1034:34–39

119. Celik M, Gokmen N, Erbayraktar S, Akhisaroglu M, Konakc S, Ulukus C, Genc S,

Genc K, Sagiroglu E, Cerami A, Brines M (2002) Erythropoietin prevents motor neuron apoptosis and neurologic disability in experimental spinal cord ischemic injury. Proc Natl Acad Sci U S A 99:2258–2263

120. Gorio A, Gokmen N, Erbayraktar S, Yilmaz O, Madaschi L, Cichetti C, Di Giulio AM, Vardar E, Cerami A, Brines M (2002) Recombinant human erythropoietin counteracts secondary injury and markedly enhances neurological recovery from experimental spinal cord trauma. Proc Natl Acad Sci U S A 99:9450–9455

121. Kilic E, Kilic U, Soliz J, Bassetti CL, Gassmann M, Hermann DM (2005) Brain-derived erythropoietin protects from focal cerebral ischemia by dual activation of ERK-1/-2 and Akt pathways. FASEB J 19:2026–2028

122. Sola A, Wen TC, Hamrick SE, Ferriero DM (2005) Potential for protection and repair following injury to the developing brain: a role for erythropoietin? Pediatr Res 57: 110R–117R

123. Viviani B, Bartesaghi S, Corsini E, Villa P, Ghezzi P, Garau A, Galli CL, Marinovich M (2005) Erythropoietin protects primary hippocampal neurons increasing the expression of brain-derived neurotrophic factor. J Neurochem 93:412–421

124. Weishaupt JH, Rohde G, Polking E, Siren AL, Ehrenreich H, Bahr M (2004) Effect of erythropoietin axotomy-induced apoptosis in rat retinal ganglion cells. Invest Ophthalmol Vis Sci 45:1514–1522

125. Shen J, Wu Y, Xu JY, Zhang J, Sinclair SH, Yanoff M, Xu G, Li W, Xu GT (2010) ERK- and Akt-dependent neuroprotection by erythropoietin (EPO) against glyoxal-AGEs via modulation of Bcl-xL, Bax, and BAD. Invest Ophthalmol Vis Sci 51:35–46

126. Grimm C, Wenzel A, Stanescu D, Samardzija M, Hotop S, Groszer M, Naash M, Gassmann M, Reme C (2004) Constitutive overexpression of human erythropoietin protects the mouse retina against induced but not inherited retinal degeneration. J Neurosci 24:5651–5658

127. Konishi Y, Chui DH, Hirose H, Kunishita T, Tabira T (1993) Trophic effect of erythropoietin and other hematopoietic factors on central cholinergic neurons in vitro and in vivo. Brain Res 609:29–35

128. Bocker-Meffert S, Rosenstiel P, Rohl C, Warneke N, Held-Feindt J, Sievers J, Lucius R (2002) Erythropoietin and VEGF promote neural outgrowth from retinal explants in postnatal rats. Invest Ophthalmol Vis Sci 43:2021–2026

129. Kretz A, Happold CJ, Marticke JK, Isenmann S (2005) Erythropoietin promotes regeneration of adult CNS neurons via Jak2/Stat3 and PI3K/AKT pathway activation. Mol Cell Neurosci 29:569–579

130. Ehrenreich H, Hasselblatt M, Dembowski C, Cepek L, Lewczuk P, Stiefel M, Rustenbeck HH, Breiter N, Jacob S, Knerlich F, Bohn M, Poser W, Ruther E, Kochen M, Gefeller O, Gleiter C, Wessel TC, De Ryck M, Itri L, Prange H, Cerami A, Brines M, Siren AL (2002) Erythropoietin therapy for acute stroke is both safe and beneficial. Mol Med 8:495–505

131. Ehrenreich H, Weissenborn K, Prange H, Schneider D, Weimar C, Wartenberg K, Schellinger PD, Bohn M, Becker H, Wegrzyn M, Jahnig P, Herrmann M, Knauth M, Bahr M, Heide W, Wagner A, Schwab S, Reichmann H, Schwendemann G, Dengler R, Kastrup A, Bartels C (2009) Recombinant human erythropoietin in the treatment of acute ischemic stroke. Stroke 40:e647–e656

132. Wustenberg T, Begemann M, Bartels C, Gefeller O, Stawicki S, Hinze-Selch D, Mohr A, Falkai P, Aldenhoff JB, Knauth M, Nave KA, Ehrenreich H (2011) Recombinant human erythropoietin delays loss of gray matter in chronic schizophrenia. Mol Psychiatry 16(26–36):21

133. Miskowiak K, O'Sullivan U, Harmer CJ (2007) Erythropoietin reduces neural and cognitive processing of fear in human models of antidepressant drug action. Biol Psychiatry 62:1244–1250

134. Ehrenreich H, Fischer B, Norra C, Schellenberger F, Stender N, Stiefel M, Siren AL, Paulus W, Nave KA, Gold R, Bartels C (2007) Exploring recombinant human erythropoietin in chronic progressive multiple sclerosis. Brain 130:2577–2588

135. Tseng MY, Hutchinson PJ, Richards HK, Czosnyka M, Pickard JD, Erber WN, Brown S, Kirkpatrick PJ (2009) Acute systemic erythropoietin therapy to reduce delayed ischemic deficits following aneurysmal subarachnoid hemorrhage: a Phase II randomized, double-blind, placebo-controlled trial. Clinical article. J Neurosurg 111:171–180

136. Nirula R, Diaz-Arrastia R, Brasel K, Weigelt JA, Waxman K (2010) Safety and efficacy of erythropoietin in traumatic brain injury patients: a pilot randomized trial. Crit Care Res Pract pii:209848

137. Asaumi Y, Kagaya Y, Takeda M, Yamaguchi N, Tada H, Ito K, Ohta J, Shiroto T, Shirato K, Minegishi N, Shimokawa H (2007) Protective role of endogenous erythropoietin system in nonhematopoietic cells against pressure overload-induced left ventricular dysfunction in mice. Circulation 115:2022–2032

138. Wu H, Lee SH, Gao J, Liu X, Iruela-Arispe ML (1999) Inactivation of erythropoietin leads to defects in cardiac morphogenesis. Development 126:3597–3605
139. Suzuki N, Ohneda O, Takahashi S, Higuchi M, Mukai HY, Nakahata T, Imagawa S, Yamamoto M (2002) Erythroid-specific expression of the erythropoietin receptor rescued its null mutant mice from lethality. Blood 100:2279–2288
140. Stuckmann I, Evans S, Lassar AB (2003) Erythropoietin and retinoic acid, secreted from the epicardium, are required for cardiac myocyte proliferation. Dev Biol 255:334–349
141. Hefer D, Yi T, Selby DE, Fishbaugher DE, Tremble SM, Begin KJ, Gogo P, Lewinter MM, Meyer M, Palmer BM, Vanburen P (2012) Erythropoietin induces positive inotropic and lusitropic effects in murine and human myocardium. J Mol Cell Cardiol 52:256–263
142. Kaygisiz Z, Erkasap N, Yazihan N, Sayar K, Ataoglu H, Uyar R, Ikizler M (2006) Erythropoietin changes contractility, cAMP, and nitrite levels of isolated rat hearts. J Physiol Sci 56:247–251
143. Hanlon PR, Fu P, Wright GL, Steenbergen C, Arcasoy MO, Murphy E (2005) Mechanisms of erythropoietin-mediated cardioprotection during ischemia-reperfusion injury: role of protein kinase C and phosphatidylinositol 3-kinase signaling. FASEB J 19:1323–1325
144. Wright GL, Hanlon P, Amin K, Steenbergen C, Murphy E, Arcasoy MO (2004) Erythropoietin receptor expression in adult rat cardiomyocytes is associated with an acute cardioprotective effect for recombinant erythropoietin during ischemia-reperfusion injury. FASEB J 18:1031–1033
145. Salisch S, Klar M, Thurisch B, Bungert J, Dame C (2011) Gata4 and Sp1 regulate expression of the erythropoietin receptor in cardiomyocytes. J Cell Mol Med 15:1963–1972
146. Depping R, Kawakami K, Ocker H, Wagner JM, Heringlake M, Noetzold A, Sievers HH, Wagner KF (2005) Expression of the erythropoietin receptor in human heart. J Thorac Cardiovasc Surg 130:877–878
147. Mihov D, Bogdanov N, Grenacher B, Gassmann M, Zund G, Bogdanova A, Tavakoli R (2009) Erythropoietin protects from reperfusion-induced myocardial injury by enhancing coronary endothelial nitric oxide production. Eur J Cardiothorac Surg 35:839–846, discussion 846
148. Parsa CJ, Kim J, Riel RU, Pascal LS, Thompson RB, Petrofski JA, Matsumoto A, Stamler JS, Koch WJ (2004) Cardioprotective effects of erythropoietin in the reperfused ischemic heart: a potential role for cardiac fibroblasts. J Biol Chem 279:20655–20662
149. Moon C, Krawczyk M, Paik D, Coleman T, Brines M, Juhaszova M, Sollott SJ, Lakatta EG, Talan MI (2006) Erythropoietin, modified to not stimulate red blood cell production, retains its cardioprotective properties. J Pharmacol Exp Ther 316:999–1005
150. Ahmet I, Tae HJ, Juhaszova M, Riordon DR, Boheler KR, Sollott SJ, Brines M, Cerami A, Lakatta EG, Talan MI (2011) A small nonerythropoietic helix B surface peptide based upon erythropoietin structure is cardioprotective against ischemic myocardial damage. Mol Med 17:194–200
151. Chin K, Oda N, Shen K, Noguchi CT (1995) Regulation of transcription of the human erythropoietin receptor gene by proteins binding to GATA-1 and Sp1 motifs. Nucleic Acids Res 23:3041–3049
152. Kirschner KM, Hagen P, Hussels CS, Ballmaier M, Scholz H, Dame C (2008) The Wilms' tumor suppressor Wt1 activates transcription of the erythropoietin receptor in hematopoietic progenitor cells. FASEB J 22:2690–2701
153. Zhou B, Ma Q, Rajagopal S, Wu SM, Domian I, Rivera-Feliciano J, Jiang D, von Gise A, Ikeda S, Chien KR, Pu WT (2008) Epicardial progenitors contribute to the cardiomyocyte lineage in the developing heart. Nature 454:109–113
154. Chu CY, Cheng CH, Chen GD, Chen YC, Hung CC, Huang KY, Huang CJ (2007) The zebrafish erythropoietin: functional identification and biochemical characterization. FEBS Lett 581:4265–4271
155. Lin JS, Chen YS, Chiang HS, Ma MC (2008) Hypoxic preconditioning protects rat hearts against ischaemia-reperfusion injury: role of erythropoietin on progenitor cell mobilization. J Physiol 586:5757–5769
156. Piuhola J, Kerkela R, Keenan JI, Hampton MB, Richards AM, Pemberton CJ (2008) Direct cardiac actions of erythropoietin (EPO): effects on cardiac contractility, BNP secretion and ischaemia/reperfusion injury. Clin Sci (Lond) 114:293–304
157. Li Y, Takemura G, Okada H, Miyata S, Maruyama R, Li L, Higuchi M, Minatoguchi S, Fujiwara T, Fujiwara H (2006) Reduction of inflammatory cytokine expression and oxidative damage by erythropoietin in chronic heart failure. Cardiovasc Res 71:684–694
158. Fu P, Arcasoy MO (2007) Erythropoietin protects cardiac myocytes against anthracycline-induced apoptosis. Biochem Biophys Res Commun 354:372–378

159. Burger D, Lei M, Geoghegan-Morphet N, Lu X, Xenocostas A, Feng Q (2006) Erythropoietin protects cardiomyocytes from apoptosis via up-regulation of endothelial nitric oxide synthase. Cardiovasc Res 72:51–59

160. Fliser D, Bahlmann FH, deGroot K, Haller H (2006) Mechanisms of disease: erythropoietin—an old hormone with a new mission? Nat Clin Pract Cardiovasc Med 3:563–572

161. Vogiatzi G, Briasoulis A, Tousoulis D, Papageorgiou N, Stefanadis C (2010) Is there a role for erythropoietin in cardiovascular disease? Expert Opin Biol Ther 10:251–264

162. Burger D, Xenocostas A, Feng QP (2009) Molecular basis of cardioprotection by erythropoietin. Curr Mol Pharmacol 2:56–69

163. Mihov D, Vogel J, Gassmann M, Bogdanova A (2009) Erythropoietin activates nitric oxide synthase in murine erythrocytes. Am J Physiol Cell Physiol 297:C378–C388

164. Li H, Wallerath T, Munzel T, Forstermann U (2002) Regulation of endothelial-type NO synthase expression in pathophysiology and in response to drugs. Nitric Oxide 7:149–164

165. Ziolo MT, Kohr MJ, Wang H (2008) Nitric oxide signaling and the regulation of myocardial function. J Mol Cell Cardiol 45:625–632

166. Joyeux-Faure M, Ramond A, Beguin PC, Belaidi E, Godin-Ribuot D, Ribuot C (2006) Early pharmacological preconditioning by erythropoietin mediated by inducible NOS and mitochondrial ATP-dependent potassium channels in the rat heart. Fundam Clin Pharmacol 20:51–56

167. Baker JE, Kozik D, Hsu AK, Fu X, Tweddell JS, Gross GJ (2007) Darbepoetin alfa protects the rat heart against infarction: dose–response, phase of action, and mechanisms. J Cardiovasc Pharmacol 49:337–345

168. Carraway MS, Suliman HB, Jones WS, Chen CW, Babiker A, Piantadosi CA (2010) Erythropoietin activates mitochondrial biogenesis and couples red cell mass to mitochondrial mass in the heart. Circ Res 106:1722–1730

169. Cook SA, Matsui T, Li L, Rosenzweig A (2002) Transcriptional effects of chronic Akt activation in the heart. J Biol Chem 277:22528–22533

170. Krishnan J, Suter M, Windak R, Krebs T, Felley A, Montessuit C, Tokarska-Schlattner M, Aasum E, Bogdanova A, Perriard E, Perriard JC, Larsen T, Pedrazzini T, Krek W (2009) Activation of a HIF1alpha-PPARgamma axis underlies the integration of glycolytic and lipid anabolic pathways in pathologic cardiac hypertrophy. Cell Metab 9:512–524

171. Matsui T, Tao J, del Monte F, Lee KH, Li L, Picard M, Force TL, Franke TF, Hajjar RJ, Rosenzweig A (2001) Akt activation preserves cardiac function and prevents injury after transient cardiac ischemia in vivo. Circulation 104:330–335

172. Gross AW, Lodish HF (2006) Cellular trafficking and degradation of erythropoietin and novel erythropoiesis stimulating protein (NESP). J Biol Chem 281:2024–2032

173. Moon C, Krawczyk M, Paik D, Lakatta EG, Talan MI (2005) Cardioprotection by recombinant human erythropoietin following acute experimental myocardial infarction: dose response and therapeutic window. Cardiovasc Drugs Ther 19:243–250

174. Jelkmann W, Wagner K (2004) Beneficial and ominous aspects of the pleiotropic action of erythropoietin. Ann Hematol 83:673–686

175. Andreotti F, Agati L, Conti E, Santucci E, Rio T, Tarantino F, Natale L, Berardi D, Mattatelli A, Musumeci B, Bonomo L, Volpe M, Crea F, Autore C (2009) Update on phase II studies of erythropoietin in acute myocardial infarction. Rationale and design of Exogenous erythroPoietin in Acute Myocardial Infarction: New Outlook aNd Dose Association Study (EPAMINONDAS). J Thromb Thrombolysis 28:489–495

176. Kang HJ, Kim HS (2008) G-CSF- and erythropoietin-based cell therapy: a promising strategy for angiomyogenesis in myocardial infarction. Expert Rev Cardiovasc Ther 6:703–713

177. McMurray JJ, Uno H, Jarolim P, Desai AS, de Zeeuw D, Eckardt KU, Ivanovich P, Levey AS, Lewis EF, McGill JB, Parfrey P, Parving HH, Toto RM, Solomon SD, Pfeffer MA (2011) Predictors of fatal and nonfatal cardiovascular events in patients with type 2 diabetes mellitus, chronic kidney disease, and anemia: an analysis of the Trial to Reduce cardiovascular Events with Aranesp (darbepoetin-alfa) Therapy (TREAT). Am Heart J 162(748–755):e743

178. Joyeux-Faure M, Durand M, Bedague D, Protar D, Incagnoli P, Paris A, Ribuot C, Levy P, Chavanon O (2011) Evaluation of the effect of one large dose of erythropoietin against cardiac and cerebral ischemic injury occurring during cardiac surgery with cardiopulmonary bypass: a randomized double-blind placebo-controlled pilot study. Fundam Clin Pharmacol 26:761–770

179. van der Meer P, van Veldhuisen DJ (2011) Acute coronary syndromes: the unfulfilled promise of erythropoietin in patients with MI. Nat Rev Cardiol 8:425–426

180. Moens AL, Kietadisorn R, Lin JY, Kass D (2011) Targeting endothelial and myocardial dysfunction with tetrahydrobiopterin. J Mol Cell Cardiol 51:559–563

181. Forstermann U, Sessa WC (2011) Nitric oxide synthases: regulation and function. Eur Heart J 33:829–837

182. Karbach S, Simon A, Slenzka A, Jaenecke I, Habermeier A, Martine U, Forstermann U, Closs EI (2011) Relative contribution of different L-arginine sources to the substrate supply of endothelial nitric oxide synthase. J Mol Cell Cardiol 51:855–861

183. Pacher P, Beckman JS, Liaudet L (2007) Nitric oxide and peroxynitrite in health and disease. Physiol Rev 87:315–424

184. Otani H (2009) The role of nitric oxide in myocardial repair and remodeling. Antioxid Redox Signal 11:1913–1928

185. Hein TW, Zhang C, Wang W, Chang CI, Thengchaisri N, Kuo L (2003) Ischemia-reperfusion selectively impairs nitric oxide-mediated dilation in coronary arterioles: counteracting role of arginase. FASEB J 17:2328–2330

186. Dweik RA (2005) Nitric oxide, hypoxia, and superoxide: the good, the bad, and the ugly! Thorax 60:265–267

187. Ryou MG, Flaherty DC, Hoxha B, Sun J, Gurji H, Rodriguez S, Bell G, Olivencia-Yurvati AH, Mallet RT (2009) Pyruvate-fortified cardioplegia evokes myocardial erythropoietin signaling in swine undergoing cardiopulmonary bypass. Am J Physiol Heart Circ Physiol 297:H1914–H1922

188. Allegra V, Mengozzi G, Martimbianco L, Vasile A (1996) Early and late effects of erythropoietin on glucose metabolism in maintenance hemodialysis patients. Am J Nephrol 16:304–308

189. Mak RH (1996) Correction of anemia by erythropoietin reverses insulin resistance and hyperinsulinemia in uremia. Am J Physiol 270:F839–F844

190. Tuzcu A, Bahceci M, Yilmaz E, Bahceci S, Tuzcu S (2004) The comparison of insulin sensitivity in non-diabetic hemodialysis patients treated with and without recombinant human erythropoietin. Horm Metab Res 36:716–720

191. Fenjves ES, Ochoa MS, Gay-Rabinstein C, Molano RD, Pileggi A, Mendez AJ, Inverardi L, Ricordi C (2004) Adenoviral gene transfer of erythropoietin confers cytoprotection to isolated pancreatic islets. Transplantation 77:13–18

192. Katz O, Stuible M, Golishevski N, Lifshitz L, Tremblay ML, Gassmann M, Mittelman M, Neumann D (2010) Erythropoietin treatment leads to reduced blood glucose levels and body mass: insights from murine models. J Endocrinol 205:87–95

193. Shuai H, Zhang J, Xie J, Zhang M, Yu Y, Zhang L (2011) Erythropoietin protects pancreatic beta-cell line NIT-1 cells against cytokine-induced apoptosis via phosphatidylinositol 3-kinase/Akt signaling. Endocr Res 36:25–34

194. Choi D, Schroer SA, Lu SY, Wang L, Wu X, Liu Y, Zhang Y, Gaisano HY, Wagner KU, Wu H, Retnakaran R, Woo M (2010) Erythropoietin protects against diabetes through direct effects on pancreatic beta cells. J Exp Med 207:2831–2842

195. Scully MS, Ort TA, James IE, Bugelski PJ, Makropoulos DA, Deutsch HA, Pieterman EJ, van den Hoek AM, Havekes LM, Dubell WH, Wertheimer JD, Picha KM (2011) A novel EPO receptor agonist improves glucose tolerance via glucose uptake in skeletal muscle in a mouse model of diabetes. Exp Diabetes Res 2011:910159

196. Noguchi CT, Wang L, Rogers HM, Teng R, Jia Y (2008) Survival and proliferative roles of erythropoietin beyond the erythroid lineage. Expert Rev Mol Med 10:e36

197. Masuda H, Asahara T (2003) Post-natal endothelial progenitor cells for neovascularization in tissue regeneration. Cardiovasc Res 58:390–398

198. Urbich C, Dimmeler S (2004) Endothelial progenitor cells: characterization and role in vascular biology. Circ Res 95:343–353

199. Carlini RG, Alonzo EJ, Dominguez J, Blanca I, Weisinger JR, Rothstein M, Bellorin-Font E (1999) Effect of recombinant human erythropoietin on endothelial cell apoptosis. Kidney Int 55:546–553

200. Carlini RG, Reyes AA, Rothstein M (1995) Recombinant human erythropoietin stimulates angiogenesis in vitro. Kidney Int 47:740–745

201. Ribatti D, Presta M, Vacca A, Ria R, Giuliani R, Dell'Era P, Nico B, Roncali L, Dammacco F (1999) Human erythropoietin induces a pro-angiogenic phenotype in cultured endothelial cells and stimulates neovascularization in vivo. Blood 93:2627–2636

202. Bahlmann FH, DeGroot K, Duckert T, Niemczyk E, Bahlmann E, Boehm SM, Haller H, Fliser D (2003) Endothelial progenitor cell proliferation and differentiation is regulated by erythropoietin. Kidney Int 64:1648–1652

203. Bahlmann FH, De Groot K, Spandau JM, Landry AL, Hertel B, Duckert T, Boehm SM,

Menne J, Haller H, Fliser D (2004) Erythropoietin regulates endothelial progenitor cells. Blood 103:921–926

204. Heeschen C, Aicher A, Lehmann R, Fichtlscherer S, Vasa M, Urbich C, Mildner-Rihm C, Martin H, Zeiher AM, Dimmeler S (2003) Erythropoietin is a potent physiologic stimulus for endothelial progenitor cell mobilization. Blood 102:1340–1346

205. Santhanam AV, d'Uscio LV, Peterson TE, Katusic ZS (2008) Activation of endothelial nitric oxide synthase is critical for erythropoietin-induced mobilization of progenitor cells. Peptides 29:1451–1455

206. Westenbrink BD, Lipsic E, van der Meer P, van der Harst P, Oeseburg H, Du Marchie Sarvaas GJ, Koster J, Voors AA, van Veldhuisen DJ, van Gilst WH, Schoemaker RG (2007) Erythropoietin improves cardiac function through endothelial progenitor cell and vascular endothelial growth factor mediated neovascularization. Eur Heart J 28:2018–2027

207. Klopsch C, Furlani D, Gabel R, Li W, Pittermann E, Ugurlucan M, Kundt G, Zingler C, Titze U, Wang W, Ong LL, Wagner K, Li RK, Ma N, Steinhoff G (2009) Intracardiac injection of erythropoietin induces stem cell recruitment and improves cardiac functions in a rat myocardial infarction model. J Cell Mol Med 13:664–679

208. d'Uscio LV, Katusic ZS (2008) Erythropoietin increases endothelial biosynthesis of tetrahydrobiopterin by activation of protein kinase B alpha/Akt1. Hypertension 52:93–99

209. Singh AK (2011) Is there a deleterious effect of erythropoietin in end-stage renal disease? Kidney Int 80:569–571

210. Pfeffer MA, Burdmann EA, Chen CY, Cooper ME, de Zeeuw D, Eckardt KU, Feyzi JM, Ivanovich P, Kewalramani R, Levey AS, Lewis EF, McGill JB, McMurray JJ, Parfrey P, Parving HH, Remuzzi G, Singh AK, Solomon SD, Toto R (2009) A trial of darbepoetin alfa in type 2 diabetes and chronic kidney disease. N Engl J Med 361:2019–2032

211. Zhang Y, Thamer M, Kaufman JS, Cotter DJ, Hernan MA (2011) High doses of epoetin do not lower mortality and cardiovascular risk among elderly hemodialysis patients with diabetes. Kidney Int 80:663–669

212. Weiner DE, Miskulin DC, Seefeld K, Ladik V, Zager PG, Singh AK, Johnson HK, Meyer KB (2007) Reducing versus discontinuing erythropoietin at high hemoglobin levels. J Am Soc Nephrol 18:3184–3191

213. Corwin HL, Gettinger A, Fabian TC, May A, Pearl RG, Heard S, An R, Bowers PJ, Burton P, Klausner MA, Corwin MJ (2007) Efficacy and safety of epoetin alfa in critically ill patients. N Engl J Med 357:965–976

214. Bohlius J, Wilson J, Seidenfeld J, Piper M, Schwarzer G, Sandercock J, Trelle S, Weingart O, Bayliss S, Brunskill S, Djulbegovic B, Benett CL, Langensiepen S, Hyde C, Engert E (2006) Erythropoietin or darbepoetin for patients with cancer. Cochrane Database Syst Rev. 3, CD003407

215. Tang YD, Rinder HM, Katz SD (2007) Effects of recombinant human erythropoietin on antiplatelet action of aspirin and clopidogrel in healthy subjects: results of a double-blind, placebo-controlled randomized trial. Am Heart J 154(494):e491–e497

216. Stohlawetz PJ, Dzirlo L, Hergovich N, Lackner E, Mensik C, Eichler HG, Kabrna E, Geissler K, Jilma B (2000) Effects of erythropoietin on platelet reactivity and thrombopoiesis in humans. Blood 95:2983–2989

217. Held TK, Gundert-Remy U (2010) Pharmacodynamic effects of haematopoietic cytokines: the view of a clinical oncologist. Basic Clin Pharmacol Toxicol 106:210–214

218. Arcasoy MO (2008) Erythropoiesis-stimulating agent use in cancer: preclinical and clinical perspectives. Clin Cancer Res 14:4685–4690

Chapter 3

Tissue-Protective Cytokines: Structure and Evolution

Pietro Ghezzi and Darrell Conklin

Abstract

Cytokines are important mediators of host defense and immunity, and were first identified for their role in immunity to infections. It was then found that some of them are pathogenic mediators in inflammatory diseases and much of the emphasis is now on pro-inflammatory cytokines, also in consideration of the fact that TNF inhibitors became effective drugs in chronic inflammatory diseases. The recent studies on the tissue-protective activities of erythropoietin (EPO) led to the term "tissue-protective cytokine." We discuss here how tissue-protective actions might be common to other cytokines, particularly those of the 4-alpha helical structural superfamily.

Key words Helical cytokines, Hematopoietic cytokines, 3D structure, Evolution, Host defense

1 Tissue-Protective Cytokines in the Context of Host Resistance

Surviving an infection is achieved in two ways: resistance (reducing the number of pathogens, such as by innate or adaptive immunity) and tolerance (surviving the damage produced by the pathogen), as outlined in Fig. 1. Read et al. (1) recently suggested that "damage control may be more important than pathogen control," and the underlying mechanisms can probably be implicated in tissue damage non-necessarily associated with infections (2).

Initially, cytokines were identified in an effort to characterize the effectors of "pathogen control," with a focus on the antiviral and anticancer activity of the immune system (3). The idea was that these effector cytokines could be used as drugs. This led to the discovery of interferons, interleukins 1 and 2, and TNF. However the real impact in terms of "translational medicine" came from the discovery of the pro-inflammatory actions of TNF (4) and IL-1 (5), with the finding that cytokines produced by the immune system in response to infection can also cause inflammation and tissue damage. The enormous impact of this paradigm-shifting concept is demonstrated by the fact that TNF inhibitors (anti-TNF antibodies and soluble

Pietro Ghezzi and Anthony Cerami (eds.), *Tissue-Protective Cytokines: Methods and Protocols*, Methods in Molecular Biology, vol. 982, DOI 10.1007/978-1-62703-308-4_3, © Springer Science+Business Media, LLC 2013

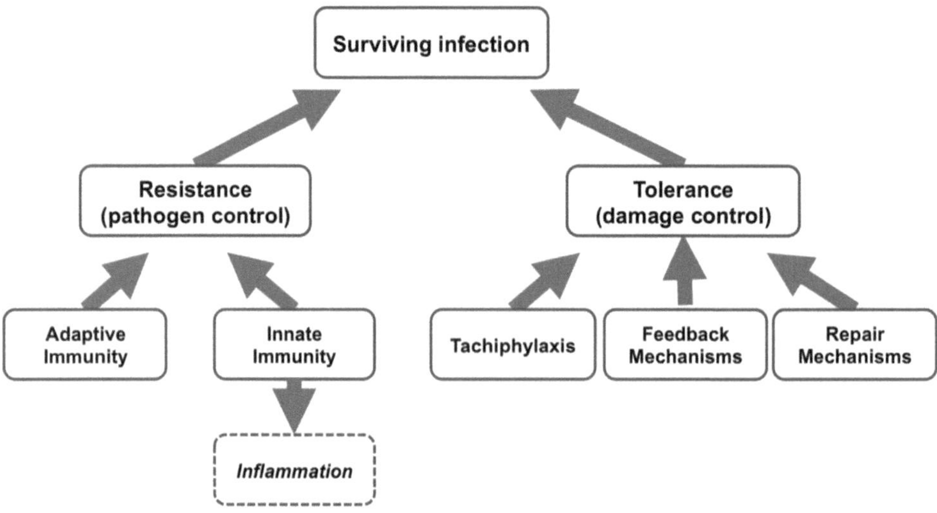

Fig. 1 Surviving infections is achieved either by killing pathogens, or controlling their growth, through adaptive or innate immunity. TNF, IFNg, IL-1 have a key role in innate immunity but also cause tissue damage. Tolerance, the ability to control damage induced by pathogens though tachiphylaxis or various protective mechanisms, is also essential. Tolerance also protects/repair from tissue damage caused by innate immunity/inflammation. Tissue-protective cytokines are important as a means of regulating inflammation and promoting repair. These protective cytokines and mechanisms are probably not only important in infections but also when tissue injury is induced, for instance, by ischemia or trauma

TNF receptors) are now top-selling biologicals in the therapy of chronic inflammatory diseases. These findings also led to the expansion of the list of pro-inflammatory cytokines, which now include, among many others, IL-6, IL-18, and IL-33. These cytokines are essential mediators in host defense against pathogens, and the role of IL-1 or TNF in innate immunity and of IL-2 or IL-6 in adaptive immunity are described in any textbook of immunology.

Looking at the right side of Fig. 1, other cytokines act, among other things, as negative regulators of inflammatory cytokines, to control excessive injury due to inflammatory cells. For instance, IL-10 can be considered as part of the feedback mechanism to regulate TNF production.

The finding that EPO has tissue-protective activities, originally identified by Brines et al. in models of ischemic or traumatic brain injury (6), was the most recent addition to the picture of the biological role of cytokines, and provided a first example of how a cytokine can be a mediator of "damage control." Probably, EPO does so not only by taming inflammation (7) but also promoting repair by various mechanisms that include angiogenesis, neurogenesis, and plasticity (8).

Induction of EPO expression by hypoxia and ischemia could well be viewed in this perspective. Transcription of EPO is regulated by HIF-1 while several cytokines implicated in inflammation and innate immunity are regulated through NF-kB. NF-kB and

HIF are often viewed as oppositely regulated (for instance, HIF is activated by hypoxia and antioxidants, NF-kB is activated by hyperoxia and inhibited by antioxidants (9, 10)). On the other hand, it has been shown that inflammatory/immune cytokines can activate HIF (11) and it is becoming more and more clear that hypoxia and inflammation are linked (12, 13). It should also be noted that HIF is activated by a number of infectious pathogens (14, 15). While the relationships between HIF-induced cytokines (mainly EPO and VEGF) in the context of inflammation are the subject of many studies, it is not always clear how the tissue-protective activity might not be exclusive of EPO but is shared by other cytokines.

The broad spectrum of tissue-protective activities of EPO, demonstrated now by over 500 papers published in the past 10 years, also prompted a number of clinical trials.

Like EPO, G-CSF is also being studied in clinical trials for its tissue- or neuro-protective effect in a variety of indications (Table 1).

The protective effects of EPO and G-CSF have led us to reevaluate the literature on tissue-protective actions of other cytokines. Some of these works, many focusing on neuroprotection, are summarized in Table 2. Although this table may not be comprehensive, it can be seen that the largest body of evidence, apart from EPO that is not listed here, is on G-CSF, GM-CSF, IL-6, and other cytokines of the IL-6 family (CNTF, LIF, CT-1, IL-11). All these belong to the family of hematopoietic cytokines (16, 17). In the case of IL-6, there is some interest in the development of a superagonist (hyper-IL-6) which is a fusion protein containing IL-6 and IL-6 receptor alpha that show a broad range of neuroprotective and reparative activities (18–21), and has been studied for its pro-myelinating properties (21). Interestingly, other proteins of this structural family have protective activities. Growth hormone (GH) is protective in various injury models (22–30). Also, Flt3 ligand (Flt3L), which has common structure with hematopoietic cytokines (31) improves recovery after spinal cord injury in rats (32) and has neurotrophic activity in vitro (33).

Some of the studies listed above have investigated the mechanism of protective activity. To be precise, in the case of G-CSF and GM-CSF, the rationale for their use in clinical trials has been mainly to promote repair through mobilization of stem cells. This is also an effect that may be important for the broad tissue-protective action of EPO (34). It should be noted that most of the studies listed here provide evidence for a direct protective effect, not mediated by immunomodulation as it could happen with some cytokines, for instance IL-13 (35). Of note, even IL-10, originally tested in models of cerebral ischemia on the ground of its anti-inflammatory activity mediated by inhibition of TNF synthesis, was found to be directly neuroprotective in vitro (36, 37). The same holds true for EPO that, while inhibiting the induction of inflammatory cytokines in experimental stroke (7), also protects directly primary neurons

Table 1
Major clinical trials based on tissue-protective actions of cytokines

Cytokine	Indication	References
EPO	Stroke	(55–58) NCT00362414, NCT00697164, NCT00478517, NCT00938314 (with hCG), NCT00362414 (with hCG)
EPO	Forearm I/R	NCT00691613
EPO	TBI	NCT00987454, NCT00260052, NCT00260052, NCT00987454, NCT00375869
EPO	AMI	NCT00390832 (with G-CSF), NCT00449488, NCT00390832, NCT00524901, NCT00648089
EPO	SAH	(59), NCT00140010, NCT00140010
EPO	ALS	(60)
CEPO[a]	Stroke	NCT00756249, NCT00870844
HBP[b] (ARA290)	Limb ischemia	EudraCT-2010-018584-41
HBP (ARA290)	RA	EudraCT-2010-023469-22
HBP (ARA290)	Neuropathic pain	EudraCT-2010-021518-45
G-CSF	Liver failure	NCT01341951
G-CSF	(with Sitagliptin) AMI	NCT00650143
G-CSF	Coronary artery disease	NCT00886509
G-CSF	Stroke recovery	NCT00809549
G-CSF	AMI	NCT00394498
G-CSF	Parkinson's disease	NCT01227681
G-CSF	ALS	(61)
G-CSF + EPO	(+EPO) Liver disease	NCT01384565
G-CSF + EPO	Acute-on-chronic liver failure	NCT01383460

[a]Carbamylated EPO
[b]Helix B surface peptide derived from EPO (62)
Sources: Pubmed; www.clinicaltrials.gov; www.clinicaltrialsregister.eu

in vitro (38, 39). Studies with knockout mice also showed that deficiency in some cytokines worsened injury in animal models of diseases (Table 3).

Cytokines can be classified in different families depending on the receptors used, overall functions and 3D structure. Of the cytokines we listed in Table 2, most of them (EPO, G- and GM-CSF, IL-3, IL-6, IL-11, LIF, CT1, CNTF) belong to the

Table 2
Tissue-protective, neuro-protective, and trophic activities of cytokines in preclinical models unrelated to hematopoiesis

Cytokine	Cell target	End-point	References
CNTF	Striatal neurons	Excitotoxicity	(63)
CT-1	PC12 cells	Apoptosis	(64)
CT-1	Rats, mice	Liver ischemia/reperfusion	(65)
Flt3L	Rats and primary neurons	Spinal cord injury	(32, 33)
G-CSF	Mice	Skeletal muscle regeneration	(66)
G-CSF	Stroke and peripheral limb ischemia	Tissue plasticity	(67)
G-CSF	Mice	Myocardial infarction	(68)
G-CSF		Cardiomyocyte proliferation	(69)
G-CSF	Rat	Muscle proliferation and strength following skeletal muscle injury	(70)
G-CSF	Rats	Optic nerve crush, neuroprotection	(71)
G-CSF	Mouse, rat	Stroke	(72, 73)
GH	Hypoxic–ischemic injury	Neuroprotection	(74)
IFN-b	Adult mouse neural progenitor cells	Decreased apoptosis upon growth factor withdrawal	(75)
IL-10	Primary neurons	Excitotoxicity	(36)
IL-3	Septal cholinergic cell line SN6	Neurotrophic	(76)
IL-3	Neurons	Amyloid toxicity	(77)
IL-3	Gerbil	Cerebral ischemia (CA1 neurons)	(78)
IL-3	Spinal motor neuron in rats	Sciatic nerve axotomy-induced death	(79)
IL-3, GM-CSF	Murine sympathetic neurons	Neurotrophic	(80)
IL-3, GM-CSF	Rats	Brain lesions	(81)
IL-3, GM-CSF, G-CSF	Central cholinergic neurons	Neurotrophic	(82)
IL-6	Mice	LPS-induced TNF production	(83)
IL-6	Mice	T cell-mediated liver injury	(84)
IL-6	Diabetic rats	Neurovascular function, nerve perfusion, and vascular endothelium	(85)

(continued)

Table 2 (continued)

Cytokine	Cell target	End-point	References
IL-6	Mice, rats	Liver ischemia	(86)
IL-6		Proliferation of human muscle satellite cells following acute muscle damage	(87)
IL-6	Mice	Ethanol-induced hepatic steatosis	(88)
IL-6	Oligodendrocyte	Differentiation	(89)
IL-6	Rat	Diabetic neuropathy	(90)
IL-6	Peripheral nerves	Neurotrophic	(91)
IL-6	Astrocytes	Oxidative stress	(92)
IL-6	Mice	Pulmonary hyperoxic toxicity	(93)
IL-6	Organotypic brain culture	Regeneration following lesion	(94)
IL-6	Cerebellar granule neurons	NMDA toxicity	(95)
IL-6	Mice and vitro	Hyperoxia-induced mitochondrial damage	(96)
IL-6	Mice	Regeneration of axotomized hypoglossal nerve	(97)
IL-6	Mice	Stroke MCAO	(98, 99)
IL-6	Mice	CCl4 hepatotoxicity	(100)
IL-6	Mice	Trimethyltin neurotoxicity	(101)
IL-6	Mice	Myelination	(102)
IL-11	Mice	Acetaminophen hepatotoxicity	(103)
LIF	Wobbler mice motor neuron	Neuro-protection	(104)
SCF	Tubular epithelial cells, neurons in vitro	Antiapoptotic	(105, 106)
SCF	Mice	Acetaminophen acute liver injury	(107)

structural group of the 4-alpha-helical cytokines (40), which bind receptors of the hematopoietic receptor family.

In the case of the IL-6 family of cytokines, all using gp130 as the signal transduction subunit, it has been proposed to define them as neuropoietic cytokines (41) and, in the paper reporting CNTF cloning, the authors concluded, from expression patterns, that its neurotrophic function may be exerted only under pathological conditions (42).

Table 3
Studies with knockout mice showing a protective role of cytokines

Cytokine	Cell target	End-point	Reference
IL-6	Neuron	Trimethyltin neurotoxicity	(101)
IL-6	Brain	Cortical freeze lesion	(108)
IL-6		Wound healing	(109)
GM-CSF	Ko in non-hematopoietic cells	Increased NSAID ileal injury	(110)
GM-CSF		Impaired skin wound healing	(111)

2 4-Alpha-Helical Cytokines

Helical cytokines form a structurally conserved family with all members containing a structurally conserved four helices, by convention labeled A through D, in a four-helix bundle. As with all soluble helix bundle proteins, the helices are amphipathic, having a solvent-facing hydrophilic side and a buried hydrophobic side. The distinguishing factor of what may be called the helical cytokine fold is the particular topological layout of helices; they are always in an up–up–down–down orientation, unlike any other known helix bundle fold.

In addition to the distinctiveness of their helix bundle fold topology, further evidences for the common evolutionary origins of helical cytokines is in their gene structures, which in cases conserve the number of exons and intron phase patterns. Despite containing a common structural core and having evolutionary origins, the helical cytokines exhibit a diverse variety of features (40), including the type of receptor they bind (43), their number of exons and intron phase patterns (44), whether they are short or long chain (the long chain cytokines having longer helices than the short chain), whether they bind their receptor through monomers, homodimers, or heterodimers, and a further topoisomeric feature where the loop between helices B and C threads over or through the loop between helices A and B. Furthermore they can be grouped into four categories based on the structure of their preprotein sequence (see (40) for details): the presence (groups 1, 3, 4) or lack (group 2) of a signal peptide, the presence of C-terminal hydrophilic sequence that is released by extracellular proteases (group 3) or alternatively a transmembrane domain C-terminal of the helix bundle (group 4).

Thus the discovery or verification of new members of this family is difficult due to the lack of sequence homology between members of different homology groups. The standard BLAST method for homology identification lacks the sensitivity to recognize the homology between a significant fraction of pairs of family members. Secondary structure prediction methods may reveal predominately helical structure but with little specificity for particular topologies. Methods to predict amphipathic helical regions from sequence (45) are useful but again lack the necessary specificity for the helical cytokine fold, being easily attracted to long coiled-coil structural regions. These difficulties motivated the creation of a new approach specifically tailored to the helical cytokines (40) which achieves high sensitivity at an acceptable level of specificity. The basic operation of this method is to "thread" a sequence through a core of amphipathic helical profiles, its specificity arising from the cumulative contribution of each helix and the compliance of the loop region lengths to permissible ranges. The method was used to directly discover or support the inference of several human helical cytokines from EST and genomic sequence (40); particularly IL-19, IL-20, IL-21, IL-27, IL-29 (Fox et al. (46) describe in more detail the computational identification of the interferon lambda family), and IL-31. A list of the known human helical cytokines is given in Table 4.

3 4-Alpha-Helical Cytokines in Invertebrates

Studies on invertebrates gained high respectability in the field of innate immunity as a lot has been learned in the *Drosophila* on key molecules in this pathway, such as the Toll-like receptors and the transcription factor NF-kB. Many cytokines have been described in invertebrates (47–50). Though helical cytokines have been reported in all vertebrate genomes, in invertebrate genomes orthologous vertebrate cytokines have not been identified. Nevertheless, intriguing evidence for their existence arises from several sources, including: orthologous receptors, which presumably would bind a ligand of the helical cytokine structure, structural prediction and data mining of transcriptomes, and immunoreactivity to vertebrate cytokine antibodies. In *Drosophila* the DOME receptor belongs mediates phosphorylation of a JAK orthologue through a ligand known as UPD, which has no homology to a known helical cytokine and does not demonstrate the features of a helical cytokine in its sequence (51, 52). Data mining of Drosophila transcripts revealed a protein known as HF (helical factor) that has an amphipathic helical structure and topology consistent with the helical cytokine fold (52) though this can only be conclusively known through structure determination. In the invertebrate chordate *Ciona intestinalis* data mining of putative coding exons revealed two cytokine receptors (53), though their ligands have not been identified.

Table 4
The human helical cytokines

Gene symbol	Group	Description
BSF3	1	Cardiotrophin-like cytokine NNT-1
CNTF	2	Ciliary neurotrophic factor
CSF1	4	Macrophage colony stimulating factor
CSF2	1	Granulocyte-macrophage colony-stimulating factor
CSF3	1	Granulocyte colony-stimulating factor
CTF1	2	Cardiotrophin-1
EPO	1	Erythropoietin
FLT3LG	1,4	FLT3 ligand
GH1	1	Somatotropin
IFNA1	1	Interferon alpha-1
IFNB1	1	Interferon beta
IFNG	1	Interferon gamma
IFNK	1	Interferon kappa
IFNT1	1	Interferon tau-1
IFNW1	1	Interferon omega-1
IL2	1	Interleukin-2
IL3	1	Interleukin-3
IL4	1	Interleukin-4
IL5	1	Interleukin-5
IL6	1	Interleukin-6
IL7	1	Interleukin-7
IL9	1	Interleukin-9
IL10	1	Interleukin-10
IL11	1	Interleukin-11
IL12A	1	Interleukin-12 alpha chain
IL13	1	Interleukin-13
IL15	1	Interleukin-15
IL19	1	Interleukin-19
IL20	1	Interleukin-20
IL21	1	Interleukin-21

(continued)

Table 4 (continued)

Gene symbol	Group	Description
IL22	1	Interleukin-22
IL23	1	Interleukin-23
IL24	1	Interleukin-24
IL26	1	Interleukin-26
IL27	1	Interleukin-27
IL28A/B	1	Interleukin-28A/B (interferon lambda-2/3)
IL29	1	Interleukin-29 (interferon lambda-1)
IL31	1	Interleukin-31
IL34	1	Interleukin-34
KITLG	4	Stem cell factor
LEP	1	Leptin
LIF	1	Leukemia inhibitory factor
OSM	3	Oncostatin M
PRL	1	Prolactin
THPO	3	Thrombopoietin

Cytokines are classified into groups 1 through 4, with their official gene name. Omitted are various close homologs and isoforms; the interferon alpha homologs; and the somatotropin homologs GH2, CSH1, CSH2, and CSHL1

4 Conclusions

It is tempting to hypothesize that the tissue-protective activities of many cytokines in this structural family preceded their specific functions developed when blood cells and immune system evolved. A recent study has shown that a protein immunoreactive with anti-human EPO antibodies is present in the grasshopper and leech (though not in Drosophila), and that recombinant human EPO is neuroprotective in insects (54). Further studies in model organisms will help identifying the evolutionary relationship between the various protective roles of cytokines.

Acknowledgments

Supported by the European Regional Development Fund, Trans-Channel Neuroscience Network to P.G.

References

1. Read AF, Graham AL, Raberg L (2008) Animal defenses against infectious agents: is damage control more important than pathogen control. PLoS Biol 6(12):e4. doi:08-PLBI-P-4573 (pii)10.1371/journal.pbio.1000004
2. Medzhitov R, Schneider DS, Soares MP (2012) Disease tolerance as a defense strategy. Science 335(6071):936–941. doi:335/6071/936 (pii)10.1126/science.1214935
3. Oppenheim JJ (2001) Cytokines: past, present, and future. Int J Hematol 74(1):3–8
4. Beutler B, Cerami A (1986) Cachectin and tumour necrosis factor as two sides of the same biological coin. Nature 320(6063):584–588. doi:10.1038/320584a0
5. Dinarello CA (1996) Biologic basis for interleukin-1 in disease. Blood 87(6):2095–2147
6. Brines ML, Ghezzi P, Keenan S, Agnello D, de Lanerolle NC, Cerami C, Itri LM, Cerami A (2000) Erythropoietin crosses the blood–brain barrier to protect against experimental brain injury. Proc Natl Acad Sci U S A 97(19):10526–10531. doi:97/19/10526 (pii)
7. Villa P, Bigini P, Mennini T, Agnello D, Laragione T, Cagnotto A, Viviani B, Marinovich M, Cerami A, Coleman TR, Brines M, Ghezzi P (2003) Erythropoietin selectively attenuates cytokine production and inflammation in cerebral ischemia by targeting neuronal apoptosis. J Exp Med 198(6):971–975. doi:10.1084/jem.20021067jem.20021067 (pii)
8. Wang L, Zhang Z, Wang Y, Zhang R, Chopp M (2004) Treatment of stroke with erythropoietin enhances neurogenesis and angiogenesis and improves neurological function in rats. Stroke 35(7):1732–1737
9. D'Angio CT, Finkelstein JN (2000) Oxygen regulation of gene expression: a study in opposites. Mol Genet Metab 71(1–2):371–380. doi:10.1006/mgme.2000.3074S1096-7192(00)93074-9 (pii)
10. Michiels C, Minet E, Mottet D, Raes M (2002) Regulation of gene expression by oxygen: NF-kappaB and HIF-1, two extremes. Free Radic Biol Med 33(9):1231–1242. doi:S0891584902010456 (pii)
11. Haddad JJ, Harb HL (2005) Cytokines and the regulation of hypoxia-inducible factor (HIF)-1alpha. Int Immunopharmacol 5(3):461–483. doi:S1567-5769(04)00380-7 (pii)10.1016/j.intimp. 2004.11.009
12. Colgan SP, Taylor CT (2010) Hypoxia: an alarm signal during intestinal inflammation. Nat Rev Gastroenterol Hepatol 7(5):281–287. doi:nrgastro.2010.39 (pii)10.1038/nrgastro.2010.39
13. Nizet V, Johnson RS (2009) Interdependence of hypoxic and innate immune responses. Nat Rev Immunol 9(9):609–617. doi:nri2607 (pii)10.1038/nri2607
14. Zinkernagel AS, Johnson RS, Nizet V (2007) Hypoxia inducible factor (HIF) function in innate immunity and infection. J Mol Med (Berl) 85(12):1339–1346. doi:10.1007/s00109-007-0282-2
15. Werth N, Beerlage C, Rosenberger C, Yazdi AS, Edelmann M, Amr A, Bernhardt W, von Eiff C, Becker K, Schäfer A, Peschel A, Kempf VAJ (2010) Activation of hypoxia inducible factor 1 is a general phenomenon in infections with human pathogens. PLoS ONE 5(7):e11576
16. Taga T, Kishimoto T (1997) Gp130 and the interleukin-6 family of cytokines. Annu Rev Immunol 15:797–819. doi:10.1146/annurev.immunol.15.1.797
17. Miyajima A, Kitamura T, Harada N, Yokota T, Arai K (1992) Cytokine receptors and signal transduction. Annu Rev Immunol 10:295–331. doi:10.1146/annurev.iy.10.040192.001455
18. Gewiese-Rabsch J, Drucker C, Malchow S, Scheller J, Rose-John S (1802) Role of IL-6 trans-signaling in CCl_4 induced liver damage. Biochim Biophys Acta 11:1054–1061. doi:S0925-4439(10)00160-2(pii)10.1016/j.bbadis.2010.07.023
19. Katz A, Chebath J, Friedman J, Revel M (1998) Increased sensitivity of IL-6-deficient mice to carbon tetrachloride hepatotoxicity and protection with an IL-6 receptor-IL-6 chimera. Cytokines Cell Mol Ther 4(4):221–227
20. Schafer KH, Mestres P, Marz P, Rose-John S (1999) The IL-6/sIL-6R fusion protein hyper-IL-6 promotes neurite outgrowth and neuron survival in cultured enteric neurons. J Interferon Cytokine Res 19(5):527–532. doi:10.1089/107999099313974
21. Zhang PL, Izrael M, Ainbinder E, Ben-Simchon L, Chebath J, Revel M (2006) Increased myelinating capacity of embryonic stem cell derived oligodendrocyte precursors after treatment by interleukin-6/soluble interleukin-6 receptor fusion protein. Mol Cell Neurosci 31(3):387–398. doi:S1044-7431(05)00258-7(pii)10.1016/j.mcn.2005.10.014
22. An Y, Xiao YB (2007) The preventative role of growth hormone on acute liver injury induced by cardiopulmonary bypass in a rat model. Eur J Cardiothorac Surg 31(6):1037–1043. doi:S1010-7940(07)00335-1 (pii)10.1016/j.ejcts.2007.01.077

23. Byts N, Samoylenko A, Fasshauer T, Ivanisevic M, Hennighausen L, Ehrenreich H, Siren AL (2008) Essential role for Stat5 in the neurotrophic but not in the neuroprotective effect of erythropoietin. Cell Death Differ 15(4):783–792. doi:cdd20081 (pii)10.1038/cdd.2008.1
24. Isgaard J, Aberg D, Nilsson M (2007) Protective and regenerative effects of the GH/IGF-I axis on the brain. Minerva Endocrinol 32(2):103–113
25. Li RC, Guo SZ, Raccurt M, Moudilou E, Morel G, Brittian KR, Gozal D (2011) Exogenous growth hormone attenuates cognitive deficits induced by intermittent hypoxia in rats. Neuroscience 196:237–250. doi:S0306-4522(11)00971-7 (pii)10.1016/j.neuroscience.2011.08.029
26. Ling FA, Hui DZ, Ji SM (2007) Protective effect of recombinant human somatotropin on amyloid beta-peptide induced learning and memory deficits in mice. Growth Horm IGF Res 17(4):336–341. doi: S1096-6374(07)00067-6(pii)10.1016/j.ghir.2007.04.012
27. Saceda J, Isla A, Santiago S, Morales C, Odene C, Hernandez B, Deniz K (2011) Effect of recombinant human growth hormone on peripheral nerve regeneration: experimental work on the ulnar nerve of the rat. Neurosci Lett 504(2):146–150. doi:S0304-3940(11)01295-X(pii)10.1016/j.neulet.2011.09.020
28. Scheepens A, Sirimanne ES, Breier BH, Clark RG, Gluckman PD, Williams CE (2001) Growth hormone as a neuronal rescue factor during recovery from CNS injury. Neuroscience 104(3):677–687. doi:S0306-4522(01)00109-9 (pii)
29. Shin DH, Lee E, Kim JW, Kwon BS, Jung MK, Jee YH, Kim J, Bae SR, Chang YP (2004) Protective effect of growth hormone on neuronal apoptosis after hypoxia-ischemia in the neonatal rat brain. Neurosci Lett 354(1):64–68. doi:S0304394003011571 (pii)
30. Yi C, Cao Y, Mao SH, Liu H, Ji LL, Xu SY, Zhang M, Huang Y (2009) Recombinant human growth hormone improves survival and protects against acute lung injury in murine Staphylococcus aureus sepsis. Inflamm Res 58(12):855–862. doi:10.1007/s00011-009-0056-0
31. Verstraete K, Vandriessche G, Januar M, Elegheert J, Shkumatov AV, Desfosses A, Van Craenenbroeck K, Svergun DI, Gutsche I, Vergauwen B, Savvides SN (2011) Structural insights into the extracellular assembly of the hematopoietic Flt3 signaling complex. Blood 118(1):60–68. doi:blood-2011-01-329532 (pii)10.1182/blood-2011-01-329532
32. Urdzikova L, Likavcanova-Masinova K, Vanecek V, Ruzicka J, Sedy J, Sykova E, Jendelova P (2011) Flt3 ligand synergizes with granulocyte-colony-stimulating factor in bone marrow mobilization to improve functional outcome after spinal cord injury in the rat. Cytotherapy 13(9):1090–1104. doi:10.3109/14653249.2011.575355
33. Brazel CY, Ducceschi MH, Pytowski B, Levison SW (2001) The FLT3 tyrosine kinase receptor inhibits neural stem/progenitor cell proliferation and collaborates with NGF to promote neuronal survival. Mol Cell Neurosci 18(4):381–393. doi:10.1006/mcne.2001.1033S1044-7431(01)91033-4 (pii)
34. Heeschen C, Aicher A, Lehmann R, Fichtlscherer S, Vasa M, Urbich C, Mildner-Rihm C, Martin H, Zeiher AM, Dimmeler S (2003) Erythropoietin is a potent physiologic stimulus for endothelial progenitor cell mobilization. Blood 102(4):1340–1346. doi:10.1182/blood-2003-01-02232003-01-0223 (pii)
35. Farmer DG, Ke B, Shen XD, Kaldas FM, Gao F, Watson MJ, Busuttil RW, Kupiec-Weglinski JW (2011) Interleukin-13 protects mouse intestine from ischemia and reperfusion injury through regulation of innate and adaptive immunity. Transplantation 91(7):737–743. doi:10.1097/TP.0b013e31820c861a
36. Grilli M, Barbieri I, Basudev H, Brusa R, Casati C, Lozza G, Ongini E (2000) Interleukin-10 modulates neuronal threshold of vulnerability to ischaemic damage. Eur J Neurosci 12(7):2265–2272. doi:ejn090 (pii)
37. Bachis A, Colangelo AM, Vicini S, Doe PP, De Bernardi MA, Brooker G, Mocchetti I (2001) Interleukin-10 prevents glutamate-mediated cerebellar granule cell death by blocking caspase-3-like activity. J Neurosci 21(9):3104–3112
38. Siren AL, Fratelli M, Brines M, Goemans C, Casagrande S, Lewczuk P, Keenan S, Gleiter C, Pasquali C, Capobianco A, Mennini T, Heumann R, Cerami A, Ehrenreich H, Ghezzi P (2001) Erythropoietin prevents neuronal apoptosis after cerebral ischemia and metabolic stress. Proc Natl Acad Sci U S A 98(7):4044–4049. doi:10.1073/pnas.05160659805160 6598 (pii)
39. Digicaylioglu M, Lipton SA (2001) Erythropoietin-mediated neuroprotection involves cross-talk between Jak2 and NF-kappaB signalling cascades. Nature 412(6847):641–647. doi:10.1038/3508807435088074 (pii)

40. Conklin D (2004) Recognition of the helical cytokine fold. J Comput Biol 11(6):1189–1200. doi:10.1089/cmb.2004.11.1189
41. Bazan JF (1991) Neuropoietic cytokines in the hematopoietic fold. Neuron 7(2):197–208. doi:0896-6273(91)90258-2 (pii)
42. Stockli KA, Lottspeich F, Sendtner M, Masiakowski P, Carroll P, Gotz R, Lindholm D, Thoenen H (1989) Molecular cloning, expression and regional distribution of rat ciliary neurotrophic factor. Nature 342(6252): 920–923. doi:10.1038/342920a0
43. Hill EE, Morea V, Chothia C (2002) Sequence conservation in families whose members have little or no sequence similarity: the four-helical cytokines and cytochromes. J Mol Biol 322(1):205–233. doi:S0022283602006538 (pii)
44. Conklin D, Haldeman B, Gao Z (2005) Gene finding for the helical cytokines. Bioinformatics 21(9):1776–1781. doi:bti283 (pii)10.1093/bioinformatics/bti283
45. Eisenberg D, Weiss RM, Terwilliger TC (1984) The hydrophobic moment detects periodicity in protein hydrophobicity. Proc Natl Acad Sci U S A 81(1):140–144
46. Fox BA, Sheppard PO, O'Hara PJ (2009) The role of genomic data in the discovery, annotation and evolutionary interpretation of the interferon-lambda family. PLoS One 4(3):e4933. doi:10.1371/journal.pone.0004933
47. Beck G, Habicht GS (1991) Primitive cytokines: harbingers of vertebrate defense. Immunol Today 12(6):180–183
48. Beck G, Habicht GS (1996) Characterization of an IL-6-like molecule from an echinoderm (Asterias forbesi). Cytokine 8(7):507–512. doi:S1043-4666(96)90069-1 (pii)10.1006/cyto.1996.0069
49. Beschin A, Bilej M, Torreele E, De Baetselier P (2001) On the existence of cytokines in invertebrates. Cell Mol Life Sci 58(5–6):801–814
50. Reichhart JM, Meister M, Dimarcq JL, Zachary D, Hoffmann D, Ruiz C, Richards G, Hoffmann JA (1992) Insect immunity: developmental and inducible activity of the Drosophila diptericin promoter. EMBO J 11(4):1469–1477
51. Huising MO, Kruiswijk CP, Flik G (2006) Phylogeny and evolution of class-I helical cytokines. J Endocrinol 189(1):1–25. doi:189/1/1 (pii)10.1677/joe.1.06591
52. Malagoli D, Conklin D, Sacchi S, Mandrioli M, Ottaviani E (2007) A putative helical cytokine functioning in innate immune signalling in Drosophila melanogaster. Biochim Biophys Acta 1770(6):974–978. doi:S0304-4165(07)00057-8(pii)10.1016/j.bbagen.2007.02.008
53. Liongue C, Ward AC (2007) Evolution of Class I cytokine receptors. BMC Evol Biol 7:120. doi:1471-2148-7-120 (pii)10.1186/1471-2148-7-120
54. Ostrowski D, Ehrenreich H, Heinrich R (2011) Erythropoietin promotes survival and regeneration of insect neurons in vivo and in vitro. Neuroscience 188:95–108. doi:S0306-4522(11)00550-1(pii)10.1016/j.neuroscience.2011.05.018
55. Ehrenreich H, Hasselblatt M, Dembowski C, Cepek L, Lewczuk P, Stiefel M, Rustenbeck HH, Breiter N, Jacob S, Knerlich F, Bohn M, Poser W, Ruther E, Kochen M, Gefeller O, Gleiter C, Wessel TC, De Ryck M, Itri L, Prange H, Cerami A, Brines M, Siren AL (2002) Erythropoietin therapy for acute stroke is both safe and beneficial. Mol Med 8(8):495–505
56. Ehrenreich H, Weissenborn K, Prange H, Schneider D, Weimar C, Wartenberg K, Schellinger PD, Bohn M, Becker H, Wegrzyn M, Jahnig P, Herrmann M, Knauth M, Bahr M, Heide W, Wagner A, Schwab S, Reichmann H, Schwendemann G, Dengler R, Kastrup A, Bartels C (2009) Recombinant human erythropoietin in the treatment of acute ischemic stroke. Stroke 40(12):e647–e656. doi:STROKEAHA.109.564872 (pii)10.1161/STROKEAHA.109.564872
57. Ehrenreich H, Kastner A, Weissenborn K, Streeter J, Sperling S, Wang KK, Worthmann H, Hayes RL, Von Ahsen N, Kastrup A, Jeromin A, Herrmann M (2011) Circulating damage marker profiles support a neuroprotective effect of erythropoietin in ischemic stroke patients. Mol Med. doi:molmed.2011.00259 (pii)10.2119/molmed.2011.00259
58. Yip HK, Tsai TH, Lin HS, Chen SF, Sun CK, Leu S, Yuen CM, Tan TY, Lan MY, Liou CW, Lu CH, Chang WN (2011) Effect of erythropoietin on level of circulating endothelial progenitor cells and outcome in patients after acute ischemic stroke. Crit Care 15(1):R40. doi:cc10002 (pii)10.1186/cc10002
59. Tseng MY, Hutchinson PJ, Richards HK, Czosnyka M, Pickard JD, Erber WN, Brown S, Kirkpatrick PJ (2009) Acute systemic erythropoietin therapy to reduce delayed ischemic deficits following aneurysmal subarachnoid hemorrhage: a Phase II randomized, double-blind, placebo-controlled trial. Clinical article. J Neurosurg 111(1):171–180. doi:10.3171/2009.3.JNS081332

60. Lauria G, Campanella A, Filippini G, Martini A, Penza P, Maggi L, Antozzi C, Ciano C, Beretta P, Caldiroli D, Ghelma F, Ferrara G, Ghezzi P, Mantegazza R (2009) Erythropoietin in amyotrophic lateral sclerosis: a pilot, randomized, double-blind, placebo-controlled study of safety and tolerability. Amyotroph Lateral Scler 10(5–6):410–415. doi:10.3109/17482960902995246 (pii)10.3109/17482960902995246

61. Duning T, Schiffbauer H, Warnecke T, Mohammadi S, Floel A, Kolpatzik K, Kugel H, Schneider A, Knecht S, Deppe M, Schabitz WR (2011) G-CSF prevents the progression of structural disintegration of white matter tracts in amyotrophic lateral sclerosis: a pilot trial. PLoS One 6(3):e17770. doi:10.1371/journal.pone.0017770

62. Brines M, Patel NS, Villa P, Brines C, Mennini T, De Paola M, Erbayraktar Z, Erbayraktar S, Sepodes B, Thiemermann C, Ghezzi P, Yamin M, Hand CC, Xie QW, Coleman T, Cerami A (2008) Nonerythropoietic, tissue-protective peptides derived from the tertiary structure of erythropoietin. Proc Natl Acad Sci U S A 105(31):10925–10930. doi:0805594105 (pii)10.1073/pnas.0805594105

63. Beurrier C, Faideau M, Bennouar KE, Escartin C, Kerkerian-Le Goff L, Bonvento G, Gubellini P (2010) Ciliary neurotrophic factor protects striatal neurons against excitotoxicity by enhancing glial glutamate uptake. PLoS One 5(1):e8550. doi:10.1371/journal.pone.0008550

64. Toth G, Yang H, Anguelov RA, Vettraino J, Wang Y, Acsadi G (2002) Gene transfer of glial cell-derived neurotrophic factor and cardiotrophin-1 protects PC12 cells from injury: involvement of the phosphatidylinositol 3-kinase and mitogen-activated protein kinase kinase pathways. J Neurosci Res 69(5):622–632

65. Iniguez M, Berasain C, Martinez-Ansó E, Bustos M, Fortes P, Pennica D, Avila MA, Js P (2006) Cardiotrophin-1 defends the liver against ischemia-reperfusion injury and mediates the protective effect of ischemic preconditioning. J Exp Med 203(13):2809–2815

66. Hara M, Yuasa S, Shimoji K, Onizuka T, Hayashiji N, Ohno Y, Arai T, Hattori F, Kaneda R, Kimura K, Makino S, Sano M, Fukuda K (2011) G-CSF influences mouse skeletal muscle development and regeneration by stimulating myoblast proliferation. J Exp Med 208(4):715–727

67. Liu S-P, Lee S-D, Lee H-T, Liu DD, Wang H-J, Liu R-S, Lin S-Z, Shyu W-C (2010) Granulocyte colony-stimulating factor activating HIF-1alpha acts synergistically with erythropoietin to promote tissue plasticity. PLoS ONE 5(4):e10093–e10093

68. Ma H, Gong H, Chen Z, Liang Y, Yuan J, Zhang G, Wu J, Ye Y, Yang C, Nakai A, Komuro I, Ge J, Zou Y (2012) Association of Stat3 with HSF1 plays a critical role in G-CSF-induced cardio-protection against ischemia/reperfusion injury. J Mol Cell Cardiol 52(6):1282–1290. doi:S0022-2828(12)00106-X (pii)10.1016/j.yjmcc.2012.02.011

69. Shimoji K, Yuasa S, Onizuka T, Hattori F, Tanaka T, Hara M, Ohno Y, Chen H, Egasgira T, Seki T, Yae K, Koshimizu U, Ogawa S, Fukuda K (2010) G-CSF promotes the proliferation of developing cardiomyocytes in vivo and in derivation from ESCs and iPSCs. Cell Stem Cell 6(3):227–237

70. Stratos I, Rotter R, Eipel C, Mittlmeier T, Vollmar B (2007) Granulocyte-colony stimulating factor enhances muscle proliferation and strength following skeletal muscle injury in rats. J Appl Physiol 103(5):1857–1863, Bethesda, MD: 1985

71. Tsai RK, Chang CH, Wang HZ (2008) Neuroprotective effects of recombinant human granulocyte colony-stimulating factor (G-CSF) in neurodegeneration after optic nerve crush in rats. Exp Eye Res 87(3):242–250. doi:S0014-4835(08)00168-1 (pii)10.1016/j.exer.2008.06.004

72. Schabitz WR, Kollmar R, Schwaninger M, Juettler E, Bardutzky J, Scholzke MN, Sommer C, Schwab S (2003) Neuroprotective effect of granulocyte colony-stimulating factor after focal cerebral ischemia. Stroke 34(3):745–751. doi:10.1161/01.STR.0000057814.70180.1701.STR.0000057814.70180.17 (pii)

73. Gibson CL, Bath PM, Murphy SP (2010) G-CSF administration is neuroprotective following transient cerebral ischemia even in the absence of a functional NOS-2 gene. J Cereb Blood Flow Metab 30(4):739–743. doi:jcbfm201012 (pii)10.1038/jcbfm.2010.12

74. Scheepens A, Sirimanne ES, Breier BH, Clark RG, Gluckman PD, Williams CE (2001) Growth hormone as a neuronal rescue factor during recovery from CNS injury. Neuroscience 104(3):677–687

75. Hirsch M, Knight J, Tobita M, Soltys J, Panitch H, Mao-Draayer Y (2009) The effect of interferon-(beta) on mouse neural progenitor cell survival and differentiation. Biochem Biophys Res Commun 388:181–186

76. Tabira T, Konishi Y, Gallyas F Jr (1995) Neurotrophic effect of hematopoietic cytokines on cholinergic and other neurons in vitro. Int J Dev Neurosci 13(3–4):241–252

77. Zambrano A, Otth C, Mujica L, Concha II, Maccioni RB (2007) Interleukin-3 prevents neuronal death induced by amyloid peptide. BMC Neurosci 8:82
78. Wen TC, Tanaka J, Peng H, Desaki J, Matsuda S, Maeda N, Fujita H, Sato K, Sakanaka M (1998) Interleukin 3 prevents delayed neuronal death in the hippocampal CA1 field. J Exp Med 188(4):635–649
79. Iwasaki Y, Ikeda K, Ichikawa Y, Igarashi O, Iwamoto K, Kinoshita M (2002) Protective effect of interleukin-3 and erythropoietin on motor neuron death after neonatal axotomy. Neurol Res 24(7):643–646
80. Kannan Y, Moriyama M, Sugano T, Yamate J, Kuwamura M, Kagaya A, Kiso Y (2000) Neurotrophic action of interleukin 3 and granulocyte-macrophage colony-stimulating factor on murine sympathetic neurons. Neuroimmunomodulation 8(3):132–141. doi:54273 (pii)
81. Nishihara T, Ochi M, Sugimoto K, Takahashi H, Yano H, Kumon Y, Ohnishi T, Tanaka J (2011) Subcutaneous injection containing IL-3 and GM-CSF ameliorates stab wound-induced brain injury in rats. Exp Neurol 229(2): 507–516. doi:S0014-4886(11)00120-8 (pii) 10.1016/j.expneurol.2011.04.006
82. Konishi Y, Chui DH, Hirose H, Kunishita T, Tabira T (1993) Trophic effect of erythropoietin and other hematopoietic factors on central cholinergic neurons in vitro and in vivo. Brain Res 609(1–2):29–35
83. Aderka D, Le JM, Vilcek J (1989) IL-6 inhibits lipopolysaccharide-induced tumor necrosis factor production in cultured human monocytes, U937 cells, and in mice. J Immunol 143(11):3517–3523
84. Cheng L, Wang J, Li X, Xing Q, Du P, Su L, Wang S (2012) Interleukin-6 Induces Gr-1+CD11b+ myeloid cells to suppress CD8+ T cell-mediated liver injury in mice. PLoS ONE 6(3):e17631
85. Cotter MA, Gibson TM, Nangle MR, Cameron NE (2010) Effects of interleukin-6 treatment on neurovascular function, nerve perfusion and vascular endothelium in diabetic rats. Diabetes Obes Metab 12(8):689–699
86. Camargo CA Jr, Madden JF, Gao W, Selvan RS, Clavien PA (1997) Interleukin-6 protects liver against warm ischemia/reperfusion injury and promotes hepatocyte proliferation in the rodent. Hepatology 26(6):1513–1520. doi:S0270913997005375 (pii)10.1002/hep.510260619
87. Toth KG, McKay BR, De Lisio M, Little JP, Tarnopolsky MA, Parise G (2011) IL-6 induced stat3 signalling is associated with the proliferation of human muscle satellite cells following acute muscle damage. PLoS ONE 6(3):e17392–e17392
88. El-Assal O, Hong F, Kim W-H, Radaeva S, Gao B (2004) IL-6-deficient mice are susceptible to ethanol-induced hepatic steatosis: IL-6 protects against ethanol-induced oxidative stress and mitochondrial permeability transition in the liver. Cell Mol Immunol 1(3):205–211
89. Zhang P, Chebath J, Lonai P, Revel M (2004) Enhancement of oligodendrocyte differentiation from murine embryonic stem cells by an activator of gp130 signaling. Stem Cells 22(3):344–354
90. Cameron NE, Cotter MA (2007) The neurocytokine, interleukin-6, corrects nerve dysfunction in experimental diabetes. Exp Neurol 207(1):23–29
91. Zhang P-L, Levy AM, Ben-Simchon L, Haggiag S, Chebath J, Revel M (2007) Induction of neuronal and myelin-related gene expression by IL-6-receptor/IL-6: a study on embryonic dorsal root ganglia cells and isolated Schwann cells. Exp Neurol 208(2):285–296
92. Fujita T, Tozaki-Saitoh H, Inoue K (2009) P2Y1 receptor signaling enhances neuroprotection by astrocytes against oxidative stress via IL-6 release in hippocampal cultures. Glia 57(3):244–257
93. Kolliputi N, Waxman AB (2009) IL-6 cytoprotection in hyperoxic acute lung injury occurs via PI3K/Akt-mediated Bax phosphorylation. Am J Physiol Lung Cell Mol Physiol 297(1):L6–16–L16–16
94. Hakkoum D, Stoppini L, Muller D (2007) Interleukin-6 promotes sprouting and functional recovery in lesioned organotypic hippocampal slice cultures. J Neurochem 100(3):747–757.doi:JNC4257(pii)10.1111/j.1471-4159.2006.04257.x
95. Wang X-Q, Peng Y-P, Lu J-H, Cao B-B, Qiu Y-H (2009) Neuroprotection of interleukin-6 against NMDA attack and its signal transduction by JAK and MAPK. Neurosci Lett 450(2):122–126
96. Waxman AB, Kolliputi N (2009) IL-6 protects against hyperoxia-induced mitochondrial damage via Bcl-2-induced Bak interactions with mitofusins. Am J Respir Cell Mol Biol 41(4):385–396
97. Hirota H, Kiyama H, Kishimoto T, Taga T (1996) Accelerated Nerve Regeneration in Mice by upregulated expression of interleukin (IL) 6 and IL-6 receptor after trauma. J Exp Med 183(6):2627–2634
98. Jung JE, Kim GS, Chan PH (2011) Neuroprotection by interleukin-6 is mediated

by signal transducer and activator of transcription 3 and antioxidative signaling in ischemic stroke. Stroke 42(12):3574–3579. doi:STROKEAHA.111.626648 (pii)10.1161/STROKEAHA.111.626648

99. Yamashita T, Sawamoto K, Suzuki S, Suzuki N, Adachi K, Kawase T, Mihara M, Ohsugi Y, Abe K, Okano H (2005) Blockade of interleukin-6 signaling aggravates ischemic cerebral damage in mice: possible involvement of Stat3 activation in the protection of neurons. J Neurochem 94(2):459–468. doi:JNC3227 (pii)10.1111/j.1471-4159.2005.03227.x

100. Bansal MB, Kovalovich K, Gupta R, Li W, Agarwal A, Radbill B, Alvarez CE, Safadi R, Fiel MI, Friedman SL, Taub RA (2005) Interleukin-6 protects hepatocytes from CCl4-mediated necrosis and apoptosis in mice by reducing MMP-2 expression. J Hepatol 42(4):548–556

101. Tran HY, Shin EJ, Saito K, Nguyen XK, Chung YH, Jeong JH, Bach JH, Park DH, Yamada K, Nabeshima T, Yoneda Y, Kim HC (2012) Protective potential of IL-6 against trimethyltin-induced neurotoxicity in vivo. Free Radic Biol Med 52(7):1159–1174. doi:S0891-5849(11)01252-4(pii)10.1016/j.freeradbiomed.2011.12.008

102. Zhang P-L, Izrael M, Ainbinder E, Ben-Simchon L, Chebath J, Revel M (2006) Increased myelinating capacity of embryonic stem cell derived oligodendrocyte precursors after treatment by interleukin-6/soluble interleukin-6 receptor fusion protein. Mol Cell Neurosci 31(3):387–398

103. Nishina T, Komazawa-Sakon S, Yanaka S, Piao X, Zheng D-M, Piao J-H, Kojima Y, Yamashina S, Sano E, Putoczki T, Doi T, Ueno T, Ezaki J, Ushio H, Ernst M, Tsumoto K, Okumura K, Nakano H (2012) Interleukin-11 Links Oxidative Stress and Compensatory Proliferation. Sci Signal 5(207):ra5. doi:10.1126/scisignal.2002056

104. Ikeda K, Iwasaki Y, Tagaya N, Shiojima T, Kinoshita M (1995) Neuroprotective effect of cholinergic differentiation factor/leukemia inhibitory factor on wobbler murine motor neuron disease. Muscle Nerve 18(11):1344–1347

105. Bengatta S, Arnould C, Letavernier E, Monge M, de Preneuf HM, Werb Z, Ronco P, Lelongt B (2009) MMP9 and SCF protect from apoptosis in acute kidney injury. J Am Soc Nephrol 20(4):787–797. doi:20/4/787 (pii)10.1681/ASN.2008050515

106. Dhandapani KM, Wade FM, Wakade C, Mahesh VB, Brann DW (2005) Neuroprotection by stem cell factor in rat cortical neurons involves AKT and NFkappaB. J Neurochem 95(1):9–19. doi:JNC3319 (pii)10.1111/j.1471-4159.2005.03319.x

107. Simpson K, Hogaboam CM, Kunkel SL, Harrison DJ, Bone-Larson C, Lukacs NW (2003) Stem cell factor attenuates liver damage in a murine model of acetaminophen-induced hepatic injury. Lab Invest 83(2):199–206

108. Penkowa M, Giralt M, Carrasco J, Hadberg H, Hidalgo J (2000) Impaired inflammatory response and increased oxidative stress and neurodegeneration after brain injury in interleukin-6-deficient mice. Glia 32(3):271–285. doi:10.1002/1098-1136(200012)32:3<271::AID-GLIA70>3.0.CO;2-5 (pii)

109. Lin ZQ, Kondo T, Ishida Y, Takayasu T, Mukaida N (2003) Essential involvement of IL-6 in the skin wound-healing process as evidenced by delayed wound healing in IL-6-deficient mice. J Leukoc Biol 73(6):713–721

110. Han X, Gilbert S, Groschwitz K, Hogan S, Jurickova I, Trapnell B, Samson C, Gully J (2010) Loss of GM-CSF signalling in non-haematopoietic cells increases NSAID ileal injury. Gut 59(8):1066–1078. doi:gut.2009.203893 (pii)10.1136/gut.2009.203893

111. Fang Y, Gong SJ, Xu YH, Hambly BD, Bao S (2007) Impaired cutaneous wound healing in granulocyte/macrophage colony-stimulating factor knockout mice. Br J Dermatol 157(3):458–465.doi:BJD7979(pii)10.1111/j.1365-2133.2007.07979.x

Chapter 4

The Regenerative Activity of Interleukin-6

Eithan Galun and Stefan Rose-John

Abstract

Interleukin-6 (IL-6) is a cytokine which is involved in many inflammatory processes and in the development of cancer. In addition, IL-6 has been shown to be important for the induction of hepatic acute-phase proteins, for the regeneration of the liver and for the stimulation of B-cells. IL-6 binds to a transmembrane IL-6 receptor (IL-6R), which is present on hepatocytes and some leukocytes. The complex of IL-6 and IL-6R associates with a second protein, gp130, which is expressed on all cells of the body. Since neither IL-6 nor IL-6R has a measurable affinity for gp130, cells, which do not express IL-6R, are not responsive to the cytokine IL-6. It could be shown, however, that a naturally occurring soluble IL-6R (sIL-6R) in complex with IL-6 can bind to gp130 on cells with no IL-6R expression. Therefore, cells shedding the sIL-6R render cells, which only express gp130, responsive to the cytokine. This process has been called trans-signaling. In the present chapter, we summarize the known activities of IL-6 with a special emphasis on regenerative activities, which often depend on the sIL-6R. A designer cytokine called Hyper-IL-6, which is a fusion protein of IL-6 and the sIL-6R, can mimic IL-6 trans-signaling responses in vitro and in vivo with considerably higher efficacy than the combination of the natural proteins IL-6 and sIL-6R. We present recent examples from animal models in which the therapeutic potential of Hyper-IL-6 has been evaluated. We propose that Hyper-IL-6 can be used to induce potent regeneration responses in liver, kidney, and other tissues and therefore will be a novel therapeutic approach in regenerative medicine.

Key words gp130, Inflammation, Interleukin-6, Interleukin-6-receptor, Regeneration, Shedding, Soluble interleukin-6-receptor, STAT3, Stem cells

1 Introduction

Interleukin-6 (IL-6) is a four helical cytokine of 184 amino acid residues (1, 2). IL-6 binds to a cell surface receptor called IL-6R, which is a type I transmembrane protein with no intrinsic signaling capacity (3). When IL-6 becomes bound to IL-6R, the complex associates with a second protein called gp130 (4). Upon binding to the IL-6/IL-6R complex, gp130 dimerizes and initiates intracellular signaling (Fig. 1) (5, 6). It turned out that gp130 is not only the signal transducing receptor subunit of IL-6R, but is also part of the receptor complexes for the cytokines Interleukin-11

Pietro Ghezzi and Anthony Cerami (eds.), *Tissue-Protective Cytokines: Methods and Protocols*, Methods in Molecular Biology, vol. 982, DOI 10.1007/978-1-62703-308-4_4, © Springer Science+Business Media, LLC 2013

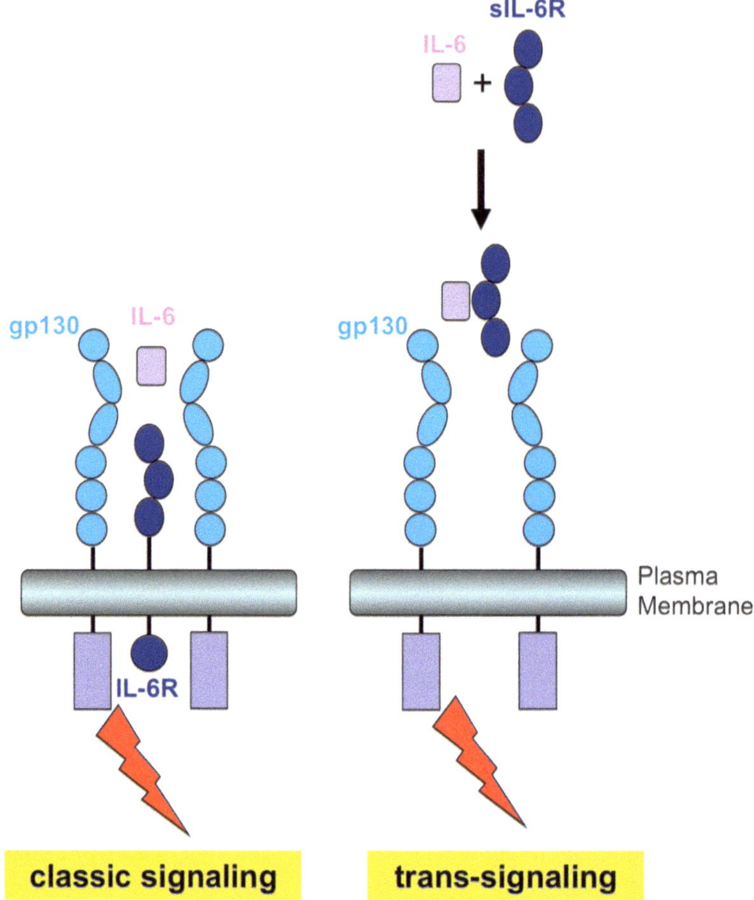

Fig. 1 The paradigm of classic- and trans-signaling of IL-6. IL-6 binds to the membrane bound IL-6R, induces dimerization of gp130 and signaling (*left*). On cells not expressing IL-6R, IL-6 can only act by binding to the sIL-6R. The complex of IL-6 and sIL-6R can bind to gp130 and induce dimerization and signaling (*right*). By this trans-signaling pathway virtually every cell in the body can respond to IL-6

(IL-11), Leukemia Inhibitory Factor (LIF), Ciliary Neurotrophic Factor (CNTF), Cardiotrophin-1 (CT-1), Cardiotrophin-like Cytokine (CLC), Oncostatin M (OSM), and Interleukin-27 (IL-27) (7, 8). These cytokines comprise the gp130 family of cytokines (8, 9). Signaling of cytokines of the gp130 family is similar, but not identical since the receptor complexes for LIF, CNTF, CT-1, CLC, OSM, and IL-27 consist of heterodimeric complexes of gp130 and LIF-R, OSM-R, or WSX-1 (5, 6).

Interestingly, gp130 is expressed on all cells of the body whereas IL-6R is only found in some cells (7). Since neither IL-6 nor IL-6R alone have a measurable affinity for gp130, it follows that cells, which do not express IL-6R cannot respond to the cytokine IL-6 (10). Cells, which do not express IL-6R include endothelial cells and smooth muscle cells. Strikingly, during inflammatory states, these cell types are known to be major target cells of IL-6 (11).

2 IL-6 Classic and Trans-signaling

Several years ago, it was found that membrane-bound IL-6R can be proteolytically released by cells thereby leading to the generation of a soluble IL-6R (sIL-6R) (12, 13). Alternatively, human cells have been shown to translate the IL-6R from an alternatively spliced mRNA lacking the coding region for the transmembrane domain, thereby also leading to the release of sIL-6R. Interestingly, in the mouse, no alternative splicing of the IL-6R mRNA has been detected. In humans, the release of sIL-6R has been shown to occur in a circadian rhythmical fashion and the increase of sIL-6R during the night has been shown not to be due to the alternative splicing of IL-6R mRNA (14). It has been concluded that the regulated release of sIL-6R is regulated by the proteolytic cleavage rather than by an alternative splicing (15). The major protease responsible for shedding of IL-6R has been identified to be the membrane-bound metalloprotease ADAM17 (15–19). Interestingly, ADAM17 has also been identified as the protease responsible for cleavage of the membrane-bound cytokine Tumor Necrosis Factor alpha (TNFα) (20, 21).

It turned out that on cells, which do not express IL-6R and which therefore cannot be stimulated by IL-6, the complex of IL-6 and sIL-6R can bind to gp130, induce dimerization and subsequent signaling (Fig. 1). This process has been called trans-signaling (10). IL-6 signaling via the membrane-bound IL-6R is called classic signaling. Conceptually, by generating sIL-6R, one cell renders a distinct cell responsive to the cytokine IL-6 (7, 8, 10, 22).

In a molecular model of the complex of IL-6 and sIL-6R, the NH_2-terminus of IL-6 is 40 Å from the COOH-terminus of sIL-6R (23, 24). Consequently, on the cDNA level, we generated a fusion protein of IL-6 and sIL-6R in which the two proteins were connected via a peptide linker of flexible amino acids, which was long enough to bridge the 40 Å. The resulting fusion protein was called Hyper-IL-6 (Fig. 2a) (25). This protein turned out to be a molecular tool that identifies cells in vitro and in vivo, which, in their response to IL-6, depended on sIL-6R. It turned out that hepatocytes (26, 27), neural cells (28, 29), neural stem cells (30), smooth muscle cells (31), hematopoietic progenitor cells (32–38), and embryonic stem cells (39, 40) require sIL-6R in their full response to IL-6.

Since IL-6 does not directly bind to gp130, a soluble version of gp130 (sgp130) was not found to interfere with classic signaling via the membrane-bound IL-6R, but to specifically inhibit IL-6

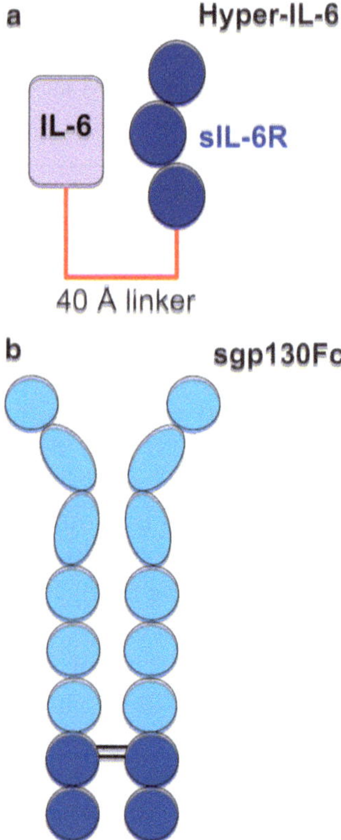

Fig. 2 Designer cytokines for the analysis of IL-6 classic- and trans-signaling. (**a**) IL-6 was fused to the sIL-6R by a flexible peptide linker. This designer cytokine was called Hyper-IL-6. (**b**) The extracellular portion of gp130 was fused to the constant portion of a human IgG1 generating the fusion protein sgp130Fc

trans-signaling via sIL-6R (41). A fusion protein of sgp130 with human IgG1 (sgp130Fc) (Fig. 2b) was shown to be about ten times more efficient than monomeric sgp130 in blocking IL-6 trans-signaling (41). Sgp130Fc was used as a molecular tool to discriminate between classic and trans-signaling in vitro and in vivo (Fig. 3). It was demonstrated in animal models of rheumatoid arthritis (42–44), peritonitis (45), sepsis (46, 47), inflammatory bowel disease (48, 49), inflammatory colonic cancer (50–52), ovarian cancer (53), and pancreatic cancer (54) that the blockade of IL-6 trans-signaling was sufficient to block the progression of the disease (Fig. 3). This has led to the concept that IL-6 trans-signaling is an emergency reaction which is not needed in everyday life, but which is switched on in situations of cellular or tissue stress (8, 55). On the other hand, IL-6 classic signaling has been shown to be needed for the acute phase protein expression in the liver (56) and for the regeneration of intestinal epithelial cells upon wounding (57, 58).

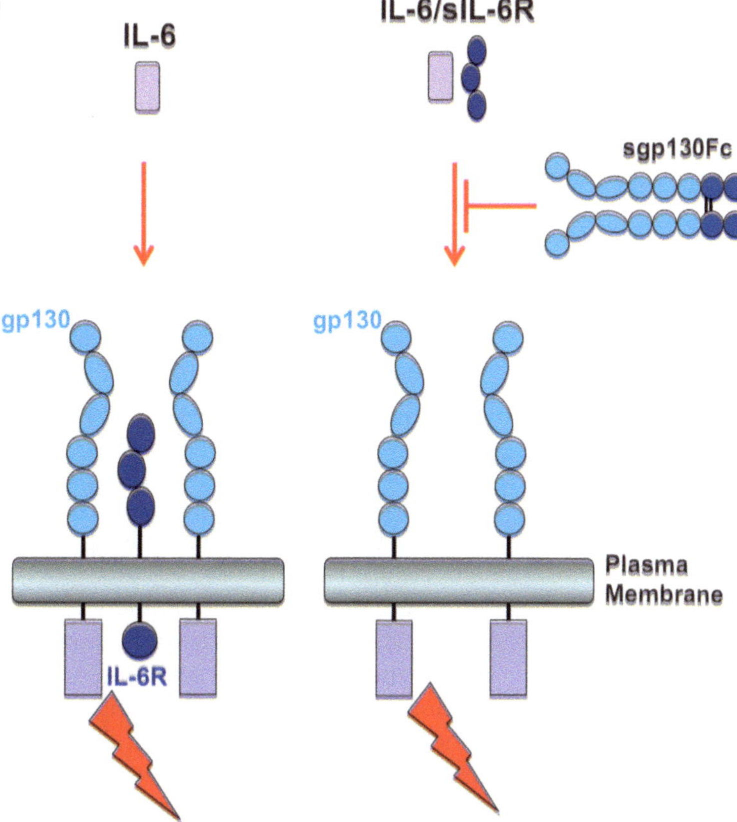

Fig. 3 The sgp130Fc fusion protein specifically blocks IL-6 trans-signaling. IL-6 stimulation of cells expressing membrane bound IL-6R is not affected by sgp130Fc (*left*), whereas IL-6 trans-signaling is efficiently blocked by sgp130Fc (*right*)

3 Signal Transduction by IL-6

As reviewed in (5, 6), signaling of IL-6 occurs upon dimerization of gp130 and activation of JAK kinases constitutively associated with the cytoplasmic portion of gp130. This leads to auto-phosphorylation of JAK kinases and subsequently to phosphorylation of five tyrosine residues within the cytoplasmic domain of gp130. The juxtamembrane tyrosine (Y759 in human gp130) leads to the activation of SHP2 and in turn to the activation of the ras/ERK and the PI3K/AKT pathways (5, 6). The other four tyrosine residues of gp130—when phosphorylated—lead to the recruitment of STAT factors, namely STAT1 and STAT3; these become phosphorylated, dimerize and translocate into the nucleus, where, as homodimers or heterodimers, they act as transcription factors and activate the transcription of STAT target genes (5, 6).

Using knock-in mouse technology, the juxtamembrane tyrosine residue of mouse gp130 was exchanged for phenylalanine. In a second knock-in mouse, the portion of gp130 containing the four membrane-distal tyrosine residues of gp130 was deleted. Thus, two mouse lines were available, one of which was incapable of activating SHP2 and subsequently the ras/ERK and the PI3K/AKT pathways, whereas the other mouse line was unable to activate the STAT pathway upon gp130 stimulation (59). These mouse lines were widely used to unravel the importance of the two signaling pathways triggered by gp130 activation.

The stimulation of embryonic stem cells with the gp130 cytokine LIF leads to a complete block of cellular differentiation, which is a prerequisite for the generation of knock-out mice, since default differentiation of these cells during the homologous recombination procedure can be prevented (60). The same block of differentiation was shown to occur in murine embryonic stem cells treated with Hyper-IL-6 indicating that only gp130 stimulation was required for this cellular response (39, 40). Activation of STAT3 was shown to be sufficient to maintain an undifferentiated state of mouse embryonic stem cells (61). It has been as, however, shown recently that the requirement for STAT3 activation can also be bypassed by the elimination of differentiation-inducing signaling from MAP kinase (62). The activity of gp130 to block differentiation has been used in the expansion of hematopoietic stem cells. Here, the cells were treated with hematopoietic cytokines such as SCF and Flt3-L in the presence of IL-6 or Hyper-IL-6. It turned out that Hyper-IL-6 led to a much more efficient expansion of hematopoietic stem cells as compared to IL-6 (25, 32–35, 37, 38).

An additional activity of IL-6 is the prevention of apoptosis in many cell types including hematopoietic cells, neural cells, epithelial cells and tumor cells (55). Due to the low or lacking IL-6R expression levels on most cells, this activity of IL-6 frequently depends on the presence of the sIL-6R (22, 55).

4 IL-6 Signaling and Trans-signaling in Tissue Regeneration

The restoration of the physiological function following injury to tissue is a complex process. The mediator of such a process and the target cells are only partially known for each and every type of tissue and injury. The cells participating in the regenerative process could originate from different populations inside and outside of any specific organ. The cell types potentially involved in tissue regeneration include the following: (1) Tissue mature parenchymal cells, e.g., in the kidney, tubular epithelium cells. (2) Tissue progenitor/stem cells, e.g., cardiac stem cells. (3) Endothelial cells, which could originate both from the injured tissue or migrate from outside of the injured organ through the bloodstream. (4) Stromal cells, e.g., liver stellate cells, which could also migrate to the tissue from outside of the organ or proliferate upon injury in the tissue/organ itself. (5) Immune cells, e.g., macrophages, which could migrate to the injured organ from the bone marrow through the blood. It is also important to know that there is evidence that trans-differentiation is a phenomenon, which is possibly participating upon regeneration, and could also contribute to the restoration of tissue function and structure. But again, the importance of trans-differentiation in regeneration is currently under investigation. Kinetics, function and significance of the different cells in tissue regeneration are also under intense investigation. During the last decade, studies have unveiled the presence, characterizations and importance of tissue progenitors. The role of these "adult tissue-specific" stem cells is now beginning to be unraveled. In particular, the role of these tissue-specific progenitors in tissue injury and regeneration is beginning to be investigated and debated. The complexity of the regenerative process is not only due to the need to understand the response to injury of the tissue parenchymal cells. Also the interaction, cross talk and dependence of regeneration on the non-parenchymal cells like endothelial cells, tissue macrophages, and the injured cells themselves needs to be addressed. The regenerative process is dependent on a network of extracellular, cell membrane components and intracellular pathways. The molecular mediators of regeneration include small molecules, e.g., serotonin delivered by circulating platelets, secreted factors as chemokines and cytokines, cellular membrane factors as receptors and cell-bound ligands, e.g., TNFα and intracellular networks of signaling molecules and scaffold proteins. This complex molecular and cellular environment needs to be coordinated upon injury to navigate the regeneration process and also terminate it upon completion. The significant role of IL-6 in tissue regeneration was reported almost two decades ago (63). The initial study from the group of Rebecca Taub, who knew about the initial early activation of STAT3 and NFκB as reported by others upon liver resection, has shown that IL-6 is upstream to STAT3 activation and is essential for liver regeneration following hepatectomy. Today, IL-6 signaling and trans-signaling effects are apparent in tissue regeneration. Below, we will discuss the current understanding of the contribution of the IL-6-gp130 pathways to tissue regeneration in different organs.

Our groups have reported that the IL-6 signaling and trans-signaling and the gp130 membrane protein are essential for tissue regeneration in various organs including hematopoietic stem cells (9); liver (26, 64, 65); kidney (66) and salivary function preservation (unpublished results). The reason for the broad tissue regeneration effect has been recently unfolded. IL-6 produced by monocytes and macrophages following tissue injury, anchors to IL-6R, which is expressed mostly on leukocytes and hepatocytes.

All three IL-6 signaling pathways (STAT3, AKT, and ERK1/2) are involved in the expression of pro-survival and cell proliferation genes. Under these conditions, sIL-6R is generated leading to the formation of IL-6/sIL-6R complexes (18, 67, 68). These can act on any tissue due to the fact that gp130 is ubiquitously expressed on all mammalian cells. Leukocytes that are instantly migrating to an acute injured tissue, be it an ischemic tissue in the heart or a liver following hepatectomy, undergo apoptosis at the site of necrosis (69). While undergoing apoptosis, leukocytes shed their IL-6R. This local event enables the local complex formation of IL-6/sIL-6R that signals through gp130 to initiate tissue regeneration. To simulate this "late" event of the IL-6/sIL-6R complex generation, we used the designer protein hyper IL-6. Administration of hyper IL-6 to mice and rats has shown to potently induce regeneration in the liver (65), kidney (66), and salivary gland (unpublished results). Furthermore, we have been recently able to show (unpublished results) that on cardiomyocytes, hyper IL-6 induces a >100-fold stronger phosphorylation of STAT 3 than of IL-6 itself. This result is in agreement with a previous report showing a reduced infarct size following the administration of the IL-6/sIL-6R complex (70).

5 Liver Regeneration: The Role of the IL6-gp130 Pathway

Liver regeneration is a complex physiological process in which both networks of molecular and cellular factors participate. The regenerative process includes both parenchymal and non-parenchymal cells. Although it is becoming clearer in recent years that tissue stem cells harbor many organs, their role in tissue regeneration is still debatable. In the liver, the role of liver stem cells/progenitors in the regenerative process might be even more convoluted. On the one hand, there is growing evidence that STAT3 activation is important in various stem cell physiology and propagation and on the other hand, it is apparent that in an acute regeneration model, liver stem cells probably contribute minimally to the liver regenerative process (71). One more important point which should be mentioned is that most of our data on the role of IL-6-gp130 signaling in liver regeneration is deduced from 2/3 partial hepatectomy (72), and not all investigators are using similar protocols, animal or animal house environments, which might have a major influence on results and conclusions. Consequently not all observations overlap.

Following partial hepatectomy in a healthy mouse, the liver regenerates during a week to ten days and thereby returns to its original size. The peak of cell DNA synthesis is about 40 h following hepatectomy. Priming of regeneration is dependent on TNFα and IL-6 produced and released from non-parenchymal cells including Kupffer cells and endothelial cells. This priming is responsible for the transition of hepatocytes from G_0 to G_1 phase. The importance of TNFα and IL-6 was shown in the relevant knock-out (KO) animals (TNFR1 and IL-6 KO mice). TNFα signals to the Kupffer cells through NFκB to produce and secrete IL-6, 3-6 h after hepatectomy. This peak of IL-6 is followed almost at the same time by the peak of activated (phosphorylated) STAT3. Following the initial priming, cells undergo G_1 to S phase transition. The downstream targets of gp130 are not only mediated through STAT3, but also through AKT which is also activated at early stages following hepatectomy (73). The transition into the S phase is dependent on three factors: HGF, TGFα, and EGF. The signal for activation of STAT3—leading to its phosphorylation and dimerization—is mediated through the activation of Jaks that activates the YxxQ motif on gp130 and the SH2 domain of STAT3. STAT3 is released from gp130 by binding to its dimerization counterpart. STAT3 is diminished in IL-6 KO mice. In mice with hepatocyte-specific deletion of STAT3 following hepatectomy, regeneration is reduced significantly. The G_1 cyclins, including cyclin D and cyclin E, are regulated by STAT3. The transition of G_1/S is regulated by STAT3 through the control of cyclin D1 and with the expression of cyclin E, culminating in E2F activation. This process controls the proliferation of the liver cells.

There is growing evidence that IL-6 trans-signaling is important for liver regeneration. This was shown in different animal models. The initial observations were in IL-6 and sIL6R double transgenic mice (74, 75). In these animals, there was an enhanced hepatocyte proliferation, without any induction of injury to the liver. Hepatocyte proliferation induced liver adenomas and regenerative hyperplasia. The administration of the designer protein hyper-IL6 to mice induced a prolonged and an enhanced effect on hepatocytes due to the higher affinity of the trans-signaling effect and to less efficient internalization (76). In a liver regeneration model in which mice underwent 70% hepatectomy, hyper-IL-6 induced a stronger and a much earlier hepatocyte regeneration effect (27). Upon inducing severe liver damage in rats with D-gal, the administration of hyper-IL-6 to induce trans-signaling rescued animals through an enhanced hepatocyte regeneration and the restoration of liver damage (26). This was repeatedly shown upon the administration of the viral vector expressing hyper-IL-6 in a similar liver damage model in which trans-signaling of IL-6 induced a robust proliferation effect on hepatocytes (64). The proliferation of hepatocytes following D-gal induced liver damage and was dependent on IL-6 trans-signaling (77). Recently, we have shown that the early response of trans-signaling in the hepatectomy model indeed induces AKT activation (65). In a hepatocyte-specific deletion of SOCS3, the cellular feedback inhibitor of gp130 signaling, STAT3 activation and ERK1/2 activity are prolonged and enhanced upon partial hepatectomy (78). Additional circulating factors affect liver regeneration. Deficiency in complement C3 and C5 is associated with impaired liver regeneration following hepatectomy. This slower regeneration process is due to the suppressed activation of STAT3 and AKT (79). These results again highlight the importance of these signaling pathways in liver regeneration.

As mentioned above, hepatocytes are among the few cell types, which express IL-6R and are therefore responsive to the cytokine IL-6. Therefore, it was puzzling that IL-6/sIL-6R double transgenic mice showed massive hepatocyte proliferation even without hepatectomy, whereas IL-6 single transgenic mice did not show such a phenotype (74, 75, 80). Although these findings were in line with the fact that injection of IL-6 alone did not accelerate liver regeneration, the application of Hyper-IL-6 resulted in a drastic amplification of the liver regeneration response (26, 27, 64). This apparent contradiction is explained by the fact that hepatocytes, like most IL-6R expressing cells, express far more gp130 on the cell surface than IL-6R (Fig. 4). Stimulation with IL-6 alone leads to STAT3 activation and a subsequent IL-6 response. In the presence of Hyper-IL-6, however, far more gp130 molecules are stimulated leading to a more profound STAT3 activation and response. We have shown that the presence of sIL-6R renders IL-6 responsive cells far more sensitive to IL-6 in vitro and in vivo (81, 82). We therefore speculate that for the onset of hepatocyte proliferation, a threshold of activated STAT3 proteins has to be overcome and that this threshold is not reached when stimulation with IL-6 alone occurs (26, 27, 64, 75, 80).

6 The Effect of IL6-gp130 on Stem Cells

The spontaneous differentiation of mouse embryonic stem cells can be blocked by the gp130 cytokine LIF. Upon the activation of gp130 by LIF—the STAT3 pathway is activated. The role of STAT3 in stemness was previously reported (61). STAT3 promotes the expression of KLF4 and Nanog, although not all investigations agree with this observation. Recent reports further indicate the importance of STAT3 signaling in pluripotency. STAT3 was shown to be responsible for the conversion of EpiSC back to embryonic stem cells (83). The role of signaling and trans-signaling in murine and human embryonic stem cells showed that gp130 stimulation activated the three downstream pathways of ERK, AKT and STAT3, although it did not prevent the loss of pluripotent markers in human cells (39). These conflicting results and the exact role of IL-6 still awaits further investigation to better determine the role of IL-6 signaling and trans-signaling in controlling stemness–differentiation–proliferation of embryonic stem cells. We have shown that in murine embryonic stem cells, a prolonged gp130 stimulation—mimicking the effect of hyper IL6/IL6 trans-signaling, prevented differentiation and supported stemness through STAT3 activation (84).

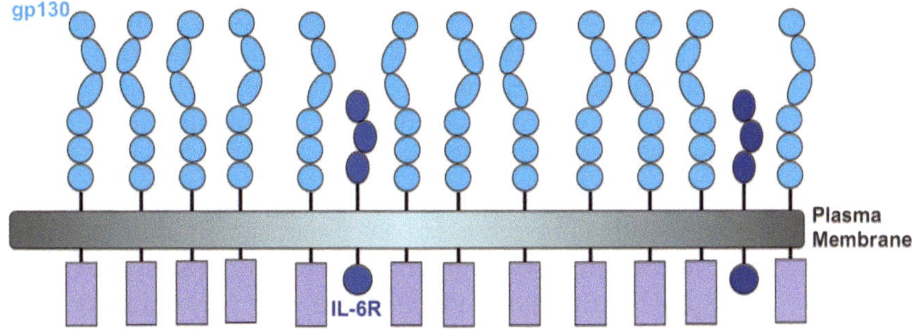

Fig. 4 Cellular expression of IL-6R and gp130. Most IL-6R expressing cells such as hepatocytes express far more gp130 molecules on the cell surface than IL-6R. These cells are responsive to IL-6 but become far more responsive to the complex of IL-6 and sIL-6R

The role of gp130 signaling is also most important in adult stem cell physiology. Muscle satellite cells are activated upon tissue injury. The proliferative response of satellite cells upon injury is dependent on IL-6 and STAT3 activation (85). The significance of mesenchymal stem cells is becoming apparent as a therapeutic platform (86). Upon integration into damaged tissue, these cells differentiate and function as parenchymal cells. We have shown that IL-6 trans-signaling is essential for the hepatic differentiation of mesenchymal stem cells in the damaged liver (87). Hyper-IL-6 through trans-signaling is important for neurogenesis and gliogenesis of neuronal stem cells, through MAPK/CREB and STAT3, respectively (30). Oval cells are liver progenitors. In IL-6 knockout mice, oval cell population is significantly reduced. IL-6 trans-signaling induces an enhanced STAT3 activation signal leading to an increased proliferation of oval cells (88). In the hematopoietic system, IL-6 trans-signaling supports the maintenance and proliferation of hematopoietic stem cells (32, 80, 89, 90).

7 The Role of the IL6-gp130 in Tissue Protection

In addition to its role in tissue regeneration, the IL-6-gp130 pathway was shown to be important for the protection from tissue injury. Interestingly, the α-cells in the pancreas express functional IL-6R. IL-6 is essential to preserve the α-cell mass under a high-fat diet stress (91). The importance of IL-6 in liver protection was shown in numerous examples in which the liver was under stress. The plasma level of IL-6 in acute and chronic liver injury is increased, possibly serving to signal through IL-6-gp130 to coordinate protective measures (92–94). In an acute inflammation model of bile duct ligation, the signaling through IL-6-gp130 is essential to protect from liver failure (95). Signaling through IL-6-gp130 protects from CCl_4-mediated liver injury (93) as well as in immune-mediated events modeled by Con-A induced hepatitis (96). In rats receiving CCl_4 that causes severe liver fibrosis (in a way modeling cirrhosis), IL-6 treatment before partial hepatectomy had a protective effect associated with an increased survival and STAT3 and AKT activation (97, 98). In the CCl_4 model, the importance of IL-6 trans-signaling was also shown. Blocking IL-6 trans-signaling increased liver damage in this model (99). The mechanism of IL-6 protection was also investigated. It was shown that IL-6 inhibits oxidative tissue injury by inducing anti-oxidative proteins including ref-1 and GPX1 (100). Many of the pro-survival, anti apoptotic (Bcl2, Bcl-XL, survivin, and FLIP) and cell cycle progression genes (cyclin A, cyclin D1-3, cyclin E, c-myc, cdc25A, and c-fos; and downregulation of p21 and p27) are STAT3 targets. The IL-6-gp130 signaling is, however, not tissue protective in all tissues. In the skin, enhanced IL-6-gp130 signaling slows wound healing due to hyperproliferation of keratinocytes (101).

8 Heart Regeneration: The Role of IL-6-gp130 Signaling

In recent years, the pivotal role of STAT3 signaling in the heart has been explored. The role of STAT3 in tissue protection and regeneration has been reported (102). These effects were apparent through specific gene regulation, prevention of injury, attenuated under clinical risk conditions for coronary heart disease and reduction of reactive oxygen species (ROS). These effects include the following: (1) The transcription factor Pim-1 was reported to be essential for cardiac progenitor cell proliferation; in addition, this factor is important in mediating myocardium contractility. Pim-1 is regulated positively by STAT3 and AKT, which both lie downstream of IL-6-gp130 signaling. Recently, it has been shown that Pim-1 is down-regulated in diabetic cardiomyopathy. The expression of Pim-1 in a diabetic heart improves cardiac function (103). Although the role of the IL-6-gp130 pathway was not specifically investigated, it is appealing to envision that Hyper-IL-6 could at least induce the same protective effect. (2) It is becoming clear that mammalians as well as fish hearts have the potential for cardiac regeneration. In the Zebrafish, it was shown that the heart could regenerate both following cryoinjury and coronary occlusion (104–106). The mouse heart was also shown to be able to regenerate at the very early stages of the postnatal period (107). Based on the role of STAT3 in heart regeneration, this effect could possibly induce both progenitor proliferation (0.25–1% of cardiomyocytes) and mature cardiomyocytes (108, 109). In humans, it has been recently estimated that about 1% of cardiomyocytes proliferate during a one year period at the age of 25 years (110). (3) In cardiomyocyte-restricted STAT3 knockout mice, STAT3 was demonstrated to be required for a preconditioned phenotype by up-regulating cardioprotective proteins including HO-1 and COX-2 and the anti-apoptotic proteins Bcl-x_L, Mcl-1, c-$FLIP_L$, and c-$FLIP_S$ (111). Interestingly, we have previously also shown that Hyper-IL-6 trans-signaling through STAT3 encounters a tissue protective effect in the kidney (66). These observations further support previous reports stating that the conditional cardiomyocyte-specific STAT3 KO mice were susceptible to cardiac injury caused by myocardial ischemia. Cardiomyocyte-specific STAT3-KO mice were also more prone to age-related heart failure. The cardioprotective effect of STAT3 is also induced through the SDF/CXCR4 signaling in an ischemic/reperfusion model (112). (4) Mice with a cardiomyocyte-restricted deletion of gp130 develop massive cardiomyocyte apoptosis and dilated cardiomyopathy when subjected to biomechanical stress (113). These same mice are susceptible to myocardial ischemia/reperfusion injury and infarction. This was shown by increased cardiomyocyte apoptosis, larger infarct sizes, reduced cardiac function and a reduced long-term survival after infarction (114). Interestingly, it was also reported that gp130 signaling in the myocardium is essential to prevent transition to congestive failure (113). (5) Diabetes mellitus is a major risk factor for coronary heart disease. It has been recently shown by the group of Jonathan Axelrod that STAT3 signaling is attenuated in the diabetic rat at a preconditioning state (115). This effect explains the cardiomyopathy associated with diabetes mellitus. This phenomenon could explain the abrogation of cardioprotection of preconditioning. (6) AKT is activated upon stimulation of gp130 in addition to STAT3 phosphorylation. In a number of studies it was also shown that AKT activation has protective properties and possibly also a regenerative effect on the heart upon acute injury (116, 117). Interestingly, these studies show that decreasing PTEN levels by microRNA-1 allowed the activation of AKT and prevented apoptosis of cardiomyocytes encountering a protective effect. The protective effect of the Granulocyte colony-stimulating factor (G-CSF) is also mediated through AKT activation (118).

9 Nerve Growth and Regeneration: The Role of IL-6-gp130

Axonal regeneration is a process dependent on SOCS3 depletion. SOCS3 is a feedback inhibitor of gp130 signaling. Upon IL-6 interaction with gp130 and the induction of signaling, SOCS3 is activated and suppressed the gp130 downstream effects, resulting in the cessation of its activation. A prolonged and sustained activation of IL-6 signaling could overcome SOCS3 effects and enable a prolonged AKT activation. Interestingly, both, PTEN suppression, leading to AKT activation and SOCS3 inhibition are two essential components for axonal regeneration (119, 120). However, although IL-6 can increase axonal growth in vitro and in vivo (121, 122), IL-6 knockout mice show normal axon regeneration (122) indicating that additional investigations are warranted to further understand the effect of gp130 signaling on axonal regeneration.

In addition to the effect of the IL-6-gp130 signaling through the activation of AKT and SOCS3, this signaling imposes an effect on sensory neuronal growth upon activation of STAT3. In primary sensory neurons, transduction and activation of STAT3 enhanced neurite growth, transduction with SOCS3 reduced neurite outgrowth, and transduction with mutant SOCS3 enhanced neurite growth (123).

10 Kidney Regeneration and IL6-gp130 Signaling

There is growing evidence that neonephrogenesis—kidney regeneration—is not only a feature of fish, but also of mammals (124). There is also evidence suggesting that there exists a multipotent stem cell in the kidney and that the adult mammal kidney could regenerate. IL-6 was previously shown to induce kidney tubular regeneration in rats (125) and it is also produced by the tubular epithelium in the kidney (126). Based on the cumulating data on the role of IL6 trans-signaling in tissue regeneration and repair, we assessed the role of hyper-IL6 in kidney protection and regeneration. The administration of hyper-IL6 to mice before inducing kidney injury prevented severe acute kidney injury and death of animals (66). This effect was mediated through STAT3 activation and subsequent upregulation of genes responsible for tissue protection as heme oxygenase 1. It is becoming apparent that also the signaling through ERK and AKT, both downstream to gp130, are essential for the protection and regeneration following acute kidney injury (127).

11 Outlook: Possible Therapeutic Applications

The biology of IL-6 acting via membrane-bound and soluble receptors is complex. As indicated above, different physiologic signals are mediated by these two pathways called classic- and trans-signaling. The activation of gp130 is central to all our phenotypic observations in regeneration, self-renewal, survival, and anti-apoptosis. Investigating the structural molecular interactions of gp130 with its family of cytokine receptors (128), e.g., OSM or CNTF and the analysis of tissue specific gp130 KO mice (129) had taught us that there could be redundancy of ligands, e.g., OSM or CT-1 and in downstream mediators—STAT3, AKT, and ERK1/2. This overlap could explain some of the controversies in the field, while different groups propose a protective vs. a regenerative potential to the

IL-6-gp130 signaling. However, the direct interaction of the chimeric protein Hyper-IL-6 with gp130, independent of the heteroreceptor component inducing its prolonged dimerization and signaling, explains the unique phenotype observed upon administration of Hyper-IL-6 to induce tissue regeneration. The generation of the designer cytokine Hyper-IL-6 made it possible to better define cell types that, in their IL-6 response, depend on sIL-6R. Since Hyper-IL-6 is between 100 to 1,000 times more active than the separate proteins IL-6 and sIL-6R, the stimulation achieved with Hyper-IL-6 is considerably impressive. Such treatment has been shown so far to lead to life-saving therapy of animals with liver and kidney damage with further experiments on the way to explore the use of HIL-6 in heart and salivary gland regeneration.

References

1. Hirano T et al (1986) Complementary DNA for a novel human interleukin (BSF-2) that induces B lymphocytes to produce immunoglobulin. Nature 324:73–76
2. Bazan JF (1990) Haemopoietic receptors and helical cytokines. Immunol Today 11:350–354
3. Yamasaki K et al (1988) Cloning and expression of the human interleukin-6 (BSF-2/IFN beta 2) receptor. Science 241:825–828
4. Hibi M et al (1990) Molecular cloning and expression of an IL-6 signal transducer, gp130. Cell 63:1149–1157
5. Heinrich PC et al (2003) Principles of interleukin (IL)-6-type cytokine signalling and its regulation. Biochem J 374:1–20
6. Scheller J et al (2006) Updating IL-6 classic- and trans-signaling. Signal Transduct 6:240–259
7. Rose-John S et al (2006) Interleukin-6 biology is coordinated by membrane-bound and soluble receptors: role in inflammation and cancer. J Leukoc Biol 80:227–236
8. Jones S et al (2011) Therapeutic strategies for the clinical blockade of IL-6/gp130 signaling. J Clin Invest 121:3375–3383
9. Campard D et al (2006) Multilevel regulation of IL-6R by IL-6-sIL-6R fusion protein according to the primitiveness of peripheral blood-derived CD133+ cells. Stem Cells 24:1302–1314
10. Rose-John S, Heinrich PC (1994) Soluble receptors for cytokines and growth factors: generation and biological function. Biochem J 300(Pt 2):281–290
11. Romano M et al (1997) Role of IL-6 and its soluble receptor in induction of chemokines and leukocyte recruitment. Immunity 6:315–325
12. Müllberg J et al (1993) The soluble interleukin-6 receptor is generated by shedding. Eur J Immunol 23:473–480
13. Müllberg J et al (1992) Protein kinase C activity is rate limiting for shedding of the interleukin-6 receptor. Biochem Biophys Res Commun 189:794–800
14. Dimitrov S et al (2006) Sleep enhances IL-6 trans-signaling in humans. FASEB J 20(12):2174–2176
15. Scheller J et al (2011) ADAM17: a molecular switch to control inflammation and tissue regeneration. Trends Immunol 32:380–387
16. Althoff K et al (2001) Recognition sequences and structural elements contribute to shedding susceptibility of membrane proteins. Biochem J 353:663–672
17. Althoff K et al (2000) Shedding of interleukin-6 receptor and tumor necrosis factor alpha. Contribution of the stalk sequence to the cleavage pattern of transmembrane proteins. Eur J Biochem 267:2624–2631
18. Matthews V et al (2003) Cellular cholesterol depletion triggers shedding of the human interleukin-6 receptor by ADAM10 and ADAM17 (TACE). J Biol Chem 278:38829–38839
19. Mullberg J et al (1995) A metalloprotease inhibitor blocks shedding of the IL-6 receptor and the p60 TNF receptor. J Immunol 155:5198–5205
20. Black RA et al (1997) A metalloproteinase disintegrin that releases tumour-necrosis factor-alpha from cells. Nature 385:729–733
21. Moss ML et al (1997) Cloning of a disintegrin metalloproteinase that processes precursor tumour-necrosis factor-alpha. Nature 385: 733–736

22. Rose-John S, Schooltink H (2007) Cytokines are a therapeutic target for the prevention of inflammation-induced cancers. Recent Results Cancer Res 174:57–66
23. Grotzinger J et al (1999) IL-6 type cytokine receptor complexes: hexamer, tetramer or both? Biol Chem 380:803–813
24. Grotzinger J et al (1997) The family of the IL-6-type cytokines: specificity and promiscuity of the receptor complexes. Proteins 27:96–109
25. Fischer M et al (1997) I. A bioactive designer cytokine for human hematopoietic progenitor cell expansion. Nat Biotechnol 15:142–145
26. Galun E et al (2000) Liver regeneration induced by a designer human IL-6/sIL-6R fusion protein reverses severe hepatocellular injury. FASEB J 14:1979–1987
27. Peters M et al (2000) Combined interleukin 6 and soluble interleukin 6 receptor accelerates murine liver regeneration. Gastroenterology 119:1663–1671
28. Marz P et al (1998) Sympathetic neurons can produce and respond to interleukin 6. Proc Natl Acad Sci U S A 95:3251–3256
29. Marz P et al (1999) Neural activities of IL-6-type cytokines often depend on soluble cytokine receptors. Eur J Neurosci 11:2995–3004
30. Islam O et al (2009) Interleukin-6 and neural stem cells: more than gliogenesis. Mol Biol Cell 20:188–199
31. Klouche M et al (1999) Novel path to activation of vascular smooth muscle cells: up-regulation of gp130 creates an autocrine activation loop by IL-6 and its soluble receptor. J Immunol 163:4583–4589
32. Audet J et al (2001) Distinct role of gp130 activation in promoting self-renewal divisions by mitogenically stimulated murine hematopoietic stem cells. Proc Natl Acad Sci U S A 98:1757–1762
33. Baiocchi M et al (2000) An IL-6/IL-6 soluble receptor (IL-6R) hybrid protein (H-IL-6) induces EPO-independent erythroid differentiation in human CD34(+) cells. Cytokine 12:1395–1399
34. Hieronymus T et al (2008) The transcription factor repertoire of Flt3+ hematopoietic stem cells. Cells Tissues Organs 188:103–115
35. Hieronymus T et al (2005) Progressive and controlled development of mouse dendritic cells from Flt3+CD11b+ progenitors in vitro. J Immunol 174:2552–2562
36. Ding X et al (2012) Polycomb group protein bmi1 promotes hematopoietic cell development from embryonic stem cells. Stem Cells Dev 21:121–132
37. Hacker C et al (2003) Transcriptional profiling identifies Id2 function in dendritic cell development. Nat Immunol 4:380–386
38. Sere KM et al (2011) Dendritic cell lineage commitment is instructed by distinct cytokine signals. Eur J Cell Biol 91:515–523
39. Humphrey RK et al (2004) Maintenance of pluripotency in human embryonic stem cells is STAT3 independent. Stem Cells 22:522–530
40. Viswanathan S et al (2002) Ligand/receptor signaling threshold (LIST) model accounts for gp130-mediated embryonic stem cell self-renewal responses to LIF and HIL-6. Stem Cells 20:119–138
41. Jostock T et al (2001) Soluble gp130 is the natural inhibitor of soluble interleukin-6 receptor transsignaling responses. Eur J Biochem 268:160–167
42. Nowell MA et al (2003) Soluble IL-6 receptor governs IL-6 activity in experimental arthritis: blockade of arthritis severity by soluble glycoprotein 130. J Immunol 171:3202–3209
43. Nowell MA et al (2009) Therapeutic targeting of IL-6 trans signaling counteracts STAT3 control of experimental inflammatory arthritis. J Immunol 182:613–622
44. Richards PJ et al (2006) Functional characterization of a soluble gp130 isoform and its therapeutic capacity in an experimental model of inflammatory arthritis. Arthritis Rheum 54:1662–1672
45. Hurst SM et al (2001) Il-6 and its soluble receptor orchestrate a temporal switch in the pattern of leukocyte recruitment seen during acute inflammation. Immunity 14:705–714
46. Barkhausen T et al (2011) Selective blockade of interleukin-6 trans-signaling improves survival in a murine polymicrobial sepsis model. Crit Care Med 39:1407–1413
47. Greenhill CJ et al (2011) IL-6 trans-signaling modulates TLR4-dependent inflammatory responses via STAT3. J Immunol 186:1199–1208
48. Atreya R et al (2000) Blockade of interleukin 6 trans signaling suppresses T-cell resistance against apoptosis in chronic intestinal inflammation: evidence in crohn disease and experimental colitis in vivo. Nat Med 6:583–588
49. Mitsuyama K et al (2006) STAT3 activation via interleukin 6 trans-signalling contributes to ileitis in SAMP1/Yit mice. Gut 55:1263–1269

50. Becker C et al (2004) TGF-beta suppresses tumor progression in colon cancer by inhibition of IL-6 trans-signaling. Immunity 21:491–501
51. Becker C et al (2005) IL-6 signaling promotes tumor growth in colorectal cancer. Cell Cycle 4:217–220
52. Matsumoto S et al (2010) Essential roles of IL-6 trans-signaling in colonic epithelial cells, induced by the IL-6/soluble-IL-6 receptor derived from lamina propria macrophages, on the development of colitis-associated premalignant cancer in a murine model. J Immunol 184:1543–1551
53. Lo CW et al (2011) IL-6 trans-signaling in formation and progression of malignant ascites in ovarian cancer. Cancer Res 71:424–434
54. Lesina M et al (2011) Stat3/Socs3 activation by IL-6 transsignaling promotes progression of pancreatic intraepithelial neoplasia and development of pancreatic cancer. Cancer Cell 19:456–469
55. Scheller J et al (2011) The pro- and anti-inflammatory properties of the cytokine interleukin-6. Biochim Biophys Acta 1813:878–888
56. Heinrich PC et al (1990) Interleukin-6 and the acute phase response. Biochem J 265:621–636
57. Bollrath J et al (2009) gp130-mediated Stat3 activation in enterocytes regulates cell survival and cell-cycle progression during colitis-associated tumorigenesis. Cancer Cell 15:91–102
58. Grivennikov S et al (2009) IL-6 and STAT3 signaling is required for survival of intestinal epithelial cells and colitis associated cancer. Cancer Cell 16:103–113
59. Ernst M, Jenkins BJ (2004) Acquiring signalling specificity from the cytokine receptor gp130. Trends Genet 20:23–32
60. Smith AG et al (1988) Inhibition of pluripotential embryonic stem cell differentiation by purified polypeptides. Nature 336:688–690
61. Matsuda T et al (1999) STAT3 activation is sufficient to maintain an undifferentiated state of mouse embryonic stem cells. EMBO J 18:4261–4269
62. Ying QL et al (2008) The ground state of embryonic stem cell self-renewal. Nature 453:519–523
63. Cressman DE et al (1996) Liver failure and defective hepatocyte regeneration in interleukin-6-deficient mice. Science 274: 1379–1383
64. Hecht N et al (2001) Hyper-IL-6 gene therapy reverses fulminant hepatic failure. Mol Ther 3:683–687
65. Nechemia-Arbely Y et al (2011) Early hepatocyte DNA synthetic response posthepatectomy is modulated by IL-6 trans-signaling and PI3K/AKT activation. J Hepatol 54:922–929
66. Nechemia-Arbely Y et al (2008) IL-6/IL-6R axis plays a critical role in acute kidney injury. J Am Soc Nephrol 19:1106–1115
67. Chalaris A et al (2010) Critical role of the disintegrin metalloprotease ADAM17 for intestinal inflammation and regeneration in mice. J Exp Med 207:1617–1624
68. Garbers C et al (2011) Species specificity of ADAM10 and ADAM17 proteins in interleukin-6 (IL-6) trans-signaling and novel role of ADAM10 in inducible IL-6 receptor shedding. J Biol Chem 286:14804–14811
69. Chalaris A et al (2007) Apoptosis is a natural stimulus of IL6R shedding and contributes to the proinflammatory trans-signaling function of neutrophils. Blood 110: 1748–1755
70. Matsushita K et al (2005) Interleukin-6/soluble interleukin-6 receptor complex reduces infarct size via inhibiting myocardial apoptosis. Lab Invest 85:1210–1223
71. Malato Y et al (2011) Fate tracing of mature hepatocytes in mouse liver homeostasis and regeneration. J Clin Invest 121:4850–4860
72. Mitchell C, Willenbring H (2008) A reproducible and well-tolerated method for 2/3 partial hepatectomy in mice. Nat Protoc 3:1167–1170
73. Jackson LN et al (2008) PI3K/Akt activation is critical for early hepatic regeneration after partial hepatectomy. Am J Physiol Gastrointest Liver Physiol 294:G1401–G1410
74. Maione D et al (1998) Coexpression of IL-6 and soluble IL-6R causes nodular regenerative hyperplasia and adenomas of the liver. EMBO J 17:5588–5597
75. Schirmacher P et al (1998) Hepatocellular hyperplasia, plasmacytoma formation, and extramedullary hematopoiesis in interleukin (IL)-6/soluble IL-6 receptor double-transgenic mice. Am J Pathol 153:639–648
76. Peters M et al (1998) In vivo and in vitro activities of the gp130-stimulating designer cytokine Hyper-IL-6. J Immunol 161:3575–3581
77. Drucker C et al (2009) Interleukin-6 trans-signaling regulates glycogen consumption after D-galactosamine-induced liver damage. J Interferon Cytokine Res 29:711–718
78. Riehle KJ et al (2008) Regulation of liver regeneration and hepatocarcinogenesis by suppressor of cytokine signaling 3. J Exp Med 205:91–103

79. Markiewski MM et al (2009) The regulation of liver cell survival by complement. J Immunol 182:5412–5418
80. Peters M et al (1997) Extramedullary expansion of hematopoietic progenitor cells in interleukin (IL)-6-sIL-6R double transgenic mice. J Exp Med 185:755–766
81. Peters M et al (1996) The function of the soluble interleukin 6 (IL-6) receptor in vivo: sensitization of human soluble IL-6 receptor transgenic mice towards IL-6 and prolongation of the plasma half-life of IL-6. J Exp Med 183:1399–1406
82. Schooltink H et al (1991) Structural and functional studies on the human hepatic interleukin-6 receptor. Molecular cloning and overexpression in HepG2 cells. Biochem J 277(Pt 3):659–664
83. Hanna JH (2010) The STATs on naive iPSC reprogramming. Cell Stem Cell 7:274–276
84. Stuhlmann-Laeisz C et al (2006) Forced dimerization of gp130 leads to constitutive STAT3 activation, cytokine-independent growth, and blockade of differentiation of embryonic stem cells. Mol Biol Cell 17:2986–2995
85. Serrano AL et al (2008) Interleukin-6 is an essential regulator of satellite cell-mediated skeletal muscle hypertrophy. Cell Metab 7:33–44
86. Williams AR, Hare JM (2011) Mesenchymal stem cells: biology, pathophysiology, translational findings, and therapeutic implications for cardiac disease. Circ Res 109:923–940
87. Lam SP et al (2010) Activation of interleukin-6-induced glycoprotein 130/signal transducer and activator of transcription 3 pathway in mesenchymal stem cells enhances hepatic differentiation, proliferation, and liver regeneration. Liver Transpl 16:1195–1206
88. Yeoh GC et al (2007) Opposing roles of gp130-mediated STAT-3 and ERK-1/2 signaling in liver progenitor cell migration and proliferation. Hepatology 45:486–494
89. Gotze KS et al (2001) gp130-stimulating designer cytokine Hyper-interleukin-6 synergizes with murine stroma for long-term survival of primitive human hematopoietic progenitor cells. Exp Hematol 29:822–832
90. Kimura T et al (2000) Signal through gp130 activated by soluble interleukin (IL)-6 receptor (R) and IL-6 or IL-6R/IL-6 fusion protein enhances ex vivo expansion of human peripheral blood-derived hematopoietic progenitors. Stem Cells 18:444–452
91. Ellingsgaard H et al (2008) Interleukin-6 regulates pancreatic alpha-cell mass expansion. Proc Natl Acad Sci U S A 105: 13163–13168
92. Dierssen U et al (2008) Molecular dissection of gp130-dependent pathways in hepatocytes during liver regeneration. J Biol Chem 283:9886–9895
93. Streetz KL et al (2003) Lack of gp130 expression in hepatocytes promotes liver injury. Gastroenterology 125:532–543
94. Wuestefeld T et al (2003) Interleukin-6/glycoprotein 130-dependent pathways are protective during liver regeneration. J Biol Chem 278:11281–11288
95. Wuestefeld T et al (2005) Lack of gp130 expression results in more bacterial infection and higher mortality during chronic cholestasis in mice. Hepatology 42:1082–1090
96. Klein C et al (2005) The IL-6-gp130-STAT3 pathway in hepatocytes triggers liver protection in T cell-mediated liver injury. J Clin Invest 115:860–869
97. Tiberio GA et al (2008) IL-6 Promotes compensatory liver regeneration in cirrhotic rat after partial hepatectomy. Cytokine 42:372–378
98. Tiberio GA et al (2007) Interleukin-6 sustains hepatic regeneration in cirrhotic rat. Hepatogastroenterology 54:878–883
99. Gewiese-Rabsch J et al (2010) Role of IL-6 trans-signaling in CCl induced liver damage. Biochim Biophys Acta 1802:1054–1061
100. Jin X et al (2007) Interleukin-6 inhibits oxidative injury and necrosis after extreme liver resection. Hepatology 46:802–812
101. Zhu BM et al (2008) SOCS3 negatively regulates the gp130-STAT3 pathway in mouse skin wound healing. J Invest Dermatol 128:1821–1829
102. Hilfiker-Kleiner D et al (2005) Many good reasons to have STAT3 in the heart. Pharmacol Ther 107:131–137
103. Katare R et al (2011) Intravenous gene therapy with PIM-1 via a cardiotropic viral vector halts the progression of diabetic cardiomyopathy through promotion of prosurvival signaling. Circ Res 108:1238–1251
104. Chablais F et al (2011) The zebrafish heart regenerates after cryoinjury-induced myocardial infarction. BMC Dev Biol 11:21
105. Jopling C et al (2010) Zebrafish heart regeneration occurs by cardiomyocyte dedifferentiation and proliferation. Nature 464:606–609
106. Schnabel K et al (2011) Regeneration of cryoinjury induced necrotic heart lesions in zebrafish is associated with epicardial activation and cardiomyocyte proliferation. PLoS One 6:e18503

107. Porrello ER et al (2011) Transient regenerative potential of the neonatal mouse heart. Science 331:1078–1080
108. Cai CL et al (2008) A myocardial lineage derives from Tbx18 epicardial cells. Nature 454:104–108
109. Zhou B et al (2008) Epicardial progenitors contribute to the cardiomyocyte lineage in the developing heart. Nature 454:109–113
110. Bergmann O et al (2009) Evidence for cardiomyocyte renewal in humans. Science 324:98–102
111. Bolli R et al (2011) A murine model of inducible, cardiac-specific deletion of STAT3: its use to determine the role of STAT3 in the upregulation of cardioprotective proteins by ischemic preconditioning. J Mol Cell Cardiol 50:589–597
112. Huang C et al (2011) SDF-1/CXCR4 mediates acute protection of cardiac function through myocardial STAT3 signaling following global ischemia/reperfusion injury. Am J Physiol Heart Circ Physiol 301:H1496–H1505
113. Hirota H et al (1999) Loss of a gp130 cardiac muscle cell survival pathway is a critical event in the onset of heart failure during biomechanical stress. Cell 97:189–198
114. Hilfiker-Kleiner D et al (2004) Signal transducer and activator of transcription 3 is required for myocardial capillary growth, control of interstitial matrix deposition, and heart protection from ischemic injury. Circ Res 95:187–195
115. Drenger B et al (2011) Diabetes blockade of sevoflurane postconditioning is not restored by insulin in the rat heart: phosphorylated signal transducer and activator of transcription 3- and phosphatidylinositol 3-kinase-mediated inhibition. Anesthesiology 114: 1364–1372
116. Glass C, Singla DK (2011) MicroRNA-1 transfected embryonic stem cells enhance cardiac myocyte differentiation and inhibit apoptosis by modulating the PTEN/Akt pathway in the infarcted heart. Am J Physiol Heart Circ Physiol 301:H2038–H2049
117. Rajesh KG et al (2005) Hydrophilic bile salt ursodeoxycholic acid protects myocardium against reperfusion injury in a PI3K/Akt dependent pathway. J Mol Cell Cardiol 39:766–776
118. Takahama H et al (2006) Granulocyte colony-stimulating factor mediates cardioprotection against ischemia/reperfusion injury via phosphatidylinositol-3-kinase/Akt pathway in canine hearts. Cardiovasc Drugs Ther 20:159–165
119. Smith PD et al (2009) SOCS3 deletion promotes optic nerve regeneration in vivo. Neuron 64:617–623
120. Sun F et al (2011) Sustained axon regeneration induced by co-deletion of PTEN and SOCS3. Nature 480:372–375
121. Cafferty WB et al (2001) Leukemia inhibitory factor determines the growth status of injured adult sensory neurons. J Neurosci 21:7161–7170
122. Cao Z et al (2006) The cytokine interleukin-6 is sufficient but not necessary to mimic the peripheral conditioning lesion effect on axonal growth. J Neurosci 26:5565–5573
123. Miao T et al (2006) Suppressor of cytokine signaling-3 suppresses the ability of activated signal transducer and activator of transcription-3 to stimulate neurite growth in rat primary sensory neurons. J Neurosci 26:9512–9519
124. Benigni A et al (2010) Kidney regeneration. Lancet 375:1310–1317
125. Homsi E et al (2002) Interleukin-6 stimulates tubular regeneration in rats with glycerol-induced acute renal failure. Nephron 92: 192–199
126. Boswell RN et al (1994) Interleukin 6 production by human proximal tubular epithelial cells in vitro: analysis of the effects of interleukin-1 alpha (IL-1 alpha) and other cytokines. Nephrol Dial Transplant 9: 599–606
127. Feliers D, Kasinath BS (2011) Erk in kidney diseases. J Signal Transduct 2011:768512
128. Skiniotis G et al (2008) Structural organization of a full-length gp130/LIF-R cytokine receptor transmembrane complex. Mol Cell 31:737–748
129. Müller P et al (2008) Identification of JAK/STAT pathway regulators–insights from RNAi screens. Semin Cell Dev Biol 19:360–369

Chapter 5

Brain Ischemic Injury in Rodents: The Protective Effect of EPO

Annelise Letourneur, Edwige Petit, Simon Roussel, Omar Touzani, and Myriam Bernaudin

Abstract

Animal models constitute an indispensable tool to investigate human pathology. Here we describe the procedure to induce permanent and transient cerebral ischemia in the mouse and the rat. The model of transient occlusion of the middle cerebral artery (MCA) is performed by the insertion of an occlusive filament until the origin of the MCA while the permanent occlusion described in the mice is performed by a distal electrocoagulation of the MCA. Those models allow evaluating the efficiency of therapeutic strategy of ischemia from tissular aspect to behavioral and cognitive impairment assessment. They were widely used in the literature to evaluate the efficiency of different drugs including the cytokines and especially erythropoietin (EPO) or its derivatives.

Key words Ischemia, Neuroprotection, Middle cerebral artery occlusion, Neurological deficits, Stroke, Cytokines

1 Introduction

Stroke is the third most common cause of death after heart diseases and cancer and is the first cause of adult long-term disability in developed countries. The very high prevalence of stroke, the lack of an appropriate therapy, and the high cost of stroke burden make this pathology a social concern (1, 2). Furthermore, the burden of this pathology is growing worldwide, typically related to an increase in cardiovascular risk factors and also because of the increasing age of the population (3–5).

A stroke corresponds to a disturbance in the blood supply to the brain and can be classified as hemorrhagic (15% of cases) or ischemic (85% of cases). The latter can be thrombotic or embolic.

Stroke results, in the worst case, in death, but most often it causes motor, sensorial, and cognitive impairments and patients surviving stroke often require institutionalization.

Pietro Ghezzi and Anthony Cerami (eds.), *Tissue-Protective Cytokines: Methods and Protocols*, Methods in Molecular Biology, vol. 982, DOI 10.1007/978-1-62703-308-4_5, © Springer Science+Business Media, LLC 2013

Although stroke occurs mostly in elderly people, many other modified factors, such as arterial hypertension, atrial fibrillation, atherosclerosis, diabetes, and smoking are well known to constitute risk factors (6–8). Today, the management of those risk factors remains the best primary prevention against stroke onset (6, 8).

Most of what we know about stroke relied on studies conducted on animal models of stroke. These models are an important tool to better understand this pathology and to propose therapeutic strategies (9–11). The ideal animal stroke model does not exist because just one model cannot correlate with the heterogeneity encountered in human pathology. Each stroke model has its particular strengths and weaknesses and the experimental model should be carefully selected depending on the purpose of the study (especially the clinical situation under investigation) and different models should be considered to investigate a given hypothesis. The model should mimic the human condition under investigation as closely as possible. This implies to take into account different aspects of the pathology ranging from the condition of the onset to the outcomes. It is also important to take into account the feasibility, the cost, and the strength of the statistical evaluation.

Experimental stroke models include models of global and focal cerebral ischemia. The former mimics brain ischemia induced by cardiac arrest or sustained severe hypotension, while the latter mimics the occlusion of a brain vessel. Ischemic focal stroke is the most frequently encountered in human clinic. Another aspect to take into consideration is the duration of the occlusion. Focal brain ischemia models may mimic transient or permanent ischemia. Most human patients suffering from ischemic stroke experience reperfusion, so that experimental models including a reperfusion phase are commonly used. The duration of the occlusion has important consequences since the lesion size is highly dependent on the duration of the occlusion.

Although animal models of ischemic stroke are essential tools to study the pathophysiology of ischemic brain injury, they are also the key to develop therapeutic approach. Major approaches developed to treat acute ischemic stroke fall into two categories: thrombolysis and neuroprotection. This issue has been addressed in many recent reviews describing the effect of many drugs and the actual concept evolves towards a combination of thrombolysis and neuroprotection (12, 13). More than 1,000 candidate neuroprotective drugs were tested in preclinical experiments (focal ischemia, global ischemia, and cell culture) and many of them were proven efficient and were subsequently tested in more than 100 clinical trials (12, 14–17). Unfortunately, despite this impressive number of drugs proven effective in animal studies, no one has really been proven efficient in human. This failure of translation from the bench to the bedside has been extensively discussed in the recent literature. It brought several reports aiming to improve this translation

(11, 18–25). Many explanations have been suggested for this failure and several recommendations were published aiming to improve the quality of both preclinical and clinical studies (18, 20). Among them, the main criticism relies on the animal model used, its relevance to the problem studied and the fact that several models should be used to prove the efficiency of a drug.

Based on the above, in this chapter, we focus on two animal models of focal cerebral ischemia. There are different approaches to induce a focal ischemic lesion. The models of permanent ischemia induced by electrocoagulation originally developed by Tamura and colleagues (26) and the permanent or transient intraluminal filament models initially developed by Longa and colleagues (27) are the most widely used. They were performed on different species and were adapted or slightly modified since their first description. Therefore, in the following sections we describe in detail the procedure to induce brain focal ischemia through proximal transient occlusion of the MCA (middle cerebral artery) in the rat (27) (intraluminal filament model) and through permanent distal occlusion of the MCA in the mouse (26) (electrocoagulation model).

2 Material

2.1 Rats

Animals need to be acclimated before any experiment (animals should be kept in an appropriate animal facility at least 5 days prior to the experiment).

The choice of the animal to be used in the experiment is of primary importance. Indeed, it is well known that the response to ischemia depends on several factors related to age (see Note 1) and strain (see Note 2). Comorbidity factors (see Note 3) and gender (see Note 4) may also influence the results. In the present example, we use male Sprague-Dawley rats (weighing from 280 to 320 g).

2.2 Mice

Male mice, 25–30 g are acclimated to the animal house for at least 5 days before experiments. All animals are housed in a 12 h day–night cycle (from 8 to 87) at a temperature of 22°C and humidity of 60%. They have access to water and food (commercial pellets) at libitum.

For obvious reasons, animals need to be housed in individual cages, at least at the acute phase after brain ischemia induction. The evolution of animal's weight and its general status should be monitored every day after surgery.

All the surgical procedure should be performed in aseptic conditions using sterilized instrument and materials. During the surgical procedure, put surgical tools on a sterile place (usually on the back of the surgical tool following its sterilization). If the surgical tools have been put by error on a non-sterilized space, re-sterilize it quickly using the bench sterilizer (few seconds).

During surgical procedure, check regularly the degree of anesthesia. The animal is well anesthetized when a loss of sensory/reflex response is observed, i.e., nonresponse to tail pinching, or paw pinching.

2.3 Equipment

1. Stereomicroscope.
2. Homeothermic blanket system with rectal probe.
3. Laser Doppler monitor with 0.7 mm diameter probe (e.g., Laser Doppler Perfusion Monitor Moor Instruments) and a probe holder.
4. Cranial headbolt (Kopf Instruments).
5. Bench Sterilizer.
6. Bench Hoover.
7. Blood gazes analyzer.
8. Computerized blood pressure monitoring (e.g., Blood Pressure Display Unit, Stoelting).
9. Microcapnograph (e.g., Micro-CapnoGraph CI240, Columbus Instruments).
10. Surgical Bone Drill System.
11. Electrosurgical unit (Harvard Apparatus).
12. Autoclave.
13. Animal shaver/hair clipper.
14. Common surgical tools set (fine biology tips forceps, hemostats, retractor, alm retractor, mouth retractor, needle holder, scalpel handle and blades, straight fine scissors, micro-scissors).
15. Cotton swabs.
16. Tape.
17. Syringes.
18. Needles.
19. Capillary tubes.
20. Nonabsorbable 5-0 silk black suture.
21. Nonabsorbable 3-0 silk black suture.
22. Carbon still burrs.
23. Polyethylene catheters.

2.4 Reagents

1. Isoflurane.
2. 30% oxygen/70% nitrous oxide.
3. Saline (0.9% sodium chloride).
4. Surgical betadine.
5. Heparin.

6. Bupivacaine.
7. Ketoprofen.
8. Xylocaine 2% jelly.
9. Lidocaine spray.
10. Artificial tears.
11. Antibiotic ointment (e.g., Neosporin or Polysporin).

3 Methods

3.1 Model of Intraluminal Middle Cerebral Artery Occlusion in the Rat

This model was first reported by Longa et al. (27), and since then, different variants have been reported (27–29). This model is commonly named the thread model or the intraluminal suture model. This procedure induces a permanent or transient proximal occlusion of the MCA. Here, we describe the method to induce transient MCA in the rat. This procedure can be applied to other species such as mice (30, 31).

3.1.1 Preparation of the Occlusive Filament

MCAO could be performed through the use of different filaments. The most employed method consists in the use of a silicon-rubber-coated, glue-coated, resin-coated or nail polish-coated filament. Flame blunted filaments should be avoided as they result in an exacerbated damage and hemorrhage. Uncoated filaments would not procure constant and reproducible occlusion (44).

The optimal diameter of the filament, for a certain body weight and strain of rat, should be determined in pilot experiments. Usually, the optimal diameter for rats weighing from 280 to 320 g is 0.380 mm (45). The length of the coated filament should also be determined with the same procedure because if too long, it could block the flow from other arteries, while a too short one would not produce optimal MCA occlusion.

In Note 5, we explain how to prepare a thermofusible glue-coated filament of 0.380 mm diameter and 2.5 mm length (Fig. 1). These threads were used in numerous published papers from our laboratory (40, 46–49). If you want to avoid this step you can also purchase the occlusive filament. Some commercials produce occluders available in different size and diameters.

3.1.2 Laser Doppler Flowmetry Measurements (Optional)

The use of laser Doppler flowmetry (LDF) is important since it allows to confirm the success and the persistence of the MCA occlusion.

Indicative cerebral blood flow (CBF) is measured in the territory supplied by the MCA of the right hemisphere. To this end:

1. Position the head of the rat into a stereotaxic frame (David Kopf Instruments).

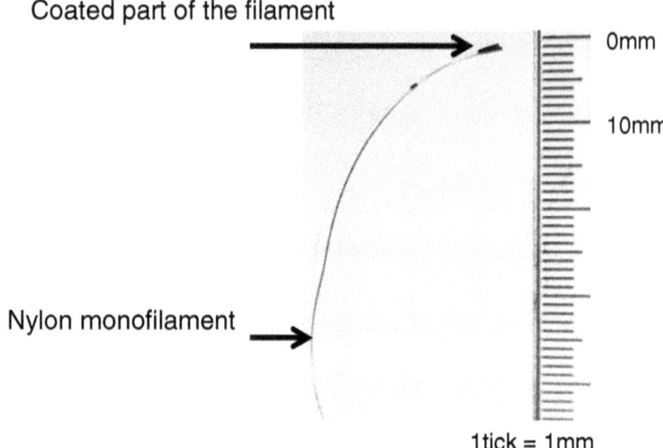

Fig. 1 Thermofusible homemade glue-coated filament of 0.380 mm diameter and 2.5 mm length

2. Perform a medial incision in the scalp with scalpel after injection of a drop of bupivacaine.
3. Dissect the temporalis muscle to expose the area of interest (coordinates 1.5 mm posterior, 5.5 mm lateral to the bregma). Refine the parietal bone with a saline-cooled dental drill on a 2 mm circle diameter.
4. Position the calibrated LDF probe on the thinned parietal bone. The LDF probe holder is positioned on the stereotaxic frame and so even when this one is tilted to induce ischemia, the LDF probe can't move.
5. Relative CBF (rCBF) is then recorded before and after the introduction of the filament into the carotid artery.

3.1.3 Physiological Parameters Monitoring

During the whole procedure, rectal temperature, blood pressure, heart rate, end tidal CO_2 and blood gazes are monitored, especially when the experimentation is of a long duration as it is well know that the evolution of the ischemic lesion is influenced by those parameters especially the temperature. Here, we describe a full monitoring of physiological parameters in the case of the animal is maintained anesthetized during hours following MCAO. This involves artificial ventilation and invasive measurement of blood pressure, heart rate and blood samples collection.

1. Rats are anesthetized with isoflurane (5%) in a mixture of 30% oxygen and 70% nitrous oxide during 5 min in a specific box under the hood.
2. Rats are intubated and artificially ventilated. If possible, continuously monitor end tidal CO_2.

3. The respirator is set to provide artificial ventilation with a respiratory volume of 2 mL at 80 breath rates per minutes. Those parameters should be adjusted regarding the End-tidal CO_2, and the blood gazes measurements (paO_2, $paCO_2$) and pH.
4. Body temperature is monitored continuously during anesthesia with a rectal probe and maintained around 37.5°C with a thermostatically controlled heating pad. Rectal temperature could be monitored after ischemia to evaluate if the animals experience fever.

3.1.4 Induction of MCAO

The rat is subjected to MCAO by insertion of the 0.380 mm diameter coated nylon monofilament previously described. The MCA occlusion is confirmed by the reduction of CBF as measured by LDF probe (Fig. 2).

1. Under anesthesia, the rat is placed on its back on the heating pad. Rectal temperature is maintained at 37.5 ± 0.5°C.
2. Apply artificial tear on the eyes.
3. Shave the fur on the ventral neck region and disinfect the surgical site with betadine.
4. Under the dissecting microscope, a ventral right median incision of about 2 cm long is performed on the neck.
5. Using forceps, the external and internal carotid arteries (ECA-ICA) and the common carotid artery (CCA) are carefully exposed under the operating microscope. The CCA and the ICA are carefully separated from the vagus nerve. The pterygopalatine artery needs to be exposed too.
6. Once the CCA, the ECA and the ICA are isolated and are free from surrounding tissue, a permanent ligature is placed on the distal portion of the ECA. Furthermore, a knot is prepared on the proximal portion of the ECA near the permanent suture but is not tight. All knots are performed using 5.0 silk suture.
7. Transient ligatures are placed on the CCA, and the ICA to temporarily block blood flow coming through when the incision in the ECA is performed.
8. A small incision is performed in the ECA between the previously installed knots.
9. The filament is introduced into the ECA, with the coated extremity directed towards the CCA. Then the proximal knot is tightened to avoid bleeding from the ECA but allowing the filament to move easily.
10. The ECA is cut and tilted to be oriented in the direction of the ICA.
11. The knot previously placed on the ICA is removed.

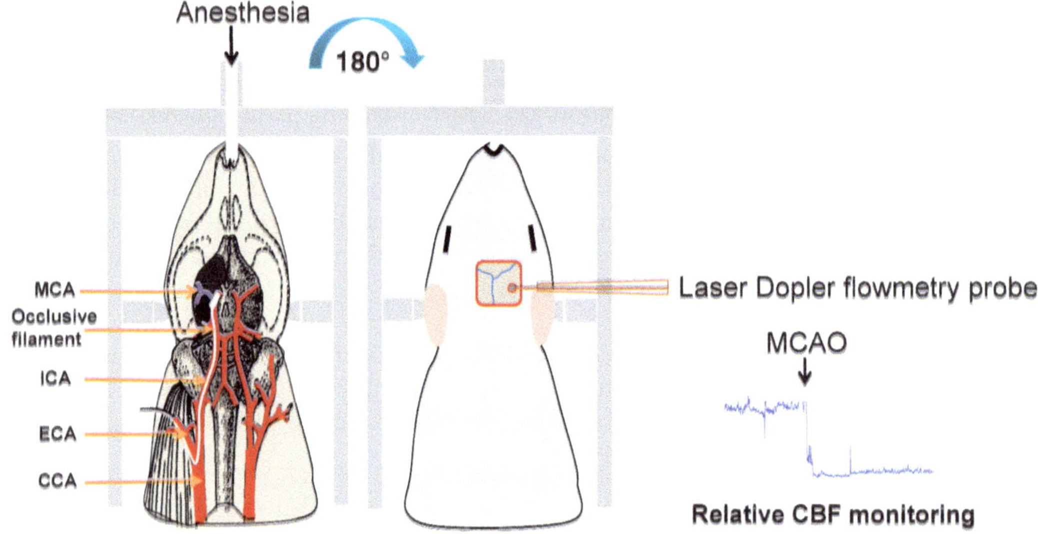

Fig. 2 Schematization of the procedure of MCAO induction

12. Then the occlusive filament is gently advanced through the ICA. Verify that the filament does not go into the pterygopalatine artery.
13. Finally, the filament is introduced along about 1.5 mm after the tympanic bubble until a resistance is felt and until the laser Doppler signal decreased to ~80% of baseline.
14. The filament is secured at the ECA by a tight ligature. The knot on the CCA is removed. The evolution of laser Doppler signal is followed for several minutes to check any modifications of blood flow. The portion of the thread remaining outside of the ECA is cut.
15. Then after about 10 min the skin incision is sutured.
16. At the end of the experiment, all the incisions are sutured with 3.0 silk nonabsorbable suture. Neosporin is applied on the wounds.
17. An administration of ketoprofen is performed (5 mg/mL IP, dose: 0.1 mL/100 g body weight as an analgesic) and animal receives wet food in the cage.

In the case of transient occlusion, the incision previously made in the neck is opened again and after a predetermined period, the suture is carefully removed and brought back to the stump of the ICA where the knot is.

3.2 Model of Permanent MCAO by Electrocoagulation in the Mice

All the points listed above regarding the choice of the strain and the age also apply to this part. This model of brain ischemia is frequently employed to investigate the effects of potential neuroprotective drugs on the extent of infarction and is also commonly performed in transgenic/knock-out mice. This model results mainly in a cortical infarction of the ipsilateral hemisphere since the occlusion affects the distal part of the MCA (50–53).

In the present example, we use male mice (weighing from 25 to 30 g).

3.2.1 Experimental Procedure

1. Induce anesthesia by exposing the mouse for 3 min to 5% isoflurane in 30% oxygen/70% nitrous oxide in a box under the hood.
2. Place the mouse in a lateral position (on its left side) on the heating blanket with the face in a mask delivering anesthesia. Place the tip of the bench hoover (vacuum cleaner) near the facial mask to avoid experimenter to inhale isoflurane.
3. A temperature probe connected to the homeothermic blanket and covered with xylocaine 2% jelly is inserted into the rectum. This allows to measure and maintain body temperature at 37.5°C.
4. The anesthesia is maintained at approximately 1.5% isoflurane and adjusted as needed.
5. Apply artificial tear on the mouse's eyes.
6. Shave the area between the right lateral part of the orbit and the right lateral auditory meatus. Clean with surgical betadine.
7. The MCA is exposed using the transtemporal approach. Realize a vertical skin incision of 0.5 cm at midpoint between the right eye and the right ear.
8. Perform an incision in the upper part of the temporalis muscle and move it further apart.
9. Use a hand drill with a 1 mm drill burr to remove parts of the skull and expose the MCA 1 mm rostral to the fusion of the zygoma and squamosal bone and about 3.5 mm ventral to the dorsal surface of the brain (50).
10. Leave intact the zygomatic arch.
11. Remove the dura mater covering the MCA.
12. Use bipolar coagulation forceps to occlude the MCA proximal to the inferior cerebral vein.
13. Transect the MCA with mini scissors where it was electrocoagulated.
14. Close the incision with 5-0 silk suture.
15. Apply antibiotic ointment on the incision (Neosporin).
16. An administration of ketoprofen is performed (5 mg/mL i.p., dose: 0.1 mL/100 g body weight as an analgesic).
17. The procedure does not last more than 30 min.
18. Give the animal wet food in the cage.

3.2.2 Evaluation of Ischemic Brain Injury

Magnetic Resonance Imaging

Magnetic resonance imaging (MRI) is the method of choice to evaluate brain ischemic lesion for three main reasons: first, it allows a noninvasive follow-up of the lesion evolution in the same animal along time, second, it allows to evaluate the lesion from minutes to months post-ischemia; third, several imaging parameters, to be functional or structural, can be acquired almost simultaneously. In fact standard T2-weighted magnetic resonance imaging (T2WI) is helpful and widely used in experimental research as in the evaluation of patients with ischemic stroke. However, these imaging modalities are useful at the chronic stage and should be used for longitudinal evaluation. Diffusion-weighted magnetic resonance imaging (DWI) that was validated using animal stroke models can demonstrate ischemic brain regions within minutes after ischemia. Furthermore, perfusion weighted imaging (PWI), is able to determine regions with reduced blood perfusion. The mismatch between the area of diffusion abnormality and the area of perfusion abnormality defines the ischemic penumbra. From a therapeutic point of view, the ischemic penumbra might be defined as a target for neuroprotective and neuroregenerative treatments. If the apparent diffusion coefficient evolves over time after stroke, the anisotropy also experience changes. Diffusion tensor imaging allows analyzing the evolution of anisotropy in the tissue. Since their development in experimental research, these techniques have been extensively applied to clinical investigations and clinical practice. Furthermore the acquisition of T2*-weighted magnetic imaging and angiography respectively allow the evaluation of potential hemorrhage and the visualization of the occluded vessel. Those different approaches are summarized in Fig. 3.

Neurologic Deficits Evaluation

As the ultimate goal for any therapeutic intervention in patients is the recovery of functional deficits, to be sensorial, motor or cognitive, the use of behavioral tests to assess the evolution of brain ischemia-induced deficits in animals is crucial (54). In general, these behavioral tests need to be sensitive enough to detect weak deficits, especially in the chronic stage as rodents are known to display a rapid and pronounced functional recovery compared to humans.

Different tests could be used to evaluate post-ischemic impairments. At the acute phase (i.e., during the first week following stroke), one can assess the neurologic deficits through the use of available neurological scales which include a battery of neurological tests comprising the spontaneous activity, the symmetry in the movement of limbs, the forepaw outstretching, the climbing, the body proprioception and the response to vibrissae touch (55). Other sensorimotor tests (accelerated rotarod test, corner test, adhesive removal test, staircase test, beam walking test, etc.) and cognitive tests (passive avoidance test, Morris water maze test, object recognition test, etc.) can be employed to assess both short- and long-term deficits. These tests have been described elsewhere (54).

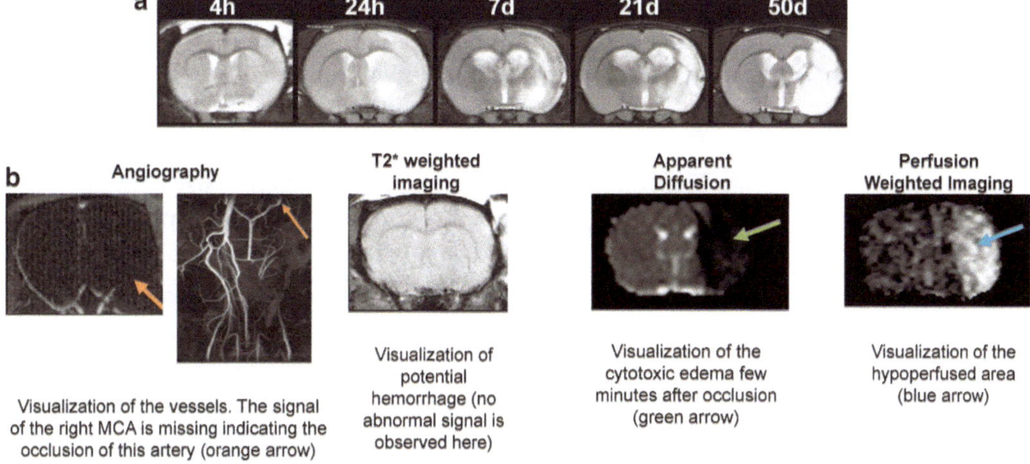

Fig. 3 MRI as a tool of choice for the evaluation of ischemic lesion. (**a**) Longitudinal evolution of the ischemic lesion in a rat subjected to 90 min of intraluminal transient MCAO. T2 weighted imaging shows clearly the vasogenic oedema at 24 h post-occlusion and the cystic cavity at latter stages. (**b**) Multiparametric evaluation of the ischemic lesion using MRI

3.3 Post-mortem Histology

3.3.1 Histological Staining Which Does Not Required Post-fixation

1. Animals are sacrificed by decapitation following overdose of anesthesia (deep isoflurane anesthesia).
2. The brain is removed from the skull and quickly frozen by rapid immersion in isopentane or liquid nitrogen (few seconds). Store it at −80°C until sectioning.

3.3.2 Histological or Immunohistological Procedures Requiring Intracardiac Perfusion and Fixation

Prepare 4% paraformaldehyde in 0.1 M PB solution as described in Note 6.

Intracardiac Perfusion

Use a perfusion pump, either a peristaltic pump or a homemade system made with air bottle connected to mercury column with tubes (see Note 7).

Wear protective eye goggles, respirator mask, and appropriate gloves during the whole perfusion process. Perform the procedure under the hood.

1. Set up perfusion pump; attach perfusion set and perfusion cannula (use the appropriate gauge size, 16–24 G, corresponding to the aorta). Have on hands the 4% paraformaldeyde in 0.1 M Phosphate Buffer, pH 7.4 and the heparinized saline solution (10 UI heparin/mL saline) in separate bottles and at room temperature or a little bit more (no more than 37°C).

2. Then place open end of perfusion tube in a beaker filled with heparinized solution. Open valve (turn on the pump) and adjust the speed you need. Fill the tube with the solution. You never want to inject air bubbles during perfusion procedure. Then turn off the pump for now.
3. The volume of solution should be scaled to size of animal (usually 150–200 mL for a rat of 250 g, about 30 mL for an adult mouse for PFA and less for heparinized saline).
4. Set up surgery site with scissors, forceps and clamps.
5. The animal to be perfused is weighed and anesthetized with either an i.p. injection ofketamine/xylazine (Xylazine (100 mg/mL)—Ketamine (100 mg/mL) solution: mix 1:8 (v/v), inject 50–100 μL/100 g i.p. (the recommended dose is 50 μL/100 g but it takes a too long wait before the rat gets deeply anesthetized). The mix can be stored at 4°C ≤6 months) or induced deeply anesthesia with isoflurane 5% for 5 min in 30% oxygen/70% nitrous oxide and maintain at 3–4% during surgery. In the first case allow 10–15 min for anesthesia to occur. Check for nonresponse to tail pinching, or paw pinching.
6. Place the animal on the operating table under the hood with its back down. Securely fix the animal with tape on each paw. Use pinch-response method to determine depth of anesthesia. Animal must be unresponsive before proceeding with the following steps.
7. Make incision with scissors just under the sternum. With sharp scissors, cut through the tissue at the bottom of diaphragm to allow access to rib cage.
8. Cut carefully the diaphragm maintaining the end of the sternum in forceps to avoid damaging the lungs or the heart.
9. Make two end horizontal cuts through the rib cage (on each side) and open up the thoracic cavity.
10. Flip the sternum using safety-lock forceps (hemostats).
11. Expose the heart dissecting the thymus tissue and visualize the aorta.
12. While holding heart steady with forceps (the heart have to still be beating, if not the perfusion will not be successful) make a small incision at the bottom tip of the left ventricle using micro-scissors.
13. Insert the cannula directly into the left ventricle and go up until you can see the cannula into the aorta.
14. Place a clamp (for example; Dietrich Bulldog Clamp, FST) near the entry point into the heart to secure the position of the cannula in the right place.
15. (Optional: clamp the descending aorta, which is visible along the spinal cord, if interested exclusively in brain perfusion.)

Comment: If fluid is coming from animal's nostrils or mouth, reposition the cannula.

16. Turn on the pump to allow perfusion with the heparinized saline and at the same time perform an incision in the right atrium.
17. When blood has been cleared from the body, change to fixative. Take care not to introduce air bubbles while transferring from one solution to the other. To avoid air bubble coming through turn off the peristaltic pump while moving the tube to the other solution and then turn it on again. The perfusion with PFA solution will induce muscle spasms ("formalin or death dance") of the whole body attesting a good perfusion.
18. Perfusion is almost complete when those spontaneous movements and lightened color of the liver are observed.
19. At the end or before perfusing another animal clean the tube and fill it again with heparinized saline. Avoid injection of PFA residue first.
20. Dispose of paraformaldehyde waste into specific containers, and wash glassware and all tools extensively.
21. Perform a decapitation and put the whole head (open the back of the skull to allow a good post-fixation of the brain) in 4% PFA in 0.1 M PB for post-fixation during 24–72 h or extract just the brain, remove the dura mater, and post-fix it in the same way. After that you can put the brain in sucrose solution and then slice it.

Cryosectioning and Staining

Section the brains coronally in 20–40 μm slices using a cryostat (Leica 3050 S), a freezing microtome, a microtome or a vibratome depending of the sample method preparation.

Mount sections on gelatin coated or positive charged plus slides (SuperFrost Ultra Plus®, SuperFrost Plus®; Menzel-Glaser). Air-dry sections during 2 or 3 days or bake slides on a slide warmer overnight.

Stain as follows:

Staining Protocol for Fresh-Frozen Sections

1. 96% Ethanol: 1′
2. 70% Ethanol: 1′
3. 50% Ethanol: 1′
4. Distilled H_2O: 1′
5. Thionin Stain: 5′ (see Note 8)
6. Distilled H_2O: 30″
7. Distilled H_2O: 30″
8. 50% Ethanol: 30″
9. 70% Ethanol: 30″
10. 96% Ethanol: 30″

11. 100% Ethanol (absolute ethanol): 30
12. Xylene: 30″
13. Xylene: 3′ = time required to coverslip slides in Eukitt.

Staining Protocol for Fixed Tissue

1. 1:1 Ethanol–chloroform 100°: 30′. This step removes lipids and facilitates cells thionin penetration.
2. 1:1 Ethanol–chloroform 100°: 30′. This step removes lipids and facilitates cells thionin penetration.
3. 70% Ethanol: 2′
4. Distilled H_2O: 2′
5. Thionin Stain: 2′
6. Distilled H_2O: 2′
7. 70% Ethanol: 2′
8. 96% Ethanol: 2′
9. 96% Ethanol + acetic acid (2 mL in 250 mL): 2′. This step improves signal differentiation.
10. 100% Ethanol: 2′
11. 100% Ethanol: 2′
12. Xylene: 2′
13. Xylene: time required to coverslip slides in Eukitt.

3.4 Analysis and Considerations on the Use of Animal Models of MCAO to Test the Neuroprotective Effects of EPO

3.4.1 Analysis

After staining and drying, the slides are scanned with high resolution. The infarct volume can be then delineated under ImageJ (Rasband W, National Institute of Health, USA). The infarcted tissue is manually delineated on each slide as is the volume of healthy tissue of each hemisphere from the beginning to the end of the lesion (Fig. 4).

The infarct volume can be calculated multiplying the sum of the delineated surfaces by the inter-slide spaces. To this purpose different equations can be used:

Infarctus volume (mm^3) = (contralateral hemisphere volume × infarct volume)/total ipsilateral volume

This formula includes edema correction.

Or:

Infarct volume (mm^3) = (healthy contralateral hemisphere volume – healthy ipsilateral hemisphere volume)

3.4.2 Use of Animal Models of MCAO to Test the Neuroprotective Effects of EPO

We described above two procedures to induce experimental ischemia. Whereas, the model of intraluminal filament occlusion and the model of permanent electrocoagulation are widely used in the literature to study the pathophysiology of stroke, they are also pertinent models to test the efficiency of therapeutic strategies.

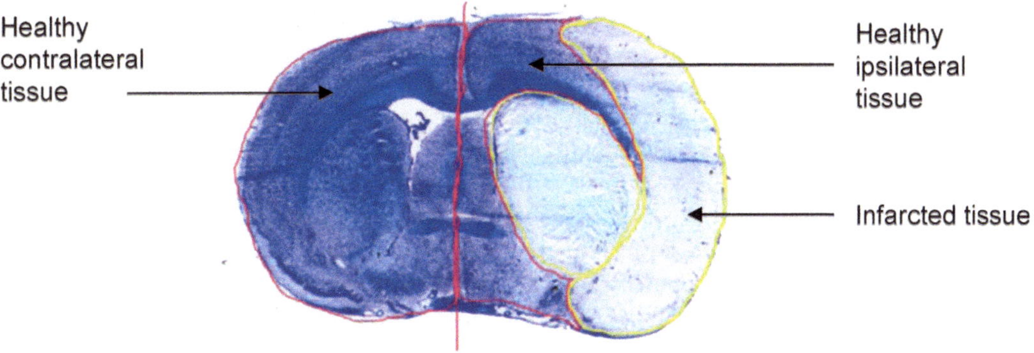

Fig. 4 Thionin histological staining on a slice of rat brain previously subjected to intraluminal MCAO

Among all the neuroprotective strategies developed and tested, several tissue-protective cytokines have been pointed out in the literature. Erythropoietin (EPO) is one of the neuroprotective cytokines but other strategies involving other cytokines, like the inhibition of interleukin-1, showed promising results. The rational for a neuroprotective effect of EPO was sustained by the existence of an endogenous expression of EPO and its receptor (EPOR) in the brain with an increase in expression rate of EPO and EPOR after a stroke. Accordingly, since the last decade, a neuroprotective effect of EPO has first been suggested in rats subjected to global ischemia (56) and in permanent focal ischemia in mice (57). Since those determining investigations, EPO treatment has shown promising features for acute treatment of brain ischemia (58–62). Local cerebral administration of rhEPO prevents ischemia-induced learning disability and neuronal death (57, 63). Those studies suggest different mechanisms underlying the neuroprotective actions of EPO independently of its hematopoietic action (56, 64–67) such as a reduction of the formation of free radicals, an anti-apoptotic activity, reduction of intracellular Ca^{2+} in the excitotoxic condition.

Since the realization that EPO has neuroprotective properties in numerous preclinical studies, it has been investigated as treatments for acute ischemic stroke. The first clinical trial was initiated in 2002. In this trial, 13 patients suffering from MCA occlusion and arriving less than 8 h after onset, received EPO intravenously daily for 3 days following stroke onset. This trial showed the absence of side effects and suggested a slight improvement of the functional recovery (using two different neurological scores), but without impact on the lesion size (61, 68). Following this pioneer clinical study, another randomized German Multicenter EPO Stroke Trial was designed on more than 500 patients to evaluate efficacy and safety of EPO in stroke. However, this time, the results published in 2009 were disappointing, showing inefficiency of EPO treatment or even an increase in the mortality rate when combined to with thrombolysis therapy (i.v. rt-PA injection) (60). Anyhow due to the adverse effects associated with long-term rhEPO administration, such as hypertension or thrombosis, experimental research focused on the dissociated effect of EPO and the future clinical application of EPO may be based on the use of new EPO derivatives (63, 69–71). Therefore, several engineered EPO derivatives have cytoprotective effects, but do not stimulate significant erythropoiesis (72). All the hematopoietic effects engendered by EPO on the bone marrow are mediated by the homodimeric EPOR (69). Desialylated EPO (or asialo EPO, generated by total enzymatic desialylation of rhEPO) can bind to the mature EPOR with the same EPOR affinity, but its very shorter plasma half-life allows reducing the hematopoietic response while maintaining its neuroprotective effects. In contrast, the second generation of EPO molecule, carbamylated EPO (CEPO, with all lysines transformed to homocitrulline by carbamylation) does not bind to the homodimeric EPOR and lacks erythropoietic activity but could confer neuroprotection. It has been suggested that EPO and CEPO neuroprotective effects might be mediated by a common beta receptor (βR), which can form a dimer with EPOR (EPOR-βR). At the moment, those non-erythropoietic EPO derivatives were tested in preclinical studies and show some promising results (71, 73–77). The neuroprotective effects of EPO and non-hematopoietic derivatives of EPO have been shown in different studies thanks to the use of these experimental stroke models. Table 1 summarizes the results coming from those studies. In parallel, two clinical trials (phase I) have been conducted to access the safety and the pharmacokinetic of CEPO administration to patients who have suffered an ischemic stroke.

3.4.3 Conclusion

In this methodological section we describe two procedures to induce experimental ischemia. The model of intraluminal filament occlusion and the model of permanent electrocoagulation are widely used in the literature to study the physiopathology of stroke and the efficiency of therapeutic strategies. They have been used to evaluate the potential neuroprotective effect of cytokines on cerebral ischemia. Table 1 summarizes the results coming from those studies. The neuroprotective effect of EPO and non-hematopoietic derivatives of EPO has been shown in different studies thanks to the use of experimental models of stroke in animals.

4 Notes

1. The age of the animal is important. To mimic stroke occurrence in adult human, it is required to work with at least adult animals (3 months or more for rats). This even if the impact of

Table 1
Studies describing the use of EPO and its analogues in experimental stroke models

MCAOt (rat)	MCAOp (rat)	MCAOt (mouse)	MCAOp (mouse)
Erythropoietin (EPO)			
Sirén et al. (64) Yu et al. (80) Aluclu et al. (81) Esneault et al. (48)	Sadamoto et al. (82) Brines et al. (56) Ding et al. (86) Wang et al. (66, 91) Villa et al. (83) Li et al. (85)	Fletcher et al. (79)	Bernaudin et al. (57) Wakida et al. (87) Li et al. (84)
Carbamylated EPO (CEPO)			
Leist et al. (89)	Wang et al. (91) Villa et al. (83)		
IgG-EPO fusion protein			
	Fu et al. (88)		Fu et al. (92)
Asialoerythropoietin			
Price et al. (77) Erbayraktar et al. (63)			
Carnasep (darbepoetin alfa)			
Belayev et al. (90)	Villa et al. (83)		
EPO-S100E			
SIOG-I-EPO	Villa et al. (83)	Gan et al. (78)	

MCAO Middle cerebral artery occlusion, *MCAOt* (temporary), *MCAOp* (permanent)

the age of rat on the brain ischemic damage is still controversial (32, 33). In any case, aging is associated with reduction in angiogenesis and degradation of some neurological function that could impact stroke evolution and outcomes. Furthermore, stroke remains a disease of the elderly in human; therefore, the use of young animals should be discarded from experimental studies.

2. The strain should be carefully chosen. Different strains could have a different intrinsic response to ischemic brain injury, due to differences in genetic background (34, 35). For example, the stroke-prone rat derived from the SHR is susceptible to develop larger and much less variable infarcts following MCAO (36, 37). This strain of rat has a genetic susceptibility to stroke independent of hypertension (38). This unique strain may provide a model to investigate genetic susceptibility to stroke. Furthermore, differences in intracranial and intracerebral vascular anastomoses could result in a larger infarct volume in some strains (32). If the genetic background

of a strain can influence the final infarct size, it could also impact the dynamic of evolution of the lesion at the acute stage. Bardutsky and colleagues showed, using MRI, that the spatiotemporal evolution of the ischemic penumbra is different between two strains of rats of the same age and without other comorbidity (39).

3. Several concomitant pathologies are known to exacerbate the ischemic brain lesion. The most often encountered comorbidities are for example arterial hypertension, diabetes and cardiovascular diseases. Hypertensive animals have been shown to display more extended lesions and earlier disappearance of the ischemic penumbra compared to normotensive ones (40, 41).
4. Almost all the experimental studies involve the exclusive use of males. The use of females adds technical complications. As female's hormones could influence per se the evolution of the ischemic lesion, the menstrual cycle of the females rats need to be monitored and the experiments performed at a specific point of the estrus cycle (42, 43).
5. How to prepare a thermofusible glue-coated filament of 0.380 mm diameter and 2.5 mm length

   Materials

   (a) Nylon monofilament (0.18 mm in diameter)
   (b) Thermofusible glue (melting temperature: 70°C)
   (c) Microscope
   (d) Homemade calibrated hole
   (e) Soldering iron
   (f) Forceps
   (g) Digital caliper (World Precision Instrument)

   Procedure

   (a) Cut the nylon monofilament into 3 or 4 cm segments
   (b) Under the operating microscope, coat the extremity of the nylon thread with the thermofusible glue (melting temperature: 70°C) by passing it through a calibrated hole made in a heated piece of metal fixed on the soldering iron.
   (c) Special attention is made to obtain a perfect cylinder without asperity.
   (d) The extremity of the filament is examined under the microscope after cooling and drying. All the shapes that are not perfectly smooth and constant are discarded.
   (e) The diameter of the tip is verified using a digital caliper.

6. Preparation of 4% paraformaldehyde (PFA) in 0.1 M phosphate buffer (PB) solution

   Materials

   (a) Paraformaldehyde (PFA) powder
   (b) Na2HPO4 dihydrate
   (c) NaH2PO4 dihydrate
   (d) Distilled water
   (e) Stirring hotplate
   (f) Filter paper
   (g) NaOH
   (h) pH meter
   (i) Pyrex bottles and beakers

   Procedure

   Prepare 1 L of 0.1 M PB, pH 7.4.

   Wear protective eye goggles, respirator mask, and appropriate gloves during the whole solution preparation process.

   (a) Put the PB solution in a flask supporting hot temperature.
   (b) Place flask on top of the stirring hotplate inside the hood.
   (c) Set the temperature of the hotplate to about 60°C and check the solution temperature with thermometer. You want to heat the solution to 55–60°C.
   (d) Add PFA powder (40 g) carefully because it is volatile. Stir the solution at mean speed. Inspect regularly to avoid overheating and consequent spilling.
   (e) Do not leave while your solution is on the heat.
   (f) When the PFA has dissolved and the solution is clear, switch off the heat but leave to stir: do not handle for safety reasons. Allow to cool.
   (g) Use filter paper to filter the whole solution when cooler.
   (h) Then put the sensor of a pH meter in the solution with slight stirring. Add NaOH as necessary to get a pH of 7.4.
   (i) Transfer the fixative to a 4°C refrigerator. Label appropriately and date. PFA 4% in 0.1 M PB can be store for 1 week before use in the refrigerator and for more than a month at -20°C. If you need huge quantity you can frozen 0.2 M 8%5 PFA solution and dissolve it into distilled water when you are ready to use it.

7. The use of a peristaltic pump is convenient but not the best choice as it does not respect the perfusion pressure and may induce vessel damages (collapse or rupture).

8. Preparation of thionin stain

   Solution A: Buffer solution

   1,000 mL distilled H_2O

   7 g Sodium Acetate Solution

   2 mL Acetic Acid Solution

   Solution B

   100 mL distilled H_2O

   1 g thionin

   Then add 25 mL solution B with 475 mL solution A Thionin is hard to dissolve. Use a glass stirring rod or a stir bar. Do not use a metal rod to stir the thionin. Filter the thionin solution. All the solution preparations and the staining are performed under the hood.

Acknowledgments

The work was funded by the Centre National de la Recherche Scientifique (CNRS), the French Ministère de l'Enseignement Supérieur et de la Recherche, and the University of Caen Lower Normandy (UCBN). This work was realized as part of the TC2N "Trans Channel Neuroscience Network" Interreg IV A 2 Mers Seas Zeeëns program, "Investing in your future" crossborder cooperation programme 2007–2013 part financed by the European Union (European Regional Development Fund).

Conflict of interest: The authors declare that they have no conflict of interest.

References

1. Demaerschalk BM, Hwang H-M, Leung G (2010) US cost burden of ischemic stroke: a systematic literature review. Am J Manag Care 16:525–533
2. Truelsen T, Bonita R (2009) The worldwide burden of stroke: current status and future projections. Handb Clin Neurol 92:327–336
3. Donnan GA et al (2008) Stroke. Lancet 371:1612–1623
4. Johnston SC, Mendis S, Mathers CD (2009) Global variation in stroke burden and mortality: estimates from monitoring, surveillance, and modelling. Lancet Neurol 8:345–354
5. Béjot Y et al (2009) Epidemiology of stroke. Med Sci (Paris) 25:727–732
6. Mostaza JM et al (2009) Patients at high risk of cerebrovascular disease: the REACH study. Cerebrovasc Dis 27(Suppl 1):77–81
7. Allen CL, Bayraktutan U (2008) Risk factors for ischaemic stroke. Int J Stroke 3:105–116
8. O'Donnell MJ et al (2010) Risk factors for ischaemic and intracerebral haemorrhagic stroke in 22 countries (the INTERSTROKE study): a case-control study. Lancet 376:112–123
9. Durukan A, Strbian D, Tatlisumak T (2008) Rodent models of ischemic stroke: a useful tool for stroke drug development. Curr Pharm Des 14:359–370
10. Durukan A, Tatlisumak T (2007) Acute ischemic stroke: overview of major experimental rodent models, pathophysiology, and therapy of focal cerebral ischemia. Pharmacol Biochem Behav 87:179–197
11. Hossmann K-A (2008) Cerebral ischemia: models, methods and outcomes. Neuropharmacology 55:257–270

12. Lapchak PA, Araujo DM (2007) Advances in ischemic stroke treatment: neuroprotective and combination therapies. Expert Opin Emerg Drugs 12:97–112
13. Khaja AM (2008) Acute ischemic stroke management: administration of thrombolytics, neuroprotectants, and general principles of medical management. Neurol Clin 26:943–961, viii
14. Auriel E, Bornstein NM (2010) Neuroprotection in acute ischemic stroke–current status. J Cell Mol Med 14:2200–2202
15. O'Collins VE et al (2006) 1,026 Experimental treatments in acute stroke. Ann Neurol 59:467–477
16. Pandya RS et al (2011) Central nervous system agents for ischemic stroke: neuroprotection mechanisms. Cent Nerv Syst Agents Med Chem. Available at: http://www.ncbi.nlm.nih.gov/pubmed/21521165. Accessed 28 Apr 2011.
17. Rosenberg N, Chen M, Prabhakaran S (2010) New devices for treating acute ischemic stroke. Recent Pat CNS Drug Discov 5:118–134
18. Fisher M et al (2009) Update of the stroke therapy academic industry roundtable preclinical recommendations. Stroke 40:2244–2250
19. Dirnagl U (2006) Bench to bedside: the quest for quality in experimental stroke research. J Cereb Blood Flow Metab 26:1465–1478
20. Macleod MR et al (2009) Good laboratory practice: preventing introduction of bias at the bench. Stroke 40:e50–e52
21. Peters JL et al (2006) A systematic review of systematic reviews and meta-analyses of animal experiments with guidelines for reporting. J Environ Sci Health B 41:1245–1258
22. Crossley NA et al (2008) Empirical evidence of bias in the design of experimental stroke studies: a metaepidemiologic approach. Stroke 39:929–934
23. DeGraba TJ, Pettigrew LC (2000) Why do neuroprotective drugs work in animals but not humans? Neurol Clin 18:475–493
24. Shuaib A (2006) Neuroprotection in acute ischemic stroke: are we there yet? Int J Stroke 1:100–101
25. Fisher M, Tatlisumak T (2005) Use of animal models has not contributed to development of acute stroke therapies: con. Stroke 36:2324–2325
26. Tamura A et al (1981) Focal cerebral ischaemia in the rat: 1 Description of technique and early neuropathological consequences following middle cerebral artery occlusion. J Cereb Blood Flow Metab 1:53–60
27. Longa EZ et al (1989) Reversible middle cerebral artery occlusion without craniectomy in rats. Stroke 20:84–91
28. Kawamura S et al (1991) Rat middle cerebral artery occlusion using an intraluminal thread technique. Acta Neurochir (Wien) 109:126–132
29. Kong LQ et al (2004) Improvements in the intraluminal thread technique to induce focal cerebral ischaemia in rabbits. J Neurosci Methods 137:315–319
30. Barber PA et al (2005) Early T1- and T2-weighted MRI signatures of transient and permanent middle cerebral artery occlusion in a murine stroke model studied at 9.4 T. Neurosci Lett 388:54–59
31. Chiang T, Messing RO, Chou W-H (2011) Mouse model of middle cerebral artery occlusion. J Vis Exp. Available at: http://www.ncbi.nlm.nih.gov/pubmed/21372780. Accessed 8 Nov 2011.
32. Duverger D, MacKenzie ET (1988) The quantification of cerebral infarction following focal ischemia in the rat: influence of strain, arterial pressure, blood glucose concentration, and age. J Cereb Blood Flow Metab 8:449–461
33. Shapira S et al (2002) Aging has a complex effect on a rat model of ischemic stroke. Brain Res 925:148–158
34. Oliff HS et al (1995) The role of strain/vendor differences on the outcome of focal ischemia induced by intraluminal middle cerebral artery occlusion in the rat. Brain Res 675:20–26
35. Cai H et al (1998) Photothrombotic middle cerebral artery occlusion in spontaneously hypertensive rats: influence of substrain, gender, and distal middle cerebral artery patterns on infarct size. Stroke 29:1982–1986, discussion 1986–1987
36. Coyle P (1986) Different susceptibilities to cerebral infarction in spontaneously hypertensive (SHR) and normotensive Sprague-Dawley rats. Stroke 17:520–525
37. Takaba H, Fukuda K, Yao H (2004) Substrain differences, gender, and age of spontaneously hypertensive rats critically determine infarct size produced by distal middle cerebral artery occlusion. Cell Mol Neurobiol 24:589–598
38. Mies G et al (1999) Hemodynamics and metabolism in stroke-prone spontaneously hypertensive rats before manifestation of brain infarcts. J Cereb Blood Flow Metab 19:1238–1246
39. Bardutzky J et al (2005) Differences in ischemic lesion evolution in different rat strains using diffusion and perfusion imaging. Stroke 36:2000–2005
40. Letourneur A et al (2011) Impact of genetic and renovascular chronic arterial hypertension on the acute spatiotemporal evolution of the ischemic penumbra: a sequential study with MRI in the rat. J Cereb Blood Flow Metab 31:504–513
41. Rewell SSJ et al (2010) Inducing stroke in aged, hypertensive, diabetic rats. J Cereb Blood Flow Metab 30:729–733
42. Brait VH et al (2010) Mechanisms contributing to cerebral infarct size after stroke: gender,

reperfusion, T lymphocytes, and Nox2-derived superoxide. J Cereb Blood Flow Metab 30:1306–1317

43. Alkayed NJ et al (1998) Gender-linked brain injury in experimental stroke. Stroke 29:159–165, discussion 166
44. Bouley J, Fisher M, Henninger N (2007) Comparison between coated vs. uncoated suture middle cerebral artery occlusion in the rat as assessed by perfusion/diffusion weighted imaging. Neurosci Lett 412:185–190
45. Spratt NJ et al (2006) Modification of the method of thread manufacture improves stroke induction rate and reduces mortality after thread-occlusion of the middle cerebral artery in young or aged rats. J Neurosci Methods 155:285–290
46. Chuquet J et al (2008) Effects of urotensin-II on cerebral blood flow and ischemia in anesthetized rats. Exp Neurol 210:577–584
47. Lecrux C et al (2007) Spontaneously hypertensive rats are highly vulnerable to AMPA-induced brain lesions. Stroke 38:3007–3015
48. Esneault E et al (2008) Combined therapeutic strategy using erythropoietin and mesenchymal stem cells potentiates neurogenesis after transient focal cerebral ischemia in rats. J Cereb Blood Flow Metab 28:1552–1563
49. Esneault E et al (2008) D-JNKi, a peptide inhibitor of c-Jun N-terminal kinase, promotes functional recovery after transient focal cerebral ischemia in rats. Neuroscience 152:308–320
50. Kuraoka M et al (2009) Direct experimental occlusion of the distal middle cerebral artery induces high reproducibility of brain ischemia in mice. Exp Anim 58:19–29
51. G-ming X et al (2004) Evaluation of murine models of permanent focal cerebral ischemia. Chin Med J 117:389–394
52. Fotheringham AP, Davies CA, Davies I (2000) Oedema and glial cell involvement in the aged mouse brain after permanent focal ischaemia. Neuropathol Appl Neurobiol 26:412–423
53. Taguchi A et al (2010) A reproducible and simple model of permanent cerebral ischemia in CB-17 and SCID mice. J Exp Stroke Transl Med 3:28–33
54. Freret T et al (2011) On the importance of long-term functional assessment after stroke to improve translation from bench to bedside. Exp Transl Stroke Med 3:6
55. Garcia JH et al (1995) Neurological deficit and extent of neuronal necrosis attributable to middle cerebral artery occlusion in rats: statistical validation. Stroke 26:627–635
56. Brines ML et al (2000) Erythropoietin crosses the blood-brain barrier to protect against experimental brain injury. Proc Natl Acad Sci U S A 97:10526–10531
57. Bernaudin M et al (1999) A potential role for erythropoietin in focal permanent cerebral ischemia in mice. J Cereb Blood Flow Metab 19:643–651
58. Eid T, Brines M (2002) Recombinant human erythropoietin for neuroprotection: what is the evidence? Clin Breast Cancer 3(Suppl 3):S109–S115
59. Digicaylioglu M (2010) Erythropoietin in stroke: quo vadis. Expert Opin Biol Ther 10:937–949
60. Ehrenreich H et al (2009) Recombinant human erythropoietin in the treatment of acute ischemic stroke. Stroke 40:e647–e656
61. Ehrenreich H et al (2004) Erythropoietin: novel approaches to neuroprotection in human brain disease. Metab Brain Dis 19:195–206
62. Jerndal M et al (2010) A systematic review and meta-analysis of erythropoietin in experimental stroke. J Cereb Blood Flow Metab 30:961–968. doi:10.1038/jcbfm.2009.267
63. Erbayraktar S et al (2003) Asialoerythropoietin is a nonerythropoietic cytokine with broad neuroprotective activity in vivo. Proc Natl Acad Sci U S A 100:6741–6746
64. Sirén AL et al (2001) Erythropoietin prevents neuronal apoptosis after cerebral ischemia and metabolic stress. Proc Natl Acad Sci U S A 98:4044–4049
65. Villa P et al (2003) Erythropoietin selectively attenuates cytokine production and inflammation in cerebral ischemia by targeting neuronal apoptosis. J Exp Med 198:971–975
66. Wang L et al (2004) Treatment of stroke with erythropoietin enhances neurogenesis and angiogenesis and improves neurological function in rats. Stroke 35:1732–1737
67. Ghezzi P, Brines M (2004) Erythropoietin as an antiapoptotic, tissue-protective cytokine. Cell Death Differ 11(Suppl 1):S37–S44
68. Ehrenreich H et al (2002) Erythropoietin therapy for acute stroke is both safe and beneficial. Mol Med 8:495–505
69. Brines M et al (2008) Nonerythropoietic, tissue-protective peptides derived from the tertiary structure of erythropoietin. Proc Natl Acad Sci U S A 105:10925–10930
70. Hermann DM (2010) Nonhematopoietic variants of erythropoietin in ischemic stroke: need for step-wise proof-of-concept studies. ScientificWorldJournal 10:2285–2287
71. Villa P et al (2007) Reduced functional deficits, neuroinflammation, and secondary tissue damage after treatment of stroke by nonerythropoietic erythropoietin derivatives. J Cereb Blood Flow Metab 27:552–563
72. Hermann DM (2009) Enhancing the delivery of erythropoietin and its variants into the ischemic brain. ScientificWorldJournal 9:967–969

73. Lapchak PA (2010) Erythropoietin molecules to treat acute ischemic stroke: a translational dilemma! Expert Opin Investig Drugs 19:1179–1186
74. Montero M et al (2007) Comparison of neuroprotective effects of erythropoietin (EPO) and carbamylerythropoietin (CEPO) against ischemia-like oxygen-glucose deprivation (OGD) and NMDA excitotoxicity in mouse hippocampal slice cultures. Exp Neurol 204:106–117
75. Doggrell SA (2004) A neuroprotective derivative of erythropoietin that is not erythropoietic. Expert Opin Investig Drugs 13:1517–1519
76. Yamashita T et al (2010) Asialoerythropoietin attenuates neuronal cell death in the hippocampal CA1 region after transient forebrain ischemia in a gerbil model. Neurol Res 32:957–962
77. Price CD et al (2010) Effect of continuous infusion of asialoerythropoietin on short-term changes in infarct volume, penumbra apoptosis and behaviour following middle cerebral artery occlusion in rats. Clin Exp Pharmacol Physiol 37:185–192
78. Gan Y et al (2012) Mutant erythropoietin without erythropoietic activity is neuroprotective against ischemic brain injury. Stroke 43: 3071–3077
79. Fletcher L et al (2009) Intranasal delivery of erythropoietin plus insulin-like growth factor-I for acute neuroprotection in stroke. Laboratory investigation. J Neurosurg 111:164–170
80. Yu YP et al (2005) Intranasal recombinant human erythropoietin protects rats against focal cerebral ischemia. Neuro sci Lett 387: 5–10
81. Aluclu MU et al (2007) Evaluation of erythropoietin effects on cerebral ischemia in rats. Neuro Endocrinol Lett 28:170–174
82. Sadamoto Y et al (1998) Erythropoietin prevents place navigation disability and cortical infarction in rats with permanent occlusion of the middle cerebral artery. Biochem Biophys Res Commun 253:26–32
83. Villa P et al (2007) Reduced functional deficits, neuroinflammation, and secondary tissue damage after treatment of stroke by nonerythropoietic erythropoietin derivatives. J Cereb Blood Flow Metab 27:552–563
84. Li Y et al (2007) Erythropoietin-induced neurovascular protection, angiogenesis, and cerebral blood flow restoration after focal ischemia in mice. J Cereb Blood Flow Metab 27: 1043–1054
85. Li L et al (2009) MRI identification of white matter reorganization enhanced by erythropoietin treatment in a rat model of focal ischemia. Stroke 40:936–941
86. Ding G et al (2010) Cerebral tissue repair and atrophy after embolic stroke in rat: a magnetic resonance imaging study of erythropoietin therapy. J Neuro sci Res 88(14):3206–3214
87. Wakida K et al (2007) Neuroprotective effect of erythropoietin, and role of metallothionein-1 and -2, in permanent focal cerebral ischemia. Neuro sci 148:105–114
88. Fu A et al (2010) Neuroprotection in experimental stroke in the rat with an IgG-erythropoietin fusion protein. Brain Res 1360:193–197
89. Leist M et al (2004) Derivatives of erythropoietin that are tissue protective but not erythropoietic. Science 305:239–242
90. Belayev L et al (2005) Neuroprotective effect of darbepoetin alfa, a novel recombinant erythropoietic protein, in focal cerebral ischemia in rats. Stroke 36:1071–1076
91. Wang Y et al (2007) Post-ischemic treatment with erythropoietin or carbamylated erythropoietin reduces infarction and improves neurological outcome in a rat model of focal cerebral ischemia. Br J Pharmacol 151: 1377–1384
92. Fu A et al (2011) Neuroprotection in stroke in the mouse with intravenous erythropoietin-Trojan horse fusion protein. Brain Res 1369: 203–207

Chapter 6

Experimental Traumatic Spinal Cord Injury

Zübeyde Erbayraktar, Necati Gökmen, Osman Yılmaz, and Serhat Erbayraktar

Abstract

Animal models are important to develope therapies for individuals suffering from spinal cord injuries. For this purpose, rats are commonly preferred. In sharp injury models, spinal cord is completely or incompletely cut to assess axonal regeneration. On the other hand, spinal cord is compressed or contused to mimic the human injury in blunt injury models for understanding as well as managing the secondary pathophysiologic processes following injury. Especially, contusions are thought to be biomechanically similar to vertebral fractures and/or dislocations and thus provide the most realistic experimental setting in which to test potential neuroprotective and regenerative strategies.

Key words Spinal cord contusion, Spinal cord injury, Spinal cord transection

1 Introduction

Experimental models of central nervous system (CNS) injury have been indispensable in the exploration of pathological mechanisms and treatment strategies. Due to the variable nature of clinical traumatic CNS injury (e.g., inconsistencies in the anatomical location of impact and the magnitude and duration of loading), experimental models must simplify the human condition in order to induce a reproducible injury that can be utilized for experimental testing. Models have enlisted a large variety of animal species (1) and have applied a broad spectrum of injury paradigms ranging from transections (2) to contusions (3, 4). Rat and mouse are the most popular animals used in such models, both because of cost and accessibility and because of the latter's transgenic potential (5). In vivo animal models allow the investigator to experimentally manipulate certain parameters that are not possible in patients and provide a more complete representation of the human spinal cord because they more closely mimic the material properties and anatomical architecture. Therefore, the load distribution and structural failure in

Pietro Ghezzi and Anthony Cerami (eds.), *Tissue-Protective Cytokines: Methods and Protocols*, Methods in Molecular Biology, vol. 982, DOI 10.1007/978-1-62703-308-4_6, © Springer Science+Business Media, LLC 2013

animal models are expected to be similar to human injury when clinically relevant biomechanical loading parameters are applied in a scale-appropriate manner. However, it is important to note that different injury paradigms address different questions, and thus each carry their own advantages and disadvantages. Sharp transection models provide valuable information about the regeneration of specific axonal tracts (6) that often is difficult to obtain from contusion models, in which many axons are invariably spared. On the other hand, contusion models are more realistic and allow the investigators to evaluate neuroprotective interventions during acute pathophysiologic processes and axonal regeneration approaches within a more representative neuropathologic milieu after injury (7).

2 Materials

2.1 Anesthesia

1. Anesthesia chamber: a fiber glass box with a lid and an inlet for anesthesia gases at one end and outlet at another end for scavenging anesthesia gases.
2. Ventilator: Rats—e.g., SAR*830*P ventilator, Life Science instruments, USA.
   Mice—e.g., Mini-Vent Type *845*, Hugo Sachs Elektronik, March-Hugstetten, Germany.
3. Anesthesia gases: Isoflurane and Oxygen (from O_2 cylinder or central supply).
4. Isoflurane vaporizer.
5. Infusion fluids: saline.

2.2 Surgery

The following sets of instruments are required for laminectomy and spinal cord injury induction.

1. Surgical instruments: scalpel handle, scalpel blades, jeweler's forceps, needle holders, Adson's tissue forceps, bone rongeurs (Friedman bone rongeurs, curved).
2. Sterile cotton tip applicators and sterile sponges (or can be sterilized along with the instruments in the autoclave).
3. Sterilization equipment: autoclave.
4. Suture materials: Nonabsorbable monofilament suture material like Nylon (Ethilon) is used for skin closure (*4-0* for rats and *5-0* for mice).
5. Material to control bleeding during craniectomy: Gelfoam (dental packs, absorbable gelatin sponge, size *4*), bone wax (Ethicon).
6. Magnifying instruments: dissecting microscope.

7. Temporary aneurysm clips with a closing force of 53 × *g*, and a clip applier.
8. Heating pad.
9. Sterile syringes and sterile needles.
10. Impactors:
    - Mice: Ohio State University Impactor.
    - Rats: New York University Impactor, Ohio State University Impactor.
11. Rectal temperature: rectal thermistor.
12. Skin disinfectant: povidone-iodine-based disinfectant.
13. Analgesic: buprenorphine.
14. Local anesthetic: 0.5% bupivacaine.
15. Antibiotic: 25 mg/ml imipenem.

3 Methods

3.1 Anesthesia

1. Rats: Anesthesia is induced by placing rats in the anesthesia chamber with 5% isoflurane in 100% oxygen flow for 3–5 min. Then, it is transferred onto a heating pad and i.p. injection of 40 mg/kg thiopental sodium was used for the maintenance of anesthesia.
2. Mice: The mouse is induced in the anesthesia chamber with 3% isoflurane in 100% oxygen for 3–5 min. It is then transferred onto a heating pad. Maintenance can be accomplished by either 1.5% isoflurane or i.p. injection of 312.5 mg/kg tribromoethanol.
3. Temperature probe is inserted into the rectum and the temperature is maintained at 36.5–37.5°C by a heating pad.

3.2 Surgery

1. Antibiotics and Analgesics: Preoperatively, imipenem is administered in the dosage of 10 mg/kg for both rats and mice by intramuscular route once daily for prophylaxis of infection. Buprenorphine analgesic is given in the dosage of 0.1 mg/kg (rats) and 1 mg/kg (mice) by subcutaneous route twice daily for 3 days starting on the day of surgery.
2. Laminectomy: The animals were positioned in the prone position and surgery performed under sterile conditions. After infiltration of the skin (bupivacaine 0.25%), paravertebral musculature was separated from the spinous processes to expose the vertebral column through a 2-cm skin incision (Fig. 1). Commonly, under microscopic magnification and illumination Th8–9 laminectomy without causing injury to the dural sac is preferred for inducing trauma, but Th3 lamninectomy is also used, especially during clip compression technique (Fig. 2).

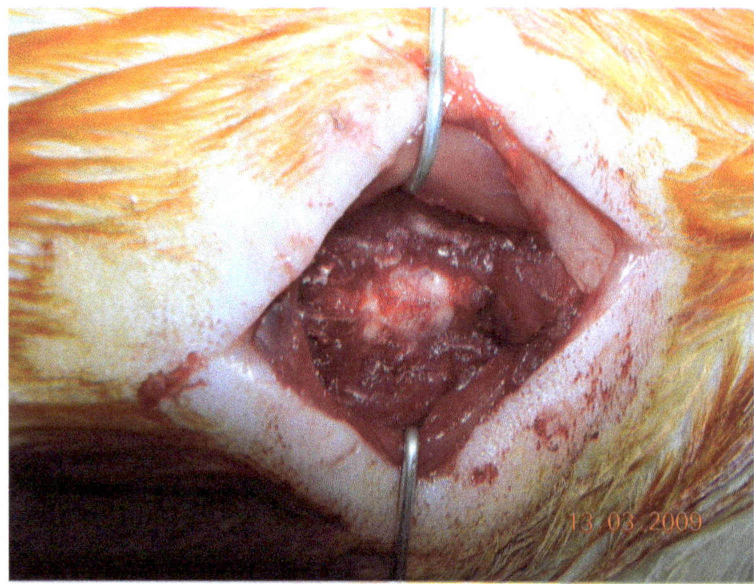

Fig. 1 Lamina of Th3 vertebra is freed from the paravertebral muscles

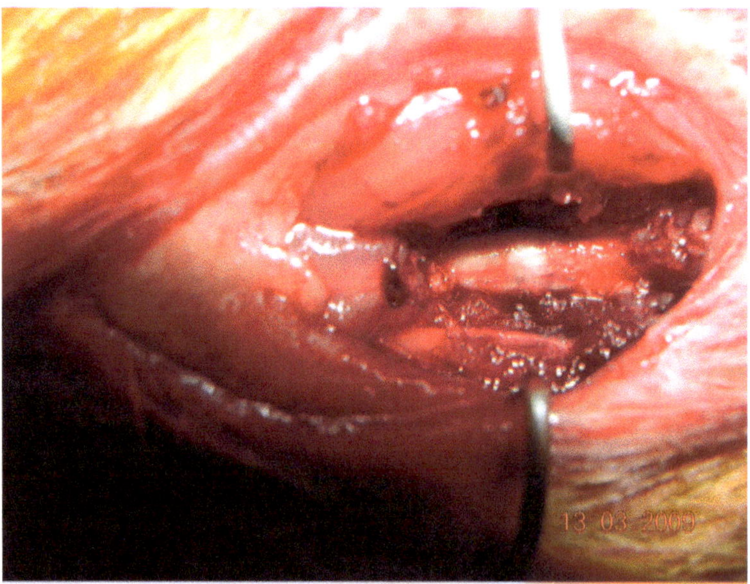

Fig. 2 After laminectomy, spinal cord is exposed

3.2.1 Spinal Cord İnjury

Spinal dura is opened with a No:11 scalpel blade and enlargened with a microscissor.

Transection Models of the Spinal Cord (See Note 1)

Complete Transection: A transverse section of the whole spinal cord is performed with a No:15 scalpel blade. Hemorrhage is controlled by using hemostatic materials (see Note 2).

Incomplete Transection: Unilateral or dorsal hemisections are performed with a No:11 scalpel blade under dissecting microscope. Bleeding is controlled by using hemostatic materials (see Note 3).

Spinal Cord Contusion and Compression Models (See Note 4)

Spinous processes of vertebrae above and below the injury site are secured in clamps to reduce motion of the spinal column during the impact. In mice, size of the laminectomy (approximately 1.5 mm in diameter) was consistent between animals to allow room for the impactor tip. Care was taken not to remove the lateral part of the vertebra at the site of laminectomy to maintain vertebral column stability. The lateral processes of Th8 and Th10 vertebrae were cleared of muscle to allow for stabilization of the vertebral column using forceps.

New York University Impactor: In this weight-drop contusion model (3), a 10-g rod is dropped from heights of *6.25, 12.5, 25, and 50*mm onto the exposed dorsal surface of the spinal cord at T9–T10, producing more severe neurologic injuries with increasing height (see Note 5).

Ohio State University Impactor: It is a computer feedback-controlled electromechanical impactor rather than a weight drop (4, 8–10). The impactor probe is slowly screwed down to the dural surface which it contacts and displaces *30*μm with a force of approximately *3,000* dynes for providing a consistent starting point from which to initiate the injury. Then, the device rapidly impacts the cord for a predetermined amount of displacement typically in the range of *0.8–1.1*mm for less than *10*ms and is held for *4–5*ms before releasing in less than 10 ms. There is no bouncing of the impactor back onto the cord, which is a potential source of variation in a weight-drop paradigm (see Note 6).

Clip Compression: A temporary aneurysm clip exerting a *53×g* closing force on the spinal cord is applied extradurally for *1*min (Fig. 3). A probable hemorrhage can be controlled by using hemostatic materials. The compressive force is applied to both the volar and dorsal aspects of the cord, making it somewhat more representative of the human injury (Fig. 4) (see Note 7).

1. Closure: The skin incision is closed in layers and the animals allowed to recover fully from anesthesia and returned to their cages.
2. Since this intervention is an important trauma to the animals, hydration with subcutaneous fluids may help them to survive. Besides, manual drainage of urine twicely in a day is a life saving maneuver for those animals prior to the establishment of reflex bladder.

3.3 Measures of Structural and Functional Outcomes

Experimental interventions in animal spinal cord injury models can be evaluated anatomically, biochemically, neurophysiologically, and/or functionally. Anatomic assessment involves the use

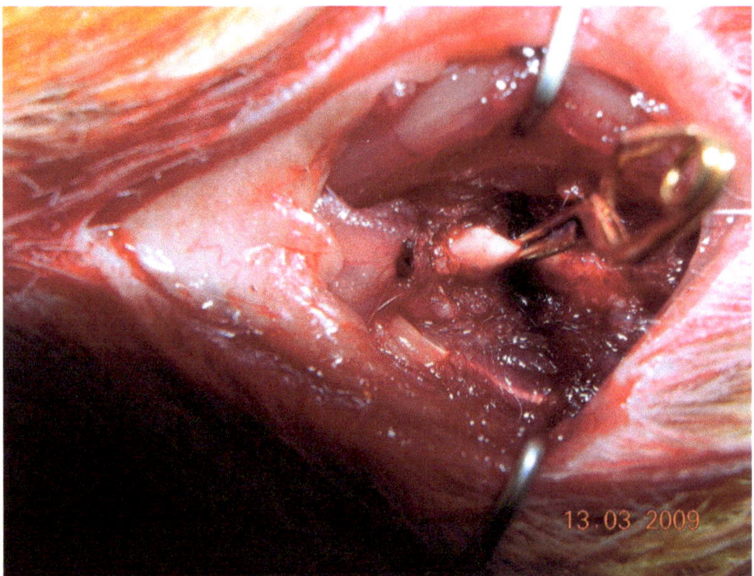

Fig. 3 Spinal cord is compressed with clip application

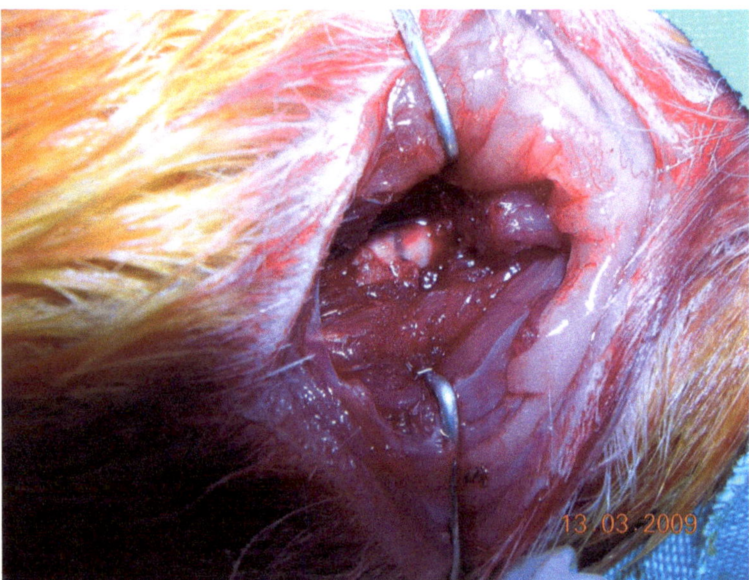

Fig. 4 After clip removal, primary spinal cord injury becomes visible

of histopathologic techniques utilizing both immunohistochemistry and axonal tracers. Neurophysiologic testing such as somatosensory-evoked potentials and motor-evoked potentials also has been found to be useful for evaluating therapies against experimental spinal cord injuries. Evoked potentials inherently tend to reflect the activity of larger fibers, which may comprise only a small part of the total axonal population at the injury site of interest (11). Although integrity of autonomic and sensory systems, particularly with regard to bowel and bladder function and pain (12) are significant for a good recovery, much attention has widely been directed to the assessment of locomotive performance. BBB locomotor rating scale described by Basso, Beattie, and Bresnahan (13) in 1995 is the most frequently used measure of motor performance in rats. Animals scored (1–8) are able to move hindlimb joint without weight support. Those scored (9–13) can make varying degrees of hindlimb weight support and forelimb—hindlimb coordination, while those score (14–22) demonstrated improvements in paw and tail position, toe clearance, and trunk stability during a fully supported and coordinated gait. Rather than remaining with a BBB score of 0, rats with complete transections can achieve a score of 3 to 4, whereas mildly injured rats can achieve the full score of 21 after a number of weeks (8, 14) (see Notes 5 and 7). Its use allows the communication and standardized comparison of results from different institutions. Moreover, a modified version of the BBB scale has been developed for the evaluation of mice after spinal cord injury. The grid walking test (15), the narrow beam test (16), and the inclined plane test (17) are other commonly used locomotor tests. During investigation of cervical spinal cord injury, skilled forelimb movement can be assessed by reaching tasks, in which the animal must grasp a small food pellet placed at varying heights and deliver it to its mouth (18). There also has been the development of sophisticated video and computerized kinematic analysis systems for animals, which may improve ability to detect small functional changes (19).

4 Notes

1. Models in which the spinal cord is sharply transected, either completely or partially, are useful for studying the anatomic regeneration of axons. They are complementary to the contusion models, which better simulate the biomechanics and neuropathology of human injury.
2. *Complete Transection*: The implication in studies using transection models is that with the ensured completeness of the lesion, anterogradely labeled axons observed distal to the lesion have indeed regenerated from above and are responsible for the functional recover. The so-called "spinalized" animal can demonstrate some native locomotor abilities of the completely transected spinal cord after a period of gravity-assisted treadmill training of the hindlimbs (20–22) or pharmacologic intervention such as intrathecally administered alpha adrenergic agonists (23, 24). Apart from this type of injury, in case the recovery is due to axonal regeneration of descending systems

and not to native spinal cord circuitry, the function is expected to be lost when after a retransection of the cord is established.

3. *Incomplete Transection*: In incomplete transection models, tracts of the spinal cord are cut selectively. Since neurologic deficit can be relatively mild, the postoperative animal care is easy, especially with regard to bladder function. Thus, comparison of the regeneration in a particular tract with its heathy partner on the contralateral hemicord is possible. Retrograde tracers are available to confirm the lack of any axon escaping from the induced injury of a particular tract. To which degree an observed functional improvement is due to true regeneration of the injured tract or to functional compensation from other systems that are spared is still waiting to be explored. Unilateral or dorsal hemisection models are commonly used to transect the ipsilateral rubrospinal tract or bilateral rubrospinal plus corticospinal tracts, respectively.
4. Blunt contusive or compressive force to the spinal cord, which is more representative of what occurs in most human injuries (5, 25–28) leads to a temporal pattern of injury maturation resulting with central cystic cavities in an athrophic cord and gliosis. Such blunt trauma leaves some normal-appearing parenchyma peripherally at the injury core and preserved function is related with the quantity of this spared rim. Due to the incomplete nature of injury as well as the complexity of the tracts and the existing discrepancies between the conditions (such as an anesthesia, or multiple drug use) under which injuries occur in patients and those in animals, much effort should be directed at optimizing the selected experimental method in order to induce reproducible and consistent neurologic injuries.
5. *New York University Impactor*: This model can induce consistent injuries in terms of the resulting neuropathology as well as the functional impairment as evaluated on the BBB scale. The average BBB scores from these four groups 6 weeks after injury were 19.0 ± 15.8, 11.4 ± 1.1, 10.6 ± 0.6, and 7.9 ± 1.8, respectively, whereas animals with full transections achieved a score of 3.3 ± 2.1. It should be noted that even in the most severe injury, caused by weight dropped from a height of *50* mm, the animals' functional recovery nearly reaches some degree of hindlimb weight support. There is an interaction between the kinetic energy of the descending impactor, its mass and its velocity at impact. The relation is linear between energy and drop height and nonlinear between impact velocity and drop height. In the attempt to compare results obtained in other weight-drop devices, it is important to be aware of the biomechanical implications of differences in drop height, impact mass, or both.
6. *Ohio State University Impactor*: While both the OSU and NYU impactors were developed for the rat, the Ohio State group

recently adapted the OSU impactor to accommodate the much smaller body of the mouse, which will allow for the evaluation of contusion injuries in transgenic animals. However, no superiority of one over the other was determined.

7. *Clip Compression*: Apart from models using impactors, this model simulates continuous cord compression secondary to residual spinal column displacement (29). As in weight-drop models, different severities of spinal cord injury can be created by adjustment of the closing force of the clip, the duration of compression, or both. With a greater closing force, fewer axons are spared at the injury site, which in turn relates to a poorer functional recovery after injury. In clip compression model rats subjected to a *53×g* closing force for *1*min (considered a "severe" injury) recovers to a BBB score of approximately *8.5* by *6*weeks after injury (7, 30). The clip compression model has provided a very valuable setting in which study acute pathophysiology after cord injury (31, 32) the timing of decompression (33), and potential therapies such as electrical stimulation (34) and neuroprotective agents (35–37). However, the current contusion and compression models inherently fail to simulate the complex biomechanical stresses of distraction, compression, bending, and shear to which the human spinal cord is subjected during trauma.

References

1. Fernandez E et al (1991) Experimental studies on spinal cord injuries in the last fifteen years. Neurol Res 13:138–159
2. Cheng H, Cao Y, Olson L (1996) Spinal cord repair in adult paraplegic rats: partial restoration of hind limb function. Science 273:510–513
3. Gruner JA (1992) A monitored contusion model of spinal cord injury in the rat. J Neurotrauma 9:123–126
4. Stokes BT (1992) Experimental spinal cord injury: a dynamic and verifiable injury device. J Neurotrauma 9:129–131
5. Jakeman LB et al (2000) Traumatic spinal cord injury produced by controlled contusion in mouse. J Neurotrauma 17:299–319
6. Kobayashi NR et al (1997) BDNF and NT-4/5 prevent atrophy of rat rubrospinal neurons after cervical axotomy, stimulate GAP-43 and Talpha1-tubulin mRNA expression, and promote axonal regeneration. J Neurosci 17:9583–9595
7. Schwartz G, Fehlings MG (2001) Evaluation of the neuroprotective effects of sodium channel blockers after spinal cord injury: improved behavioral and neuroanatomical recovery with riluzole. J Neurosurg 94:245–256
8. Blight AR (1992) Spinal cord injury models: neurophysiology. J Neurotrauma 9:147–149
9. Holmes GM et al (1998) External anal sphincter hyperreflexia following spinal transection in the rat. J Neurotrauma 15:451–457
10. Basso DM, Beattie MS, Bresnahan JC (1995) A sensitive and reliable locomotor rating scale for open field testing in rats. J Neurotrauma 12:1–21
11. Bresnahan JC et al (1987) A behavioral and anatomical analysis of spinal cord injury produced by a feedback-controlled impactiondevice. Exp Neurol 95:548–570
12. Kunkel-Bagden E, Dai HN, Bregman BS (1993) Methods to assess the development and recovery of locomotor function after spinal cord injury in rats. Exp Neurol 119:153–164
13. Soblosky JS et al (1997) Ladder beam and camera video recording system for evaluating forelimb and hindlimb deficits after sensorimotor cortex injury in rats. J Neurosci Methods 78:75–83
14. Hicks SP, D'Amato CJ (1975) Motor-sensory cortex-corticospinal system and developing locomotion and placing in rats. Am J Anat 143:1–42
15. Rivlin AS, Tator CH (1977) Objective clinical assessment of motor function after experimental

spinal cord injury in the rat. J Neurosurg 47: 577–581

16. Z'Graggen WJ et al (1998) Functional recovery and enhanced corticofugal plasticity after unilateral pyramidal tract lesion and blockade of myelin-associated neurite growth inhibitors in adult rats. J Neurosci 18:4744–4757
17. Cheng H et al (1997) Gait analysis of adult paraplegic rats after spinal cord repair. Exp Neurol 148:544–557
18. Belanger M et al (1996) A comparison of treadmill locomotion in adult cats before and after spinal transection. J Neurophysiol 76:471–491
19. de Leon RD et al (1998) Full weight-bearing hindlimb standing following stand training in the adult spinal cat. J Neurophysiol 80:83–91
20. de Leon RD et al (1998) Locomotor capacity attributable to step training versus spontaneous recovery after spinalization in adult cats. J Neurophysiol 79:1329–1340
21. Chau C, Barbeau H, Rossignol S (1998) Early locomotor training with clonidine in spinal cats. J Neurophysiol 79:392–409
22. Chau C, Barbeau H, Rossignol S (1998) Effects of intrathecal alpha 1- and alpha 2-noradrenergic agonists and norepinephrine on locomotion in chronic spinal cats. J Neurophysiol 79:2941–2963
23. Ducker TB, Hamit HF (1969) Experimental treatments of acute spinal cord injury. J Neurosurg 30:693–697
24. Ford RW (1983) A reproducible spinal cord injury model in the cat. J Neurosurg 59: 268–275
25. Griffiths IR (1976) Spinal cord blood flow after acute experimental cord injury in dogs. J Neurol Sci 27:247–259
26. Yeo JD et al (1975) The experimental contusion injury of the spinal cord in sheep. Paraplegia 12:279–298
27. Noyes DH (1987) Correlation between parameters of spinal cord impact and resultant injury. Exp Neurol 95:535–547
28. Noyes DH (1987) Electromechanical impactor for producing experimental spinal cord injury in animals. Med Biol Eng Comput 25: 335–340
29. Rivlin AS, Tator CH (1978) Effect of duration of acute spinal cord compression in a new acute cord injury model in the rat. Surg Neurol 10:38–43
30. Erbayraktar S et al (2003) Asialoerythropoietin is a nonerythropoietic cytokine with broad neuroprotective activity in vivo. Proc Natl Acad Sci U S A 27:6741–6746
31. Fehlings MG, Tator CH (1992) The effect of direct current field polarity on recovery after acute experimental spinal cord injury. Brain Res 579:32–42
32. Fehlings MG, Tator CH, Linden RD (1989) The relationships among the severity of spinal cord injury, motor and somatosensory evoked potentials, and spinal cord blood flow. Electroencephalogr Clin Neurophysiol 74:241–259
33. Leist M (2004) Derivatives of erythropoietin that are tissue protective but not erythropoietic. Science 305:239–242
34. Guha A et al (1987) Decompression of the spinal cord improves recovery after acute experimental spinal cord compression injury. Paraplegia 25:324–339
35. Celik M et al (2002) Erytropoietin prevents motor neuron apoptosis and neurological disability in experimental spinal cord ischemic injury. Proc Natl Acad Sci U S A 99:2258–2263
36. Gorio A et al (2002) Recombinant human erythropoietin counteracts secondary injury and markedly enhances neurological recovery from experimental spinal cord trauma. Proc Natl Acad Sci U S A 99:9450–9455
37. Grasso G et al (2006) Amelioration of spinal cord compressive injury by pharmacological preconditioning with erythropoietin and a nonerythropoietic erythropoietin derivative. J Neurosurg Spine 4:310–318

Chapter 7

Erythropoietin as a Neuroprotectant for Neonatal Brain Injury: Animal Models

Christopher M. Traudt and Sandra E. Juul

Abstract

Prematurity and perinatal hypoxia-ischemia are common problems that result in significant neurodevelopmental morbidity and high mortality worldwide. The Vannucci model of unilateral brain injury was developed to model perinatal brain injury due to hypoxia-ischemia. Because the rodent brain is altricial, i.e., it develops postnatally, investigators can model either preterm or term brain injury by varying the age at which injury is induced. This model has allowed investigators to better understand developmental changes that occur in susceptibility of the brain to injury, evolution of brain injury over time, and response to potential neuroprotective treatments. The Vannucci model combines unilateral common carotid artery ligation with a hypoxic insult. This produces injury of the cerebral cortex, basal ganglia, hippocampus, and periventricular white matter ipsilateral to the ligated artery. Varying degrees of injury can be obtained by varying the depth and duration of the hypoxic insult. This chapter details one approach to the Vannucci model and also reviews the neuroprotective effects of erythropoietin (Epo), a neuroprotective treatment that has been extensively investigated using this model and others.

Key words Vannucci model, Erythropoietin, Hypoxic-ischemia, Common carotid artery ligation

1 Introduction

1.1 Neonatal Brain Injury

Extreme prematurity and neonatal encephalopathy significantly impair the outcomes of affected infants. Even with optimal care, both conditions carry a 50% risk of death, or in survivors, mental retardation, cerebral palsy, hydrocephalus, and seizures. One of eight babies is born preterm, with 1 in 100 born at the extreme end of the spectrum (less than 28 weeks of gestation). Neonatal encephalopathy occurs in 3–5 of every 1,000 live births. Consequences of these conditions bring heartbreak for the child and family as well as expense. Loss of productivity, dependency, recurrent use of medical and rehabilitation services, and reduced life expectancy all exacerbate the burden (1–4). Many factors influence the outcomes of these vulnerable neonates, including the duration and severity of brain injury, brain maturity at time of

Pietro Ghezzi and Anthony Cerami (eds.), *Tissue-Protective Cytokines: Methods and Protocols*, Methods in Molecular Biology, vol. 982, DOI 10.1007/978-1-62703-308-4_7, © Springer Science+Business Media, LLC 2013

injury, as well as the general condition of the maternal–infant pair prior to the injury (e.g., nutrition, hypoxic preconditioning, infection, stress). In term infants (>36 weeks of gestation) the neurons of the cerebral cortex, hippocampus, and basal ganglia–thalamus are most often affected, while preterm infants most often develop diffuse white matter injury (5), followed by impaired development of gray as well as white matter (6).

Animal models have been used to study the pathophysiology of neonatal brain injury and how vulnerability to specific injury changes as the brain matures. Animal models of neonatal hypoxic-ischemic brain injury are essential for the development and testing of novel therapeutic approaches. Rodent models are commonly used because they are inexpensive, easy to work with, and the gene expression pathways are readily translated to human relevance. Models of neonatal brain injury include the application of prolonged hypoxia alone (7, 8), transient middle cerebral artery occlusion (9, 10), and unilateral ligation of the common carotid artery followed by exposure of the neonatal rat to hypoxia (11–15). Hyperoxia may also play an important role in brain injury, by increasing oxidative injury in vulnerable tissues (16–18). Depending on the precise nature of the investigator's question, postnatal age at the time of insult can be varied, and the severity of injury can be adjusted by varying the length of hypoxia, the degree of hypoxia, and the environmental temperature. Although rodent models are valuable, there are significant limitations to their use, including substantial morphological differences between human and rodent brains (e.g., the degree of encephalization and gray–white matter ratios). Thus, while rodent models are ideal for developing a framework for therapies, and the genetic and biochemical pathways by which they function, prior to clinical trials, it is important to test these approaches in larger mammals such as fetal sheep, piglets or nonhuman primates (19–24).

1.2 Erythropoietin Neuroprotection

Erythropoietin (Epo) is a circulating glycoprotein first identified for its role in erythropoiesis. More recently, the neurodevelopmental and neuroprotective functions of Epo, acting via cell-specific Epo receptors (EpoR) in the brain, have been the subject of extensive investigation. EpoR are expressed in brain throughout embryology and fetal development (25), and Epo has trophic effects both on the vascular and nervous systems (26, 27). As the fetal brain development proceeds, EpoR become increasingly regional and cell-specific (28). The function of Epo during brain development is thought to be both trophic and protective: Epo and EpoR knockout mice have smaller brains and decreased tolerance to hypoxic insults with increased neuronal apoptosis (26). To date, hundreds of studies have been published using Epo as a neuroprotective strategy in adult and neonatal models of brain injury. Significant neuroprotection has been identified in brain injury caused by hypoxia-ischemia, stroke, trauma, kainate-induced seizures, and subarachnoid hemorrhage (29–32).

EpoR are present on neural progenitor cells (33), select populations of mature neurons (34), astrocytes (35), oligodendrocytes (36), microglia (37), and endothelial cells (33) within the brain. Anti-apoptotic pathways are activated when Epo binds to cell surface EpoRs, which dimerize to activate phosphorylation of Janus kinase 2 (JAK2), phosphorylation and activation of the mitogen-activated protein kinase (MAPK), extracellular signal-regulated kinase (ERK1/2), as well as the phosphatidylinositol 3-kinase (PI3K)/Akt (protein kinase B) pathway, and signal transducer and activator of transcription 5 (STAT5) (38). Epo protects oligodendrocytes from interferon-γ and LPS toxicity (39), improves white matter survival in vivo (40) and promotes oligodendrocyte maturation and differentiation in culture (35).

Epo promotes neuroprotection by stimulation and interaction with other protective factors such as brain-derived neurotrophic factor (BDNF) and glial cell derived neurotrophic factor (GDNF) (33, 41). Epo actively participates in the prevention of oxidative stress with generation of antioxidant enzymes, inhibition of nitric oxide production, and decrease of lipid peroxidation. Long-term survival of injured or newly generated cells may be fostered by Epo together with vascular endothelial growth factor (VEGF) to promote angiogenesis and repair (42, 43). Enhanced migration of neural progenitor cells occurs after treatment with Epo (44). Thus, some protective effects of Epo are the result of direct neuronal receptor-mediated interaction, and others are indirect. In fact, a recent study shows there to be neuroprotective effects of Epo in the absence of neural EpoR (45).

Epo crosses the intact blood–brain barrier at high dosages (5000 U/kg i.p.) (46) and may cross at lower doses in the setting of acute brain injury associated with disruption of the blood–brain barrier (47). Neuroprotective Epo doses range from 300 U/kg/dose to 30,000 U/kg, depending on the species studied, the injury model, severity of injury, and the number of doses given (47, 48). Multiple Epo treatments provide superior neuroprotection when compared to single Epo doses following brain injury in rodent models (48, 49). This is likely due to the multiple effects of Epo that are important during the evolving injury response: Epo decreases the early inflammatory response; decreases both early and late neuronal apoptosis (50); and stimulates late repair processes such as neurogenesis, angiogenesis, and migration of regenerating neurons (51). Higher Epo dosages (20–30,000 U/kg), particularly when given repeatedly, lose protective properties, may cause harm, and are not recommended (52). Of note circulating blood concentrations in neonates given 500–1,000 U/kg/dose are similar to those achieved in rat pups given 5,000 U/kg/dose (48, 53). Complications of Epo treatment in adults include polycythemia, rash, seizures, hypertension, shortened time to death, myocardial infarction, congestive heart failure, progression of tumors, and stroke. These complications have not been observed in neonates.

Through animal experimentation, it is known that a defined injury produced by a discrete hypoxic-ischemic insult will evolve through discrete stages including acute necrosis, inflammation, apoptosis, and late repair (54). However, there is tremendous variability between individual responses to a defined injury. It is not understood why one individual will respond to an injury with an injurious cascade of inflammation and ongoing damage while another individual can curtail this response and mobilize a repair pathway, leading to healing. These responses are likely mediated by the sequential activation of specific proteins and corresponding metabolic response. Investigating these responses in a systematic manner can provide insight into new therapeutic approaches to neonatal brain injury. There is potential for therapeutic intervention at each stage in the evolution of injury; however, the efficacy of treatment is dependent on the timing of directed interventions. For example, one could target acute injury, decreasing the initial damage caused by the inflammatory response to injury, or, one could target the later repair phase. To date, hypothermia is the only proven therapy for neonatal hypoxic-ischemic brain injury, and this is only appropriate for term infants. Translational Epo studies are now underway, and there is hope that this therapy might benefit both preterm brain injury and, in conjunction with hypothermia, improve outcomes for term infants with neonatal encephalopathy.

1.3 Unilateral Brain Injury in a Neonatal Rodent Model

The most commonly used small animal model of neonatal brain injury is the Vannucci model of unilateral brain injury (11). It is modified from the Levine preparation of adult rats (55) and is created by unilateral common carotid artery ligation or electrocauterization followed by a defined hypoxic exposure, generally using 8% oxygen. Ipsilateral brain damage can include the cerebral cortex, basal ganglia, hippocampus, and periventricular white matter injury (14). The brain hemisphere contralateral to the ligated carotid artery is generally uninjured. Severity of brain injury can be modified by adjusting the duration of hypoxia (usually ranging between 30 min and 3.5 h), with the resulting injury ranging from selective neuronal necrosis to gross infarction. The postnatal (P) age at which the experiment is performed determines whether the model is targeting premature brain injury, or term brain injury. A Web-based program is available to translate brain maturity from one species to another (56). The Vannucci procedure was first described in rat pups, but has since been modified for mice. Of note, mice are generally more sensitive to hypoxia than are rats, and significant strain differences exist, and brain maturation is slightly different than in rats (57, 58).

1.4 Overview of Procedure

Neonatal rat pups are anesthetized, undergo isolation and electrocauterization of the right common carotid artery, followed by exposure to 8% oxygen for 90 min to produce a moderate unilateral hypoxic-ischemic brain injury. Depending on the length of the hypoxic episode, the timing of peak cerebral edema ranges from 4 to 24 h (23, 59). The blood–brain barrier is altered during the recovery from the hypoxic injury as evidenced by extravasation of horseradish peroxidase from the vascular space into the brain parenchyma (60). Cerebral blood flow to the ipsilateral brain is unaffected by the unilateral common carotid ligation, but is decreased during the hypoxia-ischemic episode (12) (see Note 1).

2 Materials

1. Personal protective equipment (lab coat, sterile gloves, mask).
2. Instrument sterilizer.
3. Inhalant anesthetic (e.g., Isoflurane), matching vaporizer with regulated oxygen supply and adjustable flow control.
4. Anesthesia induction chamber.
5. Anesthetic exhaust or trap.
6. Dissecting microscope with light source.
7. Plastic animal cradle. Should contain a V-shaped indentation in which to hold the rat pup supine during surgery (Fig. 1).
8. Heating pad × 2.
9. Small animal incubator (Veterinary warmer).
10. Fine rat-tooth forceps.
11. Scalpel with #11 blade.
12. Microretractor or suitable substitute.
13. Sharp fine forceps × 2.
14. Cauterizer.
15. Skin staples, skin glue, or suture material for wound closure.
16. Gauze or cotton swabs.
17. Pentobarbital (or other approved euthanizing agent, to euthanize failed surgeries).
18. Body bag.
19. Hypoxia chamber (sealed container with ports for gas delivery).
20. Thermoregulated water bath.
21. Thermometer.
22. 8% oxygen tank with regulator.
23. Oxygen meter.

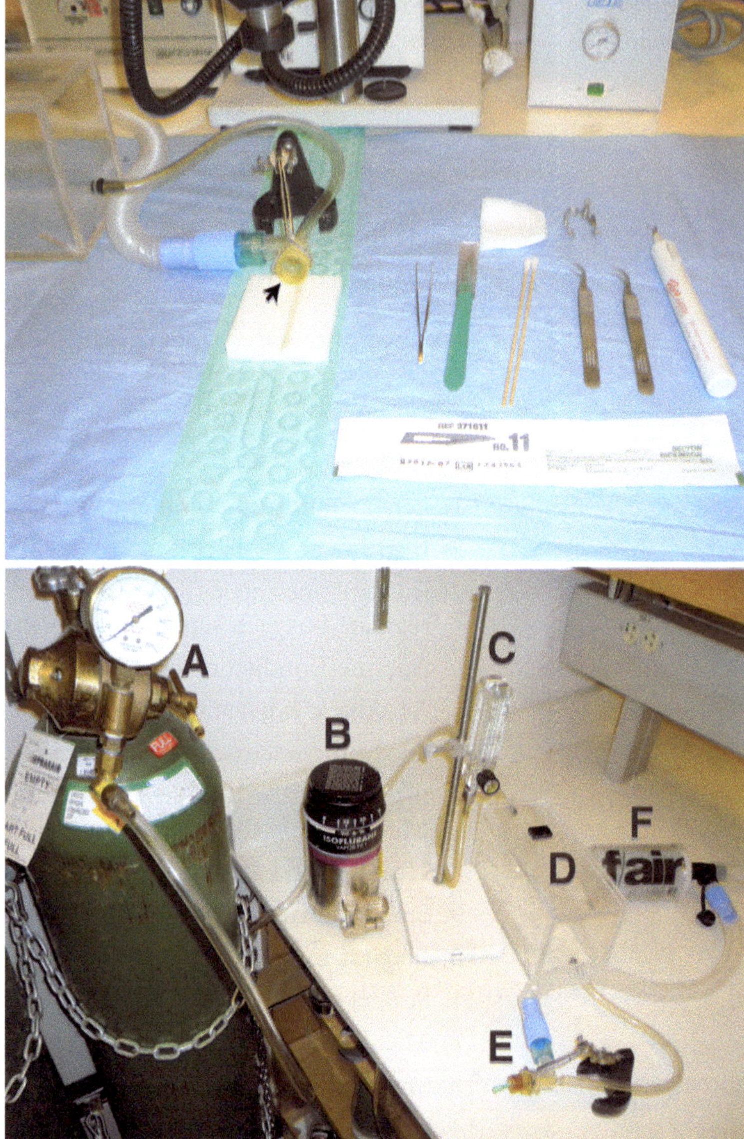

Fig. 1 Setup. The top panel illustrates the surgical setup, including the animal cradle, anesthetic outlet (*arrow*), and surgical instruments on a sterile field. The bottom panel shows the setup for anesthesia. The oxygen tank (**A**) is connected to the anesthesia vaporizer (**B**), a flow meter (**C**), the induction chamber (**D**), anesthetic outlet (**E**), and gas scavenger (**F**)

3 Methods

3.1 Setup

3.1.1 Animal Warming Station

1. Turn on heating pad or veterinary warmer.
2. Place a home cage on top of a heating pad or inside a veterinary warmer. This will provide a warmed area to keep the pups when they are not with their dam.

3.1.2 Surgical Station (Fig. 1)

1. Prepare a sterile field in the area adjacent to the dissecting microscope. Tools are autoclaved prior to use, and a hot bead sterilizer is used between animals. Surgeons wear sterile gloves, face masks, and clean lab coats.
2. Lay out sterilized surgical instruments, warming pad and animal cradle (see Fig. 1, top panel).
3. Prepare dissecting microscope by turning on light source, and checking focal range.
4. Prepare anesthesia delivery system by connecting regulated oxygen supply to the anesthetic vaporizer. Connect the vaporizer to the induction chamber. The induction chamber connects to the anesthesia outlet (Fig. 1, bottom panel, arrow) which is connected to an exhaust or trap. A rubber dam can be secured to the anesthesia outlet to decrease the escape of anesthetic gas into the environment. To do this, cut a 3 in. square from a laboratory glove, secure to the outlet with a rubber band, and cut a small hole in the taut surface. The pup's nose will be inserted into this hole.

3.1.3 Hypoxia Station

1. Obtain an 8% O_2 tank and regulator, then connect tubing to channel the gas supply through a humidifier followed by a water trap then run the line to the hypoxia chamber. The humidifier and gas trap are readily made using two Erlenmeyer flasks, the humidifier contains dH_2O and is immersed in the water bath to warm and humidify the gas supply.
2. The hypoxia chamber must have an input and one-way exhaust port to prevent pressure build up. The optimal chamber is transparent acrylic or plastic with an air-tight opening. In advance, the minimal flow rate needed to quickly flush the chamber with 8% oxygen should be determined using an oxygen meter.
3. Prior to gas exposure, heat the water bath/chamber apparatus so that the humidifier and chamber will be at the desired temperatures.

3.2 Hypoxic-Ischemic Brain Injury

Select, identify, weigh, and assign animals to treatment groups as appropriate for the experimental protocol.

3.2.1 Anesthesia

1. Fill vaporizer with anesthetic.
2. Open gas supply and adjust gas flow to 2–3 LPM, depending on size of induction chamber.
3. Turn on vaporizer to 2.5–3%.
4. Place animal in chamber. Before the start of any surgical steps, ensure adequate sedation by checking for loss of spontaneous and reflexive movements (including lack of postural/geotaxis

orientation or response to touch), increased muscle relaxation, and decreased respiration. 3–5 min of pre-anesthesia is generally sufficient.

3.2.2 Common Carotid Artery Electrocautery

1. Aseptic precautions (must include method of instrument sterilization prior to initial use and between animals, if applicable)—Tools are autoclaved prior to use, and a hot bead sterilizer is used between animals. Surgeons wear sterile gloves, face masks, and clean lab coats. Turn on heating pad under the animal cradle.
2. Place the sedated animal supine onto the animal cradle with its nose in the anesthesia outlet (Fig. 1, arrow). Adjust the focus of the dissecting scope as necessary. Place a piece of tape across the animal's chest and forelimbs to gently secure the animal to the cradle. See Note 2 for additional comments.
3. Prepare the neck with Betadine surgical scrub. After a small midline neck incision the right CCA is isolated and permanently cauterized and the incision is closed with skin adhesive or wound clips as appropriate and topical Betadine solution is applied.
4. Use a #11 scalpel blade to make a 1 cm incision in skin of the mid anterior neck. Retract skin to improve access.
5. Use blunt dissection with fine curved forceps (repeated opening of the forceps) to retract the muscle layers and expose the underlying veins and arteries. The veins are dark, avoid them. The artery will be a pulsatile structure with bright red blood, surrounded by a white muscular layer. Dissect down to expose the strap muscles, then laterally to expose the sterno-cleido-mastoid muscles, and the carotid is deep in the cleft between those muscles.
6. Gently lift the exposed common carotid artery using a curved forceps (Fig. 2a).
7. To cauterize the artery, begin by touching the cauterizer to the forceps (not the artery). The heat will be transferred from the forceps to the artery. Observe the color change in the artery as it coagulates. Once it becomes opaque, you may touch the cauterizer to the artery to completely sever the structure.
8. Align the skin and apply a skin staple to close the incision. Alternatively, the incision can be sutured closed or glued using Vetbond.
9. Return the pups to the warming station and monitor recovery. Once the animal has recovered from anesthesia, it may be placed back in the home cage with the dam for 2 h before hypoxia exposure (see Note 3).

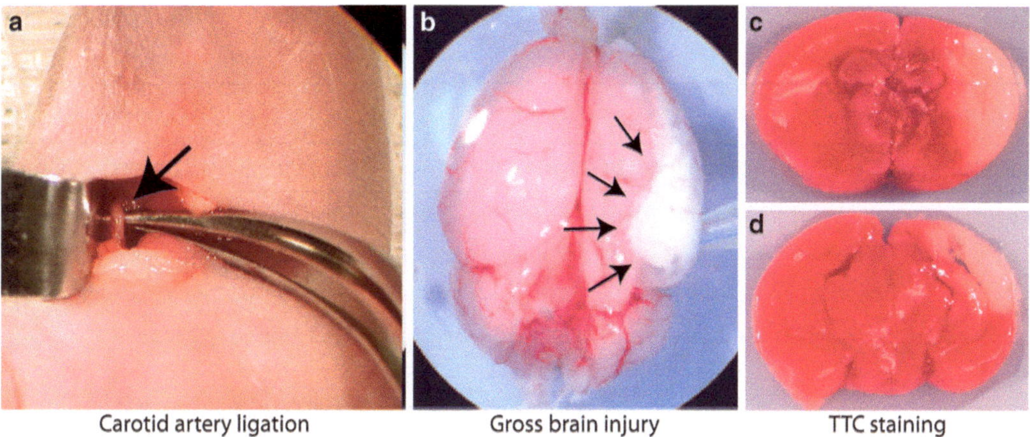

Fig. 2 Brain injury. Panel (**a**) shows the isolated common carotid artery (*arrow*). Panel (**b**) shows a brain with severe gross injury 48 h after ligation. *Arrows* outline the injured region. Panels (**c**) and (**d**) show TTC staining of injured brain from an anterior section (**c**), and posterior (**d**)

3.2.3 Hypoxia Exposure

1. Place pups in the preheated hypoxia chamber (heated to 36°C) (see Note 4).
2. Turn on the 8% oxygen tank at 15 L/min for 5 min to purge the tank. Decrease the flow to 5 L/min for duration of hypoxic event (30 min to 3.5 h, depending on severity of brain injury desired) (see Note 5).
3. When hypoxia exposure is completed and animals are stable, return pups to the dams.

3.3 Evaluation of Injury

3.3.1 Gross Evaluation of Injury

Weight gain, growth, timing, and cause of mortality and morbidity should be recorded. All brains should be scored for gross injury at the time of sacrifice. Gross brain injury is visually apparent 24 h after injury (Fig. 2b), and can be evaluated using an ordinal scale modified from Vannucci (61) as shown in Table 1. If animals are not perfusion-fixed, tetrazolium chloride (TTC) staining, which is incorporated into actively respiring tissues, can be used to clearly demarcate areas of non-vital tissue (Fig. 2c, d). The intensity of the color is proportional to the rate of respiration in the tissues, so living tissue stains red, while infarcted areas appear white.

3.3.2 Triphenyltetrazolium Chloride Staining

When TTC diffuses into *actively respiring* tissues it accepts electrons from the mitochondrial electron transport chain and the stain is reduced to the pink compound formazan. The accumulation of this pink compound stains actively metabolizing tissues red, and the intensity of the red is proportional to the rate of respiration in those tissues. This method can therefore be used to distinguish between live (stain, red) and infarcted (no stain, white) areas on 2 mm thick brain sections.

Table 1
Ordinal scale

Brain injury scale		
Scale	**Acute injury (24–48 h)**	**Chronic injury (>1 week)**
0	Normal ipsilateral hemisphere	Normal ipsilateral hemisphere
1	Mild edema with <25% ipsi/contra size difference	Mild atrophy with <25% ipsi/contra size difference
2	Moderate edema (25–50%) difference	Moderate atrophy (25–50% difference)
3	Liquefaction of 50–75% hemisphere	Cystic cavitation 50–75% hemisphere
4	Liquefaction of ≥75% hemisphere	Cystic cavitation ≥75% hemisphere

Using a mouse brain grid, rat pup brains are coronally sliced into 2 mm sections. Sections are incubated in 2% TTC in 0.9% Sodium Chloride for 30 min at room temperature. TTC is removed, and samples washed with fixative (4% Paraformaldehyde). Sections are removed from the solution for photographing.

Further investigation into specific effects of injury or neuroprotective treatments can be done using mRNA protein measurement or immunohistochemical approaches (see Note 6).

4 Notes

1. There are many confounders that may affect the extent of injury with this model. The age of the rodent changes which cell lines and brain regions are more at risk from hypoxic-ischemic injury (62, 63). Rodent pups should be sexed since gender differences have been seen in response to hypoxic-ischemic injury (64). These animals are exquisitely sensitive to temperature variation, and hypothermia can attenuate hypoxic-ischemic injury; therefore, body temperature should be constant during hypoxia-ischemia and throughout the experimental period. The hypoxic exposure should occur within 2–3 h after ligation of the carotid artery to allow suckling of the pups, thus avoiding the attenuation of injury secondary to fasting hypoglycemia (65). If larger experiments are planned, pups can be randomly assigned to dams and will be cross fostered. This decreases the maternal influence on experimental outcomes. Culling the litter size to 8 pups per litter will increase uniformity of pup growth and improve survival.
2. Anesthetic choice may affect outcomes, so a consistent anesthetic must be used across groups and when comparing groups

over time. For example, isoflurane is a vasodilator and also may have neurotoxic effects, while other anesthetics such as xenon may be neuroprotective.

3. Rat dams will eat the pups if disturbed. This occurs most commonly in the first days of life, but can still occur at P7. To minimize this, make sure there is no blood on the pups when they are returned to the dam after surgery. We have found that wiping the incision with a Betadine swab after incision closure decreases the likelihood that the dam will interfere.
4. Brain injury in this model can be highly variable, and is extremely temperature dependent. It is critical to monitor environmental temperatures closely before, during, and after the experiment. Small decreases in temperature will produce neuroprotection, and increases in temperature will increase brain injury. One solution is to use a plastic hypoxia chamber immersed in a temperature-controlled water bath. In advance, determine the optimal water bath temperature for the chosen chamber. We have also found that separating the pups during the hypoxia exposure will help ensure uniform temperature, and providing a heating pad turned to "low" after the experiment for several days will increase the uniformity of brain injury.
5. The depth and duration of hypoxia will determine the severity of brain injury. 6–12% oxygen exposure for 30 min to 3.5 h has been reported. Younger animals are more tolerant of hypoxia and require longer hypoxia exposures to create injury. Mice are more susceptible to hypoxic brain injury than are rats, and this is strain dependent (57), so caution must be used when beginning this experimental procedure in new strains. Despite this, working with mice can be advantageous because they have a more homogeneous background than rats, and transgenic strains are available.
6. For short-term experiments, animals can be identified using permanent marker numbers that are renewed daily. For long-term experiments, permanent identifiers must be used. Ear punches or paw tattoos work equally well. For long-term experiments, weaning at 21–28 days and separation of sexes is required.

References

1. Edwards AD, Brocklehurst P, Gunn AJ, Halliday H, Juszczak E, Levene M, Strohm B, Thoresen M, Whitelaw A, Azzopardi D (2010) Neurological outcomes at 18 months of age after moderate hypothermia for perinatal hypoxic ischaemic encephalopathy: synthesis and meta-analysis of trial data. BMJ 340:c363
2. Volpe JJ (2009) Cerebellum of the premature infant: rapidly developing, vulnerable, clinically important. J Child Neurol 24:1085–1104
3. Tyson JE, Parikh NA, Langer J, Green C, Higgins RD (2008) Intensive care for extreme prematurity—moving beyond gestational age. N Engl J Med 358:1672–1681

4. Wang B, Chen Y, Zhang J, Li J, Guo Y, Hailey D (2008) A preliminary study into the economic burden of cerebral palsy in China. Health Policy 87:223–234
5. Back SA (2006) Perinatal white matter injury: the changing spectrum of pathology and emerging insights into pathogenetic mechanisms. Ment Retard Dev Disabil Res Rev 12:129–140
6. Kapellou O, Counsell SJ, Kennea N, Dyet L, Saeed N, Stark J, Maalouf E, Duggan P, Ajayi-Obe M, Hajnal J, Allsop JM, Boardman J, Rutherford MA, Cowan F, Edwards AD (2006) Abnormal cortical development after premature birth shown by altered allometric scaling of brain growth. PLoS Med 3:e265
7. Stewart WB, Ment LR, Schwartz M (1997) Chronic postnatal hypoxia increases the numbers of cortical neurons. Brain Res 760:17–21
8. Ment LR, Schwartz M, Makuch RW, Stewart WB (1998) Association of chronic sublethal hypoxia with ventriculomegaly in the developing rat brain. Brain Res Dev Brain Res 111:197–203
9. Ashwal S, Cole DJ, Osborne S, Osborne TN, Pearce WJ (1995) A new model of neonatal stroke: reversible middle cerebral artery occlusion in the rat pup. Pediatr Neurol 12:191–196
10. Sola A, Rogido M, Lee BH, Genetta T, Wen TC (2005) Erythropoietin after focal cerebral ischemia activates the janus kinase-signal transducer and activator of transcription signaling pathway and improves brain injury in postnatal day 7 rats. Pediatr Res 57(4):481–487
11. Rice JE 3rd, Vannucci RC, Brierley JB (1981) The influence of immaturity on hypoxic-ischemic brain damage in the rat. Ann Neurol 9:131–141
12. Vannucci RC, Lyons DT, Vasta F (1988) Regional cerebral blood flow during hypoxia-ischemia in immature rats. Stroke 19:245–250
13. Yager JY, Brucklacher RM, Vannucci RC (1992) Cerebral energy metabolism during hypoxia-ischemia and early recovery in immature rats. Am J Physiol 262:H672–H677
14. Towfighi J, Zec N, Yager J, Housman C, Vannucci RC (1995) Temporal evolution of neuropathologic changes in an immature rat model of cerebral hypoxia: a light microscopic study. Acta Neuropathol 90:375–386
15. Vannicci RC, Connor JR, Mauger DT, Palmer C, Smith MB, Towfighi J, Vannucci SJ (1999) Rat model of perinatal hypoxic-ischemic brain damage. J Neurosci Res 55:158–163
16. Felderhoff-Mueser U, Sifringer M, Polley O, Dzietko M, Leineweber B, Mahler L, Baier M, Bittigau P, Obladen M, Ikonomidou C, Buhrer C (2005) Caspase-1-processed interleukins in hyperoxia-induced cell death in the developing brain. Ann Neurol 57:50–59
17. Saugstad OD (1990) Oxygen toxicity in the neonatal period. Acta Paediatr Scand 79:881–892
18. Munkeby BH, Borke WB, Bjornland K, Sikkeland LI, Borge GI, Halvorsen B, Saugstad OD (2004) Resuscitation with 100% O2 increases cerebral injury in hypoxemic piglets. Pediatr Res 56:783–790
19. Ment LR, Stewart WB, Fronc R, Seashore C, Mahooti S, Scaramuzzino D, Madri JA (1997) Vascular endothelial growth factor mediates reactive angiogenesis in the postnatal developing brain. Brain Res Dev Brain Res 100:52–61
20. Ment LR, Stewart WB, Duncan CC, Pitt BR, Cole JS (1986) Beagle puppy model of perinatal cerebral infarction. Regional cerebral prostaglandin changes during acute hypoxemia. J Neurosurg 65:851–855
21. Adcock LM, Yamashita Y, Goddard-Finegold J, Smith CV (1996) Cerebral hypoxia-ischemia increases microsomal iron in newborn piglets. Metab Brain Dis 11:359–367
22. Stave U (1965) Age-dependent changes of metabolism. II. Influences of hypoxia on tissue enzyme patterns of newborn and adult rabbits. Biol Neonat 8:114–130
23. Vannucci RC (1993) Experimental models of perinatal hypoxic-ischemic brain damage. APMIS Suppl 40:89–95
24. Myers RE (1975) Four patterns of perinatal brain damage and their conditions of occurrence in primates. Adv Neurol 10:223–234
25. Juul SE, Anderson DK, Li Y, Christensen RD (1998) Erythropoietin and erythropoietin receptor in the developing human central nervous system. Pediatr Res 43:40–49
26. Yu X, Shacka JJ, Eells JB, Suarez-Quian C, Przygodzki RM, Beleslin-Cokic B, Lin CS, Nikodem VM, Hempstead B, Flanders KC, Costantini F, Noguchi CT (2002) Erythropoietin receptor signalling is required for normal brain development. Development 129:505–516
27. Chen ZY, Asavaritikrai P, Prchal JT, Noguchi CT (2007) Endogenous erythropoietin signaling is required for normal neural progenitor cell proliferation. J Biol Chem 282:25875–25883
28. Juul SE, Yachnis AT, Rojiani AM, Christensen RD (1999) Immunohistochemical localization of erythropoietin and its receptor in the developing human brain. Pediatr Dev Pathol 2:148–158
29. Brines ML, Ghezzi P, Keenan S, Agnello D, de Lanerolle NC, Cerami C, Itri LM, Cerami A (2000) Erythropoietin crosses the blood–brain barrier to protect against experimental brain injury. Proc Natl Acad Sci U S A 97:10526–10531
30. Juul S (2002) Erythropoietin in the central nervous system, and its use to prevent

hypoxic-ischemic brain damage. Acta Paediatr Suppl 91:36–42
31. McPherson RJ, Juul SE (2008) Recent trends in erythropoietin-mediated neuroprotection. Int J Dev Neurosci 26:103–111
32. Xiong T, Qu Y, Mu D, Ferriero D (2011) Erythropoietin for neonatal brain injury: opportunity and challenge. Int J Dev Neurosci 29(6):583–591
33. Wang L, Zhang Z, Wang Y, Zhang R, Chopp M (2004) Treatment of stroke with erythropoietin enhances neurogenesis and angiogenesis and improves neurological function in rats. Stroke 35:1732–1737
34. Wallach I, Zhang J, Hartmann A, van Landeghem FK, Ivanova A, Klar M, Dame C (2009) Erythropoietin-receptor gene regulation in neuronal cells. Pediatr Res 65:619–624
35. Sugawa M, Sakurai Y, Ishikawa-Ieda Y, Suzuki H, Asou H (2002) Effects of erythropoietin on glial cell development; oligodendrocyte maturation and astrocyte proliferation. Neurosci Res 44:391–403
36. Nagai A, Nakagawa E, Choi HB, Hatori K, Kobayashi S, Kim SU (2001) Erythropoietin and erythropoietin receptors in human CNS neurons, astrocytes, microglia, and oligodendrocytes grown in culture. J Neuropathol Exp Neurol 60:386–392
37. Chong ZZ, Kang JQ, Maiese K (2003) Erythropoietin fosters both intrinsic and extrinsic neuronal protection through modulation of microglia, Akt1, Bad, and caspase-mediated pathways. Br J Pharmacol 138:1107–1118
38. Digicaylioglu M, Lipton SA (2001) Erythropoietin-mediated neuroprotection involves cross-talk between Jak2 and NF-kappaB signalling cascades. Nature 412:641–647
39. Genc K, Genc S, Baskin H, Semin I (2006) Erythropoietin decreases cytotoxicity and nitric oxide formation induced by inflammatory stimuli in rat oligodendrocytes. Physiol Res 55:33–38
40. Vitellaro-Zuccarello L, Mazzetti S, Madaschi L, Bosisio P, Fontana E, Gorio A, De Biasi S (2008) Chronic erythropoietin-mediated effects on the expression of astrocyte markers in a rat model of contusive spinal cord injury. Neuroscience 151:452–466
41. Dzietko M, Felderhoff-Mueser U, Sifringer M, Krutz B, Bittigau P, Thor F, Heumann R, Buhrer C, Ikonomidou C, Hansen HH (2004) Erythropoietin protects the developing brain against N-methyl-d-aspartate receptor antagonist neurotoxicity. Neurobiol Dis 15:177–187
42. Wang L, Chopp M, Gregg SR, Zhang RL, Teng H, Jiang A, Feng Y, Zhang ZG (2008) Neural progenitor cells treated with EPO induce angiogenesis through the production of VEGF. J Cereb Blood Flow Metab 28:1361–1368
43. Bocker-Meffert S, Rosenstiel P, Rohl C, Warneke N, Held-Feindt J, Sievers J, Lucius R (2002) Erythropoietin and VEGF promote neural outgrowth from retinal explants in postnatal rats. Invest Ophthalmol Vis Sci 43:2021–2026
44. Wang KK, Larner SF, Robinson G, Hayes RL (2006) Neuroprotection targets after traumatic brain injury. Curr Opin Neurol 19:514–519
45. Xiong Y, Mahmood A, Lu D, Qu C, Kazmi H, Goussev A, Zhang ZG, Noguchi CT, Schallert T, Chopp M (2008) Histological and functional outcomes after traumatic brain injury in mice null for the erythropoietin receptor in the central nervous system. Brain Res 1230:247–257
46. Juul SE, McPherson RJ, Farrell FX, Jolliffe L, Ness DJ, Gleason CA (2004) Erytropoietin concentrations in cerebrospinal fluid of nonhuman primates and fetal sheep following high-dose recombinant erythropoietin. Biol Neonate 85:138–144
47. Zhu C, Kang W, Xu F, Cheng X, Zhang Z, Jia L, Ji L, Guo X, Xiong H, Simbruner G, Blomgren K, Wang X (2009) Erythropoietin improved neurologic outcomes in newborns with hypoxic-ischemic encephalopathy. Pediatrics 124:e218–e226
48. Kellert BA, McPherson RJ, Juul SE (2007) A comparison of high-dose recombinant erythropoietin treatment regimens in brain-injured neonatal rats. Pediatr Res 61:451–455
49. Gonzalez FF, Abel R, Almli CR, Mu D, Wendland M, Ferriero DM (2009) Erythropoietin sustains cognitive function and brain volume after neonatal stroke. Dev Neurosci 31:403–411
50. Juul SE, McPherson RJ, Bammler TK, Wilkerson J, Beyer RP, Farin FM (2008) Recombinant erythropoietin is neuroprotective in a novel mouse oxidative injury model. Dev Neurosci 30:231–242
51. Tsai PT, Ohab JJ, Kertesz N, Groszer M, Matter C, Gao J, Liu X, Wu H, Carmichael ST (2006) A critical role of erythropoietin receptor in neurogenesis and post-stroke recovery. J Neurosci 26:1269–1274
52. Weber A, Dzietko M, Berns M, Felderhoff-Mueser U, Heinemann U, Maier RF, Obladen M, Ikonomidou C, Buhrer C (2005) Neuronal damage after moderate hypoxia and erythropoietin. Neurobiol Dis 20: 594–600
53. Juul SE, McPherson RJ, Bauer LA, Ledbetter KJ, Gleason CA, Mayock DE (2008) A phase I/II trial of high-dose erythropoietin in extremely low birth weight infants: pharmacokinetics and safety. Pediatrics 122:383–391

54. Fan X, Kavelaars A, Heijnen CJ, Groenendaal F, van Bel F (2010) Pharmacological neuroprotection after perinatal hypoxic-ischemic brain injury. Curr Neuropharmacol 8:324–334
55. Levine S (1960) Anoxic-ischemic encephalopathy in rats. Am J Pathol 36:1–17
56. Clancy B, Kersh B, Hyde J, Darlington RB, Anand KJ, Finlay BL (2007) Web-based method for translating neurodevelopment from laboratory species to humans. Neuroinformatics 5:79–94
57. Sheldon RA, Sedik C, Ferriero DM (1998) Strain-related brain injury in neonatal mice subjected to hypoxia-ischemia. Brain Res 810:114–122
58. Boasen JF, McPherson RJ, Hays SL, Juul SE, Gleason CA (2008) Neonatal stress or morphine treatment alters adult mouse conditioned place preference. Neonatology 95:230–239
59. Vannucci RC (1990) Experimental biology of cerebral hypoxia-ischemia: relation to perinatal brain damage. Pediatr Res 27:317–326
60. Vannucci RC, Christensen MA, Yager JY (1993) Nature, time-course, and extent of cerebral edema in perinatal hypoxic-ischemic brain damage. Pediatr Neurol 9:29–34
61. Vannucci RC, Towfighi J, Heitjan DF, Brucklacher RM (1995) Carbon dioxide protects the perinatal brain from hypoxic-ischemic damage: an experimental study in the immature rat. Pediatrics 95:868–874
62. Yager JY, Brucklacher RM, Vannucci RC (1996) Paradoxical mitochondrial oxidation in perinatal hypoxic-ischemic brain damage. Brain Res 712:230–238
63. Towfighi J, Mauger D, Vannucci RC, Vannucci SJ (1997) Influence of age on the cerebral lesions in an immature rat model of cerebral hypoxia-ischemia: a light microscopic study. Brain Res Dev Brain Res 100:149–160
64. Vannucci SJ, Hagberg H (2004) Hypoxia-ischemia in the immature brain. J Exp Biol 207:3149–3154
65. Yager JY, Heitjan DF, Towfighi J, Vannucci RC (1992) Effect of insulin-induced and fasting hypoglycemia on perinatal hypoxic-ischemic brain damage. Pediatr Res 31: 138–142

Chapter 8

Evaluating Effects of EPO in Rodent Behavioral Assays Related to Depression

Catharine H. Duman and Samuel S. Newton

Abstract

The cytokine erythropoietin (EPO) is an important regulator of hematopoesis and has well-known tissue protective properties. Neurotrophic action is implicated as mechanistically important in the treatment of depression, and neurotrophic actions of EPO suggest potential therapeutic utility of an EPO-like mechanism in depressive disorder. Rodent behavioral models that are responsive to clinically used antidepressants as well as to neurotrophic compounds can be used to assess potential antidepressant properties of EPO and EPO-like compounds. Rodent models described here are the forced-swim test (FST), a hyponeophagia test and the novel object recognition test. Each of these models provides different information and relevance to depression and each can be tested with EPO and similar compounds.

Key words EPO, CEPO, Depression, Behavior, Swim test, Object recognition, Hyponeophagia

1 Introduction

1.1 Erythropoietin, Neuroprotection, and Depression

Erythropoietin (EPO) is widely recognized for its role in elevating erythropoiesis and is used successfully to treat anemia. EPO receptors are expressed in neuronal cell lines and in brain, and in addition to its hematopoietic effects in the periphery, EPO can exert neurotrophic and angiogenic effects in the central nervous system (1–5).

A neuroprotective action of EPO suggests a variety of therapeutic applications but such applications could be limited by the erythropoietic effect during long-term administration. The separation of erythropoietic from neuroprotective mechanisms has been investigated based on specificity of receptor complexes and signaling pathways that mediate neuroprotection vs. hematopoiesis, and EPO derivatives have been developed that do not increase hematocrit but retain tissue protective properties (5–7). Non-erythropoietic EPO-like compounds can be tested in research

Pietro Ghezzi and Anthony Cerami (eds.), *Tissue-Protective Cytokines: Methods and Protocols*, Methods in Molecular Biology, vol. 982, DOI 10.1007/978-1-62703-308-4_8, © Springer Science+Business Media, LLC 2013

models with the benefit that they can be given over longer periods of time compared with EPO. Carbamylated EPO (Cepo) is an EPO-derivative that lacks hematopoietic activity and has been demonstrated to be neuroprotective (6).

The trophic hypothesis of depression states that deficits in trophic support and signaling could contribute to neuronal atrophy, dysfunction, and behavioral depression, while antidepressants are able to elevate trophic signaling and alleviate these abnormalities (8). The trophic actions of EPO in brain suggested it might have antidepressant properties and this has been supported by initial studies in rodents (9). EPO is induced in brain by an antidepressant treatment and EPO can produce antidepressant-like behavioral responses in mouse and rat models (9). These results are consistent with human studies showing that EPO improves mood (10).

Major depressive disorders are a heterogeneous group of illnesses that vary in symptomology and likely in etiology. Because the pathophysiological basis for depression is not known, re-creating the disease in animal models is not possible. In this chapter we present three rodent behavioral models that are relevant to depression research in different ways and that are each amenable to testing with EPO or related compounds.

1.2 Models of Depression

The forced-swim test (FST) has been widely used as screening test for antidepressant activity due to its ease of use, high throughput and relatively good predictive validity. The FST does not reproduce the pathophysiology of depression but induces changes that are sensitive to therapeutic agents in a manner predictive of their effects in humans (11–13). Stress exposure is an important component of many depression models and this test includes acute or sub-chronic stress. In the FST, a rat or mouse is placed in a cylinder with enough water so that it cannot touch the bottom with its hindpaws. A normal animal will show an immediate burst of activity, try to escape, and then eventually adopt an "immobile" posture, where it will make only those movements necessary to keep its head above water. The development of immobility may be facilitated by prior exposure to the test (13). Immobility is quantified during the test and classical antidepressants such as the monoamine oxidase inhibitors, tricyclics, and atypical antidepressants all decrease the duration of immobility in rats and mice in a dose-dependent manner (11, 12, 14). Neurotrophic factors have also been demonstrated to decrease immobility in the FST similarly to antidepressants (15, 16). In rats it is possible to separately quantify the predominant active behavior as either swimming or climbing which are associated with predominantly serotonergic and noradrenergic mechanisms, respectively, allowing the FST to detect this distinction (17, 18).

The FST may yield false positive results with drugs that increase motor activity and this possibility needs to be assessed in a separate test of drug effects on locomotor activity. Important limitations of the FST include that it does not distinguish acute from chronic antidepressant effects and therefore is not useful for assessing time-dependent actions of antidepressants. This combined with the selectivity for monoamine-based mechanisms may limit the ability of the FST to detect novel mechanisms (19, 20).

Hyponeophagia paradigms are anxiety-based, and compare feeding behavior in an anxiogenic vs. a non-anxiogenic environment. The stress employed in these models is very mild and consists of placing the experimental animal in a novel environment to induce anxiety during testing. The animal experiences conflict between the desire to approach and feed or drink, and the anxiety-induced avoidance of the novel environment.

The hyponeophagia paradigms are sensitive to acute and chronic anxiolytic effects of benzodiazepines (21). They detect anxiolytic effects of antidepressants only after chronic treatment, which agrees with the temporal profile for this effect in humans (22, 23). Utility of hyponeophagia paradigms is that they are bidirectionally sensitive, are selective for chronic effects of antidepressants and have no significant training requirement.

The neural mechanisms underlying anxiety and depression may be related as suggested by their comorbidity and the comparable efficacy of antidepressants in the treatment of both disorders (24–26). These considerations suggest that common neural substrates modulate anxiety and depressive mood, and validate use of anxiety components in animal depression models.

The novel object recognition test is a test of learning and memory in rodents. This test is based on the tendency of a rodent to interact with and explore a novel object to a greater extent than with a familiar object (27). The degree of the difference in novel vs. familiar object exploration is related to the strength of the memory the animal has of the familiar object. In this test, animals are exposed to two identical objects during a familiarization session. After a defined period of time, each animal is reintroduced to the test box that now contains one of the familiar objects but also a novel object. The time spent exploring the novel object relative to the time spent exploring the familiar object is a measure of the memory of the familiar object. Tests of learning and memory provide a functional measure of neural plasticity that is likely important to the actions of antidepressant treatments as well as to other classes of compounds including trophic factors. Various learning and memory tests have been useful in demonstrating improved cognitive function in humans and in rodent models following treatment with EPO or related compounds (7, 28–30).

2 Materials

2.1 Treatments

1. Rat recombinant erythropoietin.
2. Carbamylated erythropoietin: prepared according to (31) with some modifications. Vehicle: Phosphate buffered saline with 0.1% bovine serum albumin (molecular biology grade).
3. One-ml syringes.
4. 26 G-needles.

2.2 Rat Forced-Swim Test

1. Two clear (Plexiglas) cylinders, 25-cm diameter, 65-cm height (see Note 1).
2. Nearby water source and hose or tubing.
3. Sturdy cart or table.
4. White divider to place between cylinders.
5. Clean cages with tops for drying animals.
6. Paper and cloth towels.
7. Video camera and tripod.
8. Thermometer.
9. Timer.

2.3 Mouse FST

1. Two 4-l glass beakers, each with a mark 10 cm from the bottom.
2. Thermometer.
3. Large tub of water (e.g., a clean rodent chow bin).
4. Two clean cages.
5. Paper towels.
6. Video camera and tripod.
7. Timer.
8. Stopwatch.

2.4 Mouse Hyponeophagia

1. Carnation sweetened condensed milk diluted 1:3 in water.
2. Water bottles (125 ml size).
3. Bottle sipper tops (with ball-bearings).
4. Timer.
5. Clean mouse cages and tops (number of cages = number of mice to be tested).

2.5 Mouse Novel Object Recognition

1. Test cages (28 × 17 × 12 cm) (number of cages for each day = number of mice to be tested).
2. Cage tops.
3. Camera, tripod.

4. Two identical objects: (See Note 2).
5. A novel object (similar in size to familiar objects).
6. 70% ETOH.
7. Paper towels.
8. Two silent-operation stopwatches (no beeps or other noises when time is started or stopped).
9. Timer.

3 Methods

3.1 Rat FST

EPO (500 U/kg) is injected intraperitoneally (i.p.) once per day for 2 consecutive days. Vehicle is similarly administered to a second group of rats (8 rats/group) (see Notes 3 and 4).

3.1.1 Swim Day 1

1. EPO and vehicle are administered as on the prior 2 days to the same groups of rats. This is the third consecutive day of treatment. Time injections so that pairs of rats are injected every 20–25 min. Swim testing starts for the first pair of rats 4 h after they are injected.
2. The two cylinders are placed on a cart or table at one end of the room, with a plain white surface (such as a bare wall) behind them and a white divider between them. Visual stimuli should be the same on the outer sides of both cylinders. Ensure that normal overhead lighting falls evenly on both cylinders.
3. All rats to be tested are brought in their homecages into a quiet holding room near the testing room and acclimate for 1 h before the first pair is tested.
4. Attach hose or tubing (see Note 1) to faucet and fill cylinders while adjusting the hot and cold flows so that the temperature of water is 23–25 °C. Fill to a depth of 45 cm.
5. After rats acclimate for 1 h, two rats at a time are brought from the holding room into the testing room and lowered by the tail into the two cylinders (see Note 5). The experiment timer is started and runs for 15 min. The experimenter sits (equidistant from both cylinders) on the opposite side of the room. Behavior is not scored on day 1.
6. At the end of the 15 min session, the rats are removed from the water and placed in clean rat cages lined with several dry paper or cloth towels. They are dried with the towels and then remain in the drying cages for 15 min before being returned to their homecages.
7. Cylinders are emptied and refilled with clean water for the next pair of rats. The process is repeated for remaining animals.

3.1.2 Swim Day 2

1. EPO and vehicle are administered as on the prior 3 days to the same groups of rats. This is the fourth consecutive day of treatment. Time injections so that pairs of rats are injected every 15–20 min. Swim testing starts for the first pair of rats 4 h after they are injected.
2. The set-up is identical to that on day 1 (see Note 6). The two cylinders are placed on a cart or table at one end of the room, with a plain white surface (such as a bare wall) behind them and a white divider between them. Visual stimuli should be the same on the outer sides of both cylinders. The camera is set on a tripod at the other end of the room where the experimenter also sits (equidistant from both cylinders). Ensure that normal overhead lighting falls evenly on both cylinders.
3. All rats to be tested are brought in their homecages into a quiet holding room near the testing room and rats acclimate for 1 h before the first pair is tested.
4. Attach hose or tubing to faucet and fill cylinders while adjusting the hot and cold flows so that the temperature of water is 23–25 °C. Fill to a depth of 45 cm.
5. Two rats at a time are brought from the holding room into the testing room and gently lowered into the two cylinders. Each rat should be tested in the same cylinder as the prior day. The experiment timer is started and runs for 10 min.
6. Scoring of immobility can be performed after the experiment has been completed using the video recording, or one rat can be scored while the animals are being tested leaving one to be scored at a later time from the recorded video. Only one animal can be scored at a time. Sessions are videotaped from the side and scored by using a behavioral sampling technique to assess the primary behavior for each 5-s interval. Three behaviors are scored: (a) climbing, rat is thrashing in an upward motion along the sides of the tank; (b) swimming, movement across the tank into another quadrant; and (c) immobility, when the animal made the minimal amount of movement required to stay afloat.
7. At the end of the 10 min session, the rats are removed from the water and placed in clean rat cages lined with several dry paper or cloth towels. They are dried with the towels and then remain in the drying cages for 10 min before being returned to their homecages.
8. Cylinders are emptied and refilled with clean water for the next pair of rats. The process is repeated for remaining animals.
9. Scored behaviors are quantified and reported for the entire test period or as they occur over defined intervals of the test (see Fig. 1).

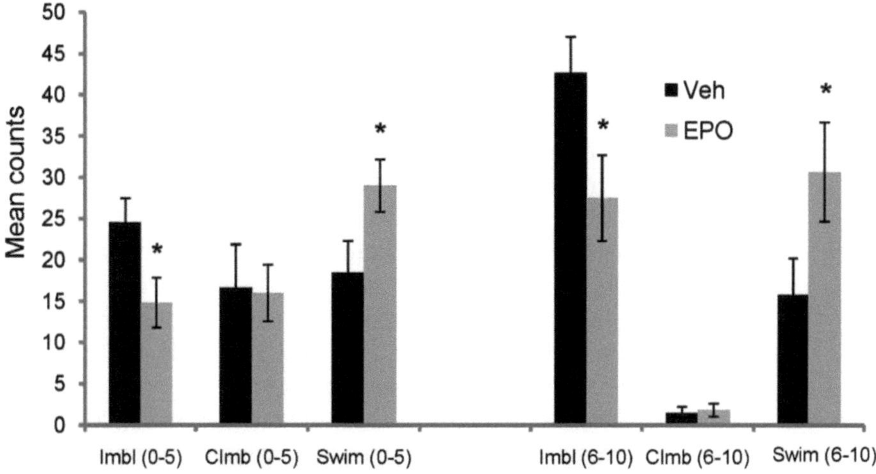

Fig. 1 Immobility in the forced-swim test (FST) is decreased by 4 days of EPO administration in rats. EPO (500 U/kg, i.p.) or vehicle (0.1% BSA in PBS) were administered to rats on 4 consecutive days, with the last dose 4 h prior to the day 2 swim test. Data are presented for two 5-min intervals of the test. $^{*}p<0.05$, Student's *t* test, $n=6$

3.2 Mouse FST

EPO (500 U/kg) is injected intraperitoneally (i.p.) once per day for 3 consecutive days. Vehicle is similarly administered to a second group of mice (8/group) (see Notes 3 and 4).

Swim Test

1. EPO and vehicle are administered as on the prior 3 days to the same groups of mice. This is the fourth consecutive day of treatment. Time injections so that pairs of mice are injected every 15 min. Swim testing starts for the first pair of mice 4 h after they are injected.
2. The two beakers are filled with the room temperature water to the 10 cm mark and beakers are placed with a white divider between them, on a cart or table at one end of the room with a plain white surface (such as a bare wall) behind them. Visual stimuli should be the same on the outer sides of both beakers. The camera is set on a tripod at the other end of the room where the experimenter also sits (equidistant from both beakers). Ensure that normal overhead lighting falls evenly on both beakers.
3. All mice to be tested are brought in their homecages into a quiet holding room near the testing room. Mice acclimate in the holding room for 1 h.
4. Two mice at a time are brought from the holding room into the testing room and lowered by the tail into the two beakers (see Note 5). The experiment timer is started.

5. Scoring of immobility begins after 2 min have elapsed. Scoring can be performed after the experiment has been completed using the recorded video, or one mouse can be scored while the animals are being tested leaving one to be scored at a later time from the recorded video. Only one mouse can be scored at a time. Each time the mouse ceases struggling and swimming movements and appears to be resting or floating (making only the minimal movements needed to stay afloat), the stopwatch is started. The watch is stopped when the bout of immobility ends. The watch is similarly started and stopped each time the mouse enters into and then ends a bout of immobility. At the end of the session the cumulative time spent in immobility is recorded off the stopwatch. Scoring continues for the remainder of the test period, which is predetermined and is typically from 4 to 8 additional minutes.
6. Mice are removed from the water at the end of the test period and placed individually into two clean mouse cages lined with several dry paper or cloth towels. They are dried with the towels and then remain in the drying cages for 5–10 min before being returned to their homecages.
7. The beakers are emptied and refilled with clean water for the next pair of mice. The process is repeated for remaining mice.
8. Cumulative immobility over the test period is reported.

3.3 Mouse Hyponeophagia

Individually house mice for at least 5 days in usual mouse cages prior to starting the habituation (see Note 4).

3.3.1 Habituation (Days 1–3)

Day 1

1. Dilute Carnation sweetened condensed milk 1:3 in water and mix well.
2. Partially fill clean water bottles with the diluted milk solution. Use at room temperature.
3. A bottle containing the milk solution is put on each cage in the place of the normal water for a 20-min period.
4. Milk bottles are then removed and normal water bottles are returned to all cages.
5. The bottles of milk can be reused for all habituation days (for the same animal) if they are refrigerated after each session and brought to room temperature before use. Discard after 3 days of use.

Days 2–3

1. EPO (500 U/kg) is injected intraperitoneally (i.p.) and vehicle is similarly administered to a second group of mice (8/group).

2. Remove milk bottles from refrigerator and allow them to come to room temperature.
3. Four hours after injections, milk bottles are placed on each cage in the place of the normal water for a 20-min period.
4. Milk bottles are then removed and normal water bottles are returned to all cages.
5. Repeat injections and milk habituation the next day.
6. Discard milk solution.

3.3.2 Home-Cage Consumption Test (Day 4)

1. EPO (500 U/kg) and vehicle are injected as on previous days.
2. Test occurs in homecages in their usual locations in the normal housing room. Prepare fresh milk solution and partially fill clean bottles.
3. Four hours after the injections, the test is performed. Water bottles are removed from the cages and are replaced with milk bottles. A timer is started and the latency it takes each mouse to drink the milk is recorded. More than one mouse can be tested at a time.
4. Label bottles and refrigerate for use in novel-cage test.

3.3.3 Novel-Cage Consumption Test (Day 5)

1. EPO (500 U/kg) and vehicle are injected as on previous days.
2. All mice to be tested are brought in their homecages into a quiet holding room near the testing room (see Note 6).
3. Remove bottles of milk from refrigerator and allow them to come to room temperature.
4. Prepare testing room. Have in the room a stack of clean mouse cages and tops equal to the number of mice to be tested. The room should have even bright lighting that falls evenly into the test area (see Note 7). Place two of the mouse cages without bedding side by side so that the observer has a clear view of the entire cage area. A white divider (cardboard or Plexiglas) is placed between the cages.
5. The novelty test is performed 4 h after the injections. Two mice at a time are brought from the holding room into the testing room and one is placed in each cage (see Notes 3 and 5). Milk bottles are placed on the cages and the timer is started (see Note 8).
6. The time when each mouse begins to drink the milk is recorded.
7. The test is terminated after 8 min for any mouse that does not drink and a cutoff latency of 480 s is given to that mouse.
8. Mice are removed from test cages and returned to homecages when both mice have completed the test.

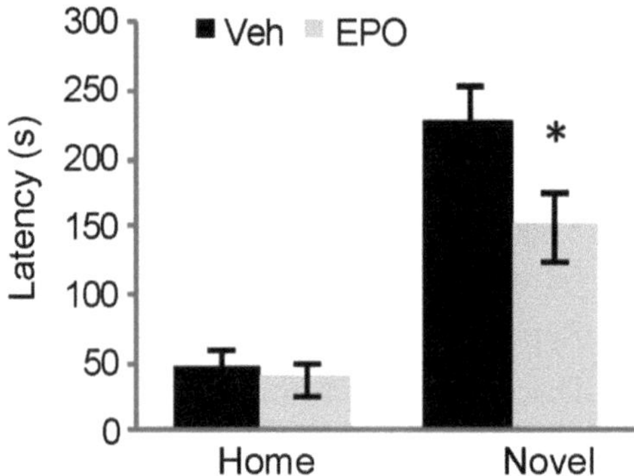

Fig. 2 Consumption latency in the novelty-induced hypophagia test is decreased in mice that received 4 days of EPO treatment (500 U/kg, i.p.). Testing was performed 4 h after the last EPO treatment. *$p=0.05$. Students *t* test, $n=7$–8

9. The procedure is repeated for remaining pairs of mice.
10. Latencies to drink in the homecage and the novel cage are the reported measures (see Fig. 2).

3.4 Novel Object Recognition in Mouse

CEPO (30 μg/kg) is injected intraperitoneally (i.p.) to a group of mice and vehicle is similarly administered to a second group of mice (8/group). This occurs on 2 consecutive days.

3.4.1 Habituation (Day 1)

1. CEPO and vehicle are administered as on the prior day to the same groups of mice. This is the 3rd consecutive day of treatment. Habituation occurs 4 h after treatments.
2. All mice to be tested are brought in their homecages into a quiet holding room near the testing room. Mice are placed into individual clean mouse cages with fresh bedding, in the experimental room which is dimly lit. Tops are placed on each cage and mice remain in these cages undisturbed for 1 h.
3. Return mice to homecages and normal housing room (see Notes 3, 4, 6).

3.4.2 Acquisition (Day 2)

1. CEPO and vehicle are administered as on the prior days to the same groups of mice. This is the fourth consecutive day of treatment. Time injections so that mice are given their object exposure sessions 4 h post-treatment.
2. All mice to be tested are brought in their homecages into a quiet holding room near the testing room. In the testing room, a clean cage with fresh bedding is set on the observation table and the cleaned objects are placed at opposite ends of the cage. The camera is set up to record the behavior during the session. The room is kept in dim lighting (see Note 7).
3. A mouse is brought to the experimental room and placed in the middle of the test cage, the top is closed and the experiment timer is started.

4. Scoring can be performed after the experiment has been completed if it is recorded on video, or can be completed during testing. The experimenter holds one stopwatch in each hand, with one watch corresponding to one of the objects. For each interval when contact/exploration occurs (direct contact or exploration with nose within 2 cm of object), the stopwatch for that object is started. The watch is stopped when the contact/exploration ends. The watches are started and stopped each time the contact occurs for the object. At the end of the session, the cumulative time of contact/exploration for each object is recorded.
5. Scoring continues for the 5-min test period.
6. The mouse is returned to its home cage.
7. The test cage is replaced by a clean test cage.
8. Objects are wiped with ETOH, dried with paper towels, and placed in the clean cage.
9. The next animal is then placed in the test cage and the procedure is repeated.

3.4.3 Novel Object Test (Day 3)

1. All mice to be tested are brought in their homecages into a quiet holding room near the testing room. In the testing room, a clean cage with fresh bedding is set on the observation table. One of the same objects from day 2 is replaced by the novel object (see Note 2) and the two objects are placed at opposite ends of the cage (see Note 9). The camera is set up to record mouse behavior during the test. The room is kept in dim lighting.
2. A mouse is brought to the experimental room and placed in the middle of the test cage, the top is closed and the experiment timer is started.
3. Scoring can be performed after the experiment has been completed if it is recorded on video, or can be completed during testing. The experimenter holds one stopwatch in each hand, with one watch corresponding to one of the objects. For each interval when contact/exploration occurs (direct contact or exploration with nose within 2 cm of object), the stopwatch for that object is started. The watch is stopped when the contact/exploration ends. The watches are started and stopped each time the contact occurs for the object. At the end of the session, the cumulative time of contact/exploration for each object is recorded.
4. Scoring continues for the 5-min test period.
5. The mouse is returned to its home cage.
6. The test cage is replaced by a clean test cage.
7. Objects are wiped with ETOH, dried with paper towels, and placed in the clean cage.

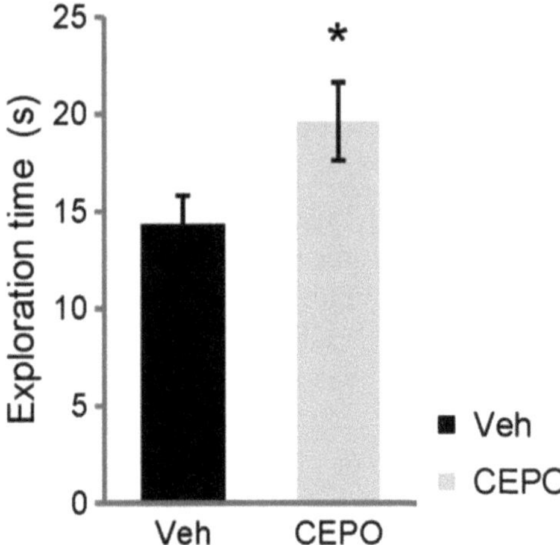

Fig. 3 Novel object exploration is increased in mice that received 4 daily administrations of CEPO (30 μg/kg, i.p.). Novel object exploration was conducted one day after the last CEPO administration. *$p = 0.05$, $n = 10$/group

8. The next animal is then placed in the test cage and the procedure is repeated.
9. Recognition index in the novel object test is generally the reported measure. It is calculated as the difference in time spent exploring the novel object and the familiar object divided by the total time spent exploring both objects. Exploration times recorded during the acquisition session (day 2) are useful controls for any side or object bias and exploration times can also be reported (see Fig. 3).

4 Notes

1. Cylinders can be custom made to have a drain port at the bottom which can be attached to PE tubing for easy draining and filling between animals.
2. For novel object recognition, it is important to run an initial pilot test to make sure that the baseline exploration of the different objects is similar. Possible objects are a Falcon tube and children's toy—Lego stacks of a similar size.
3. Animals should be handled with gloved hands in a gentle manner that does not activate or disrupt the animal.
4. Anything in the behavioral testing environment that might be a sensory stimulus should be minimized or should be kept constant over all of the days of the procedure. For example:
   (a) The same person should perform the procedure for all animals and all days.

(b) Position of the camera, test chamber, and experimenter within the testing room should be identical for each animal and on each day of a given procedure.

(c) Minimize extraneous noise that could potentially alter the animals' behavior (such as clicking or beeping of stopwatches: (cell phones have convenient silent stopwatch operation).

(d) Variables associated with the testing room should be controlled such as lighting, nearby noise and room odor.

5. Reduce side bias. For procedures that test two animals at a time, run animals of different treatment groups together and alternate sides or chambers in which animals of a given group are tested in order to minimize any bias due to the chambers or position in the room. Be sure to keep the side or chamber for a given animal the same on each day of the procedure.
6. For multiday tests, conduct the sessions at the same time on each day.
7. In tests that require some level of activity or exploration such as hyponeophagia tests or object recognition, note that the room lighting can affect the degree to which the animal explores. Bright lighting can increase anxiety in rodents and may inhibit activity and/or exploration.
8. It is important to ensure that the bottles do not drip. The mice will approach any drips on the cage floor with a relatively short latency and the data will be not useful.
9. For novel object recognition, counterbalance the sides the familiar and novel objects are placed on for the novel object test in order to minimize any side bias.

Acknowledgments

We thank Dr. Shannon Gourley and Monica Sathyanesan for their assistance in conducting the novel object recognition test. This work is supported by USHPS grant MH 078132 and the Connecticut Mental Health Center.

References

1. Masuda S et al (1993) Functional erythropoietin receptor of the cells with neural characteristics. Comparison with receptor properties of erythroid cells. J Biol Chem 268(15): 11208–11216
2. Konishi Y et al (1993) Trophic effect of erythropoietin and other hematopoietic factors on central cholinergic neurons in vitro and in vivo. Brain Res 609(1–2):29–35
3. Morishita E et al (1997) Erythropoietin receptor is expressed in rat hippocampal and cerebral cortical neurons, and erythropoietin prevents in vitro glutamate-induced neuronal death. Neuroscience 76(1):105–116

4. Brines M et al (2000) Erythropoietin crosses the blood–brain barrier to protect against experimental brain injury. Proc Natl Acad Sci U S A 97(19):10526–10531
5. Brines M, Cerami A (2005) Emerging biological roles for erythropoietin in the nervous system. Nat Rev Neurosci 6(6):484–494
6. Leist M et al (2004) Derivatives of erythropoietin that are tissue protective but not erythropoietic. Science 305(5681):239–242
7. Brines M et al (2008) Nonerythropoietic, tissue-protective peptides derived from the tertiary structure of erythropoietin. Proc Natl Acad Sci U S A 105:10925–10930
8. Duman RS, Heninger GR, Nestler EJ (1997) A molecular and cellular theory of depression. Arch Gen Psychiatry 54(7):597–606
9. Girgenti MJ et al (2009) Erythropoietin induction by electroconvulsive seizure, gene regulation, and antidepressant-like behavioral effects. Biol Psychiatry 66(3):267–274
10. Miskowiak K et al (2008) Erythropoietin improves mood and modulates the cognitive and neural processing of emotion 3 days post administration. Neuropsychopharmacology 33(3):611–618
11. Porsolt RD, Bertin A, Jalfre M (1977) Behavioral despair in mice: a primary screening test for antidepressants. Arch Int Pharmacodyn Ther 229(2):327–336
12. Porsolt RD, Le Pichon M, Jalfre M (1977) Depression: a new animal model sensitive to antidepressant treatments. Nature 266(5604):730–732
13. Porsolt RD, Bertin A, Jalfre M (1978) "Behavioural despair" in rats and mice: strain differences and the effects of imipramine. Eur J Pharmacol 51(3):291–294
14. Borsini F, Meli A (1988) Is the forced swimming test a suitable model for revealing antidepressant activity? Psychopharmacology (Berl) 94:147–160
15. Shirayama Y et al (2002) Brain-derived neurotrophic factor produces antidepressant effects in behavioral models of depression. J Neurosci 22(8):3251–3261
16. Duman CH et al (2009) Peripheral insulin-like growth factor-I produces antidepressant-like behavior and contributes to the effect of exercise. Behav Brain Res 198(2):366–371
17. Detke MJ, Rickels M, Lucki I (1995) Active behaviors in the rat forced swimming test differentially produced by serotonergic and noradrenergic antidepressants. Psychopharmacology (Berl) 121(1):66–72
18. Lucki I (1997) The forced swimming test as a model for core and component behavioral effects of antidepressant drugs. Behav Pharmacol 8:523–532
19. Willner P (1990) Animal models of depression: an overview. Pharmacol Ther 45:425–455
20. Thiébot M-H, Martin P, Puech AJ (1992) Animal behavioural studies in the evaluation of antidepressant drugs. Br J Psychiatry 160:44–50
21. Bodnoff SR et al (1988) The effects of chronic antidepressant treatment in an animal model of anxiety. Psychopharmacology (Berl) 95(3):298–302
22. Bodnoff SR et al (1989) A comparison of the effects of diazepam versus several typical and atypical anti-depressant drugs in an animal model of anxiety. Psychopharmacology (Berl) 97(2):277–279
23. Dulawa SC, Hen R (2005) Effects of chronic fluoxetine in animal models of anxiety and depression. Neurosci Biobehav Rev 29(4–5):771–783
24. Kaufman J, Charney D (2000) Comorbidity of mood and anxiety disorders. Depress Anxiety 12(Suppl 1):69–76
25. Ressler KJ, Nemeroff CB (2000) Role of serotonergic and noradrenergic systems in the pathophysiology of depression and anxiety disorders. Depress Anxiety 12(Suppl 1):2–19
26. Serretti A et al (2009) Common genetic, clinical, demographic and psychosocial predictors of response to pharmacotherapy in mood and anxiety disorders. Int Clin Psychopharmacol 24(1):1–18
27. Ennaceur A, Delacour J (1988) A new one-trial test for neurobiological studies of memory in rats. 1: Behavioral data. Behav Brain Res 31(1):47–59
28. Miskowiak K et al (2008) Differential effects of erythropoietin on neural and cognitive measures of executive function 3 and 7 days post-administration. Exp Brain Res 184(3):313–321
29. Ehrenreich H et al (2004) Erythropoietin: a candidate compound for neuroprotection in schizophrenia. Mol Psychiatry 9(1):42–54
30. Morgensen J et al (2008) Erythropoietin improves spatial delayed alternation in a T-maze in rats subjected to ablation of the prefrontal cortex. Brain Res Bull 77(1):1–7
31. Mun KC, Golper TA (2000) Impaired biological activity of erythropoietin by cyanate carbamylation. Blood Purif 18:13–17

Chapter 9

Erythropoietin and Cytoprotective Cytokines in Experimental Traumatic Brain Injury

Samson Kumar Gaddam, Jovany Cruz, and Claudia Robertson

Abstract

The various biochemical cascades that follow primary brain injury result in secondary brain injury which can adversely affect the clinical outcome. Over the last few years it has been well established that molecules like erythropoietin (Epo) have a neuroprotective role in experimental traumatic brain injury (TBI). Epo is shown to produce this effect by modulating multiple cellular processes, including apoptosis, inflammation, and regulation of cerebral blood flow. Derivatives of Epo, including asialo Epo and carbamylated Epo, have been developed to separate the neuroprotective properties from the erythropoiesis-stimulating activities of Epo which may have adverse effects in clinical situations. Peptides that mimic a portion of the Epo molecule, including Helix B surface peptide and Epotris, have also been developed to isolate the neuroprotective activities. The TBI model in rodents most commonly used to study the effect of Epo and these derivatives in TBI is controlled cortical impact injury, which is a model of focal contusion following a high velocity impact to the parietal cortex. Following TBI, rodents are given Epo or an Epo derivative vs. placebo and the outcome is evaluated in terms of physiological parameters (cerebral blood flow, intracranial pressure, cerebral perfusion pressure), behavioral parameters (motor and memory), and histological parameters (contusion volumes, hippocampus cell counts).

Key words Anesthesia, Behavior, Brain, Controlled cortical impact, Erythropoietin, Impactor, Craniectomy, Traumatic brain injury

1 Introduction

TBI is the leading cause of disabling injuries in all age groups. Apart from high mortality it can also result in significant morbidity; both short term and long term. In USA alone, an estimated 5.3 million people live with TBI related disability (1). Even mild TBI can result in cognitive problems that can affect activities of daily living (2). The primary brain injury that occurs at the time of injury is followed by several biochemical pathways in the brain that result in secondary brain injury leading to raised intracranial pressure and cerebral ischemia. Ischemia due to extracerebral causes (hypotension, anemia, and hypoxia) can also augment the secondary brain injury. The current management of head injury patients involves

Pietro Ghezzi and Anthony Cerami (eds.), *Tissue-Protective Cytokines: Methods and Protocols*, Methods in Molecular Biology, vol. 982, DOI 10.1007/978-1-62703-308-4_9, © Springer Science+Business Media, LLC 2013

monitoring and anticipating for complications of secondary brain injury and their medical and/or surgical treatment. The medical management mainly consists of controlled ventilation, anti-edema drugs, antibiotics, and antiseizure medications. Over the last few decades several biochemical pathways that are crucial in the secondary injury following TBI have been described. Several molecules like erythropoietin that affect these biochemical pathways have proved to be useful in animal TBI experiments (3, 4). Though there are no published randomized clinical trials demonstrating an effect of Epo on outcome from TBI (5), there is indirect data in the form of a retrospective study (6) and randomized trials in critically ill patients (7, 8), suggesting a beneficial role.

Cytokines are a family of polypeptides that regulate cell activation, proliferation, and differentiation. Under normal condition, cytokines are present in very low concentrations in the brain, and cytokine receptors are constitutively expressed only in low levels. After traumatic brain injury, pro-inflammatory and anti-inflammatory cytokines are rapidly upregulated. Pro-inflammatory cytokines, including interleukin-1β (IL-1β), tumor necrosis factor-α (TNF-α), and IL-6, stimulate the inflammatory response. Anti-inflammatory cytokines, including transforming growth factor-β (TGF-β) and IL-10, inhibit the expression of pro-inflammatory cytokines and reduce inflammation. The relative balance between the harmful and beneficial effects of cytokines depend on the injury conditions.

Erythropoietin (Epo) is a 34-kDa hematopoietic cytokine, synthesized predominantly by the kidney. Epo interacts with the Epo receptor (EpoR) which is expressed both in the neural and the nonneural tissues (9). Epo is shown to have a tissue protective (10) role through its influence on various pathways, including the JAK2 pathway, the MAPK signaling pathway, and the STAT-5 pathway (11–13). These pathways mainly have anti-inflammatory, anti-apoptotic, anti-excitatory, and neuroprotective roles (14, 15). Epo is also shown to increase cerebral blood flow, to maintain the integrity of blood brain barrier, and to stimulate angiogenesis (16). Studies in various experimental TBI models have demonstrated neuroprotective effects with Epo administered after TBI (3, 17–23). However, the potential for thromboembolic complications with Epo have hampered its translation as a neuroprotective agent (24, 25). Clinical trials are currently ongoing with Epo in TBI patients, but the doses being given in these trials are limited by concern for thromboembolic events and are lower than the doses found to be optimal in experimental TBI studies (26).

Studies with derivatives of Epo or peptides that mimic part of the Epo molecule have clearly shown that these neuroprotective activities of Epo can be separated from the erythropoietic effects (27–29). The receptor involved in the neuroprotective activities is less clear. Some studies have suggested that these neuroprotective activities are mediated through an Epo and CD131 heteroreceptor (30). Other studies have observed cytoprotection with Epo derivatives even in the absence of the CD131 receptor (31), and found that EpoR is necessary for neuroprotection (29).

Various chemical alterations that eliminate the erythropoietic activity of Epo, by either shortening the circulating half-time or by eliminating the ability of the derivative to interact with EpoR, have been developed (32). This has led to engineering of some non-erythropoietic derivatives of Epo like non-sialic Epo (asialoEpo), low sialic Epo (Neuro-Epo), and carbamylated Epo (CEpo). Erythropoietic activities of Epo require sustained blood levels. AsialoEpo reacts with EpoR but has a very short half-life in the circulation and does not significantly increase hematocrit although neuroprotective properties in a variety of central nervous system injury models are maintained (33). Neuro-Epo has been given intranasally to enhance the neuroprotective activity of the derivative (34, 35). CEpo, in contrast, does not interact with EpoR but is still neuroprotective in models of experimental TBI (23, 36–38).

Epo-mimetic peptides with neuroprotective activities have also been developed. Helix B surface peptide (HBSP) is a linear peptide containing the amino acids from the aqueous face of helix B (amino acids 58–85) of Epo (28). HBSP does not bind to EpoR, but is thought to activate a separate neuroprotective receptor (10). A modification of the N-terminal Q to pyroglutamate (U) for stabilization, results in an 11-aminoacid peptide called pyroglutamate Helix-B surface peptide (pHBSP) that has the cytoprotective effects with no stimulation of erythropoiesis (39, 40). Improved cerebral blood flow and better recovery of neurological function (reduced contusion volume and better performance in motor tasks) have been observed with administration of pHBSP following mild cortical impact injury complicated by hemorrhagic hypotension in rats (41).

Epotris, a 20 amino acid peptide corresponding to the C α-helix region (amino acid residues 92–111) of Epo, also has been shown to have neuroprotective properties without stimulating erythropoiesis (29). Epotris binds to the Epo receptor, and the neuroprotective activities are blocked when the EpoR is absent (29).

Studying the neuroprotective effects of these compounds requires a standardized model of experimental TBI. Human TBI is heterogeneous and complex. There are different types of TBI models that are described in literature, each of which has some features of human TBI. Fluid percussion injury, weight drop acceleration–deceleration injury, blast injury, and controlled cortical impact (CCI) are among the most commonly used experimental models (42). The CCI model mimics many aspects of human TBI, including concussion, contusion, and axonal injury (43).

Rodent TBI experiments are broadly classified into acute and chronic activities. In an acute experiment (non-survival procedure), after TBI and drug administration, the cerebral physiological and

Table 1
Impaction parameters in controlled cortical injury (CCI) for rats and mice

	Rats		Mice	
	Mild injury	Severe injury	Mild injury	Severe injury
Velocity	3 m/s	5 m/s	3 m/s	3 m/s
Duration of impact	90 ms	150 ms	100 ms	100 ms
Depth of deformation	2.5 mm	3 mm	1 mm	1.5 mm
Impactor tip (diameter)	8 mm	8 mm	3 mm	3 mm

vital parameters (blood pressure, temperature) are monitored for 2–3 h and then the animal is sacrificed (41). In chronic experiments (survival procedure) (3), after the TBI the animal is recovered from anesthesia, subjected to behavior studies, imaging studies and depending on the protocol the animal is sacrificed at a later time point (1 day, 3 days, 1 week, 2 weeks, 1 month or later). Similarly, depending on the severity of injury, TBI is classified into mild TBI or severe TBI which is mainly decided by factors like the velocity of impact, duration of impact and depth of deformation (44–47) (Table 1). TBI can also be complicated by secondary insults such as hypotension or hypoxia, which markedly worsen the resulting neurological injury. Hypotension following the TBI simulates the complex situation with the polytrauma patient (41).

2 Materials

2.1 Anesthesia

1. Anesthesia chamber: a fiber glass box (approximately 27 cm × 12 cm × 12 cm) with a lid and an inlet for anesthesia gases (closer to the floor of box) at one end and outlet (closer to the roof of box) at another end for scavenging anesthesia gases.
2. Ventilator: Rats—SAR830P ventilator that is supplied by the Life Science instruments USA. Mice—Mini-Vent Type 845 (Hugo Sachs Elektronik, March-Hugstetten, Germany) ventilator.
3. Endotracheal tubes: rats—uncuffed endotracheal tubes are made using 14/16 G intravenous catheters, No. 2.5 (ID); mice—endotracheal tubes are made using 22 G intravenous catheter or PE 60 tubes.

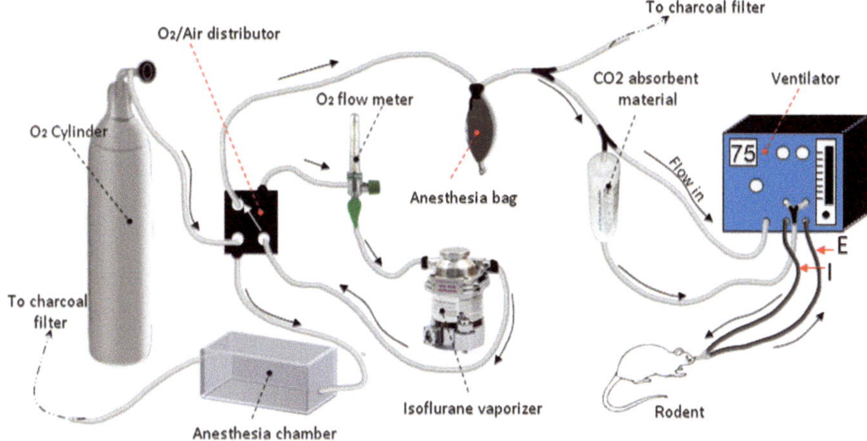

Fig. 1 Components of anesthesia circuit for rodents

4. Intravenous/Intraarterial catheters: rats—21 G intravenous catheters; mice—Micro-cannula (ID = 0.007, OD = 0.014).
5. Anesthesia gases: Isoflurane and Oxygen (from O_2 cylinder or central supply).
6. Isoflurane vaporizer and rebreathing bag (22 mm bush neck).
7. Central suction or charcoal canisters (f/air, Omnicon, Bickford Inc, NY, USA) for scavenging anesthesia gases.
8. Infusion pumps: (Harvard apparatus: Pump Dual RS-232).
9. Infusion fluids: 0.9% Saline.
10. Soda lime for CO_2 absorption (Sodasorb, WR Grace & Co. Chicago, IL).

The above parts (ventilator, O_2 cylinder, vaporizer, rebreathing bag, container with indicator soda lime on the outlet side to absorb CO_2) are assembled using plastic tubings and connectors as shown in the Fig. 1. Equipment like vaporizers requires regular calibration.

2.2 Surgery

The following sets of instruments are required for craniectomy and CCI injury induction.

1. Surgical instruments: scalpel handle (No. 3), scalpel blades (No. 10, 15), jeweler's forceps (11 cm, No. 5), needle holders (12 cm), eye scissors (10.5/11/5), iris forceps serrated (10 cm, curved points 0.7 mm), eye specula, Adson's tissue forceps (12 cm, 1 × 2 teeth), bone rongeurs (Friedman bone rongeurs, curved).
2. Microsurgical instruments: eye scissors (11/11.5 cm), micro-Adson's tissue forceps (12 cm, 1 × 2 teeth).

3. Sterile cotton tip applicators and sterile sponges (or can be sterilized along with the instruments in the autoclave).
4. Sterilization equipment: autoclave, glass beads sterilizer.
5. Suture materials: silk 5-0 (for vessels in mice) and 3-0 (for tying vessels in rats). Nonabsorbable monofilament suture material like Nylon (Ethilon) is used for skin closure (4-0 for rats and 5-0 for mice).
6. Material to control bleeding during craniectomy: Gelfoam (dental packs, absorbable gelatin sponge, size 4), bone wax (Ethicon).
7. Magnifying instruments: microscope (mice) and operating loop (rats).
8. Micro-aneurysm clips (0.8 mm width, 5 mm long) for temporary occlusion of vessel while acquiring arterial or venous access.
9. Stereotactic frame for mice or rats.
10. Micro drill; drill bits: mice (0.9 mm burr) and rats (1.4 mm burr).
11. Heating pad.
12. Sterile syringes (1, 5, 30 ml) and sterile needles (27 G, 21 G).
13. Impactors: Mice: voltage driven impactor (Bench Mark Impactor, Leica); rats: Pneumatic (N2) impactor (48).
14. Rectal thermistor (thermocouple thermometer).
15. Brain temperature: 0.003-inch-diameter type K thermocouple microprobe (Physiotemp Instrument, Inc; Clifton, NJ).
16. Microcapnometer: $EtCO_2$ monitor (Microcapnograph, CI-240, Columbus Instruments, Columbus, OH).
17. Intracranial pressure (ICP) monitor: 1.4-French ultra miniature pressure transducer, SPR-671 (Millar instruments, Houston, TX) or Codman ICP catheters (reusable).
18. Cerebral blood flow monitor: Perimed PIM3 Laser Doppler (model PF-3 with 0.25 mm fiber separation; Perimed, Piscataway, NJ) positioned in a stereotactic holder.
19. Arterial blood gases analyzer.
20. Computer software: The pneumatic impactor parameters and all physiological parameters are displayed on the monitor using the LabVIEW soft ware (National instruments corporation, New Jersey).
21. Skin disinfectant: betadine.
22. Analgesic: buprenorphine (0.3 mg/ml).
23. Antibiotic: enrofloxacin (100 mg/ml).

24. Lubricant eye ointment (artificial tears, 15% mineral oil in 83% white petrolatum).
25. Heparin to withdraw blood (2 IU/ml of heparin in 0.9% saline).
26. Citrate phosphate dextrose anticoagulant (to be used at a ratio to whole blood = 1.4:10).

2.3 Behavioral Studies

1. Beam walking: wooden beam (1 m long and 2.5 cm wide) for rats, plastic wire (50 cm long and 2 mm thick) for mice, light source, darkened goal box, noise (90 db) generator, timer, and four plastic pegs (7.5 mm long and 2.5 mm wide).
2. Beam balancing: wooden beam (1 m long and 1 cm wide for rats) and a wire of 2 mm thick for mice, large plastic container (120 cm × 40 cm and 60 cm high).
3. Morris water maze: galvanized steel pool (5 foot diameter), plexiform platform (10 × 10 cm), water, and nontoxic, white paint, heating lamp, and Chromotrack Video tracking system (Sam Diego Instruments, San Diego, CA). Computer software HVS image.

2.4 Tissue Processing

1. 4% paraformaldehyde solution; ddH_2O, 1 N NaOH, pH strip, paraformaldehyde (Sigma), sodium phosphate monobasic/monohydrate, sodium phosphate dibasic anhydrous, and hot stir plate.
2. Chemical fume hood
3. Personal protective equipment: safety glasses, sterile apron, cap, and mask.
4. 30% sucrose.

3 Methods

3.1 Anesthesia

1. Rats: Anesthesia is induced by placing rats in the anesthesia chamber with 5% isoflurane in 100% oxygen flow for 3–5 min. Then the rats are intubated with ET tubes made from 14 G/16 G catheters using an intubation stand (a plexiglass stand that allows suspending the rat by its incisors, on its sloping side using an elastic band) and a laryngoscope (infant blade). Using the laryngoscope, the vocal cords are visualized and catheter is placed into the trachea. The ET tubes are prepared using 14/16 G catheter, its stylet and the cut proximal portion of No. 2.5 (ID) ET tube as shown in Fig. 2a. The ET tube is secured by tying it to the lower jaw with an elastic tape. Anesthesia is maintained with 2% isoflurane in 100% oxygen (see Note 1).

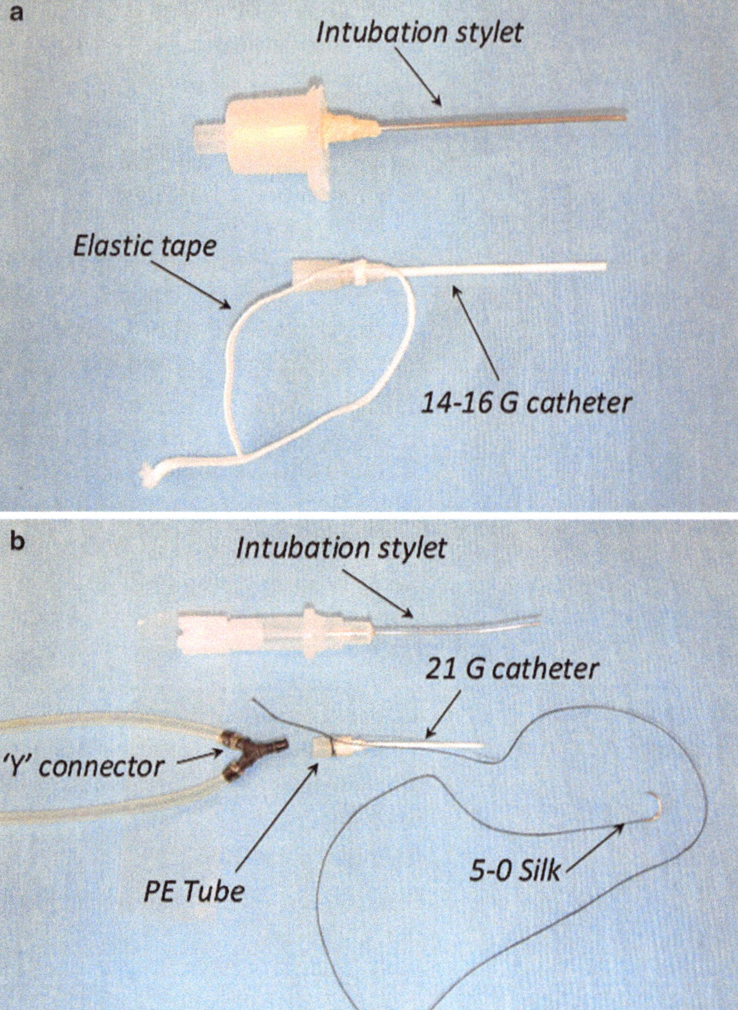

Fig. 2 Endotracheal tubes. (**a**) ET tube for rats is prepared using 14 G/16 G catheter and No. 2.5 ET tube. In rats the elastic tape that is tied around the proximal part of the catheter is secured to the lower jaw after intubation. (**b**) ET tube for mice prepared from 21 G catheter and PE tubes. Note the 5-0 silk suture knot that lodges into the superficial groove on the proximal PE tube (that is later connected to the black "Y" connector of the ventilator tubings). After intubation the ET tube is sutured to the lower jaw by taking a stitch through the skin

2. Mice: As it is difficult to visualize the vocal cords in mice we use the technique described by Michael et al. (49) for intubation (50). The mouse is induced in the anesthesia chamber with 3% isoflurane in 100% oxygen for 3–5 min. It is then

transferred onto a heating pad and secured with tapes. The nose and mouth are placed into a plastic mask (cut conical end of a 15 ml polypropylene tube) and connected to miniventilator. Small sterile drapes are placed around the mouse. A midline skin incision is made over the anterior neck below the jaw and the fat is dissected to visualize the muscles covering the trachea. With blunt iris forceps the pre-tracheal muscles are separated in the midline to visualize the trachea. Using a 5-0 silk the muscles and the skin on either side are retracted. Then under microscope or using magnifying loops, endotracheal tube is placed under direct vision into the trachea. This usually requires removal of plastic mask, gentle pull on the tongue with an iris forceps and then placing the ET tube into the mouth towards the trachea. Next step includes gentle manipulation of the tip into trachea under direct vision. An opaque ET tube or marking of the ET tube tip with a marker is helpful in visualizing the tip. The ET tube for mice is prepared as follows (Fig. 2b): cut the cannula part of 21 G catheter, place its proximal end into a polythene tube (ID 0.9 mm) which is further fit into a polythene tube (ID 14–16 mm) which should fit into the "Y" connector of the tubings from the mini ventilator. The tip of the metal stylet that comes with the cannula can be cut to adequate length (2–2.5 cm) and used as a stylet for intubation purpose. Under magnification small grooves are made on the outer surface proximal polythene tube which would later help to lodge a knot to secure it. After the ET tube is placed into the trachea, using a 5-0 silk, it is secured to the skin over lower jaw (the free end of the tie is first tied into the groove on the tube, then a stitch is taken through the skin over the lower jaw). The incision is suture closed with monofilament (4-0 or 5-0 prolene nylon). Surgical anesthesia is maintained with 2% isoflurane in 100% oxygen (see Note 2).

3. After the ET tube is secured it is connected to the ventilator. The ventilators are adjusted so that the pCO_2 is between 35 and 40 mm of Hg and pO_2 is above 100 mm of Hg. $ETCO_2$ monitor can also be used to maintain an ETCO2 of 35–40 mm of Hg. For mouse the mini-ventilator rate is usually adjusted in the range of 150–200/min with a stroke volume of 150–175 μl (0.15–0.175 ml). For rats the ventilator is set at a rate of 70–80/min, flow rate of 700–800 ml/min, and inspiratory time of 0.4 s (tidal volume of 4–5 ml). Arterial blood gases should be analyzed to check for pO_2 and pCO_2 values and correlation between pCO_2 and $ETCO_2$ (see Note 3).
4. Arterial and venous catheterization: In rats the tail artery is catheterized. A 2–4 mm incision is made in the proximal section of tail and the artery is catheterized using a 22 G arterial

cannula. This access is used to monitor systemic blood pressure and also to withdraw blood for arterial gases. A 5 mm incision is made over the left groin and the femoral vein is dissected and then cannulated using a 22 G angiocath. Venous access is used to deliver fluids, drugs and to obtain blood for testing or withdraw blood to create hypotension.

The vessel cannulation is usually done in the following steps. The vessel is dissected from surrounding structures using microinstruments (forceps, probes). The distal end of the vessel is tied with a suture (5-0 or 2-0 silk) and a clip is applied to the proximal part. A small opening is made in the vessel wall and the tip of the angiocath is introduced into the vessel. A suture is tied over the cannula to secure it. The proximal clip is removed and the catheter is advanced into the artery. Another suture is taken through the skin and the cannula is double secured (see Note 4).

5. Temperature probe is inserted into the rectum and the temperature is maintained at 36.5–37.5°C (see Note 5).

3.2 Surgery (Procedure, Monitoring, and Models)

1. Craniectomy: The animal is fixed in a stereotactic frame (3 point fixation: Bilateral ears and the upper incisors). Care should be taken while shifting and fixing the animal on the frame so connections of the ET tube to the ventilator tubings, arterial access to the intrarterial BP transducer and the venous access to the infusion pump remain intact (Fig. 3). Artificial eye ointment is applied to the eyes. The skin over the cranium is clipped. A midline incision is made over the scalp to expose the bones. The bregma (junction of fronto-parietal and interparietal sutures in the anterior part) and lambda (junction of the midline interparietal and parieto-occipital sutures) are identified (Fig. 3). The skin edges are retracted using a retractor or sutures. The periosteal layer is scraped and the boundaries of craniotomy are identified. A craniotomy is performed in the region of right parietal bone. Usually a trough is made in the shape of a rectangle within the borders of the parietal bone till the dural vessels are visible through the thinned bone. Then the entire piece of bone is lifted with a small clamp (another technique is to drill the entire surface of the parietal bone till it becomes thin enough to visualize the dural vessels and then using a small, blunt curved forceps the thinned out bone is elevated). This step should be followed carefully done to avoid dural tears. The left parietal bone is also drilled so that it is thin enough to allow cerebral blood flow measurements if required. Brain temperature probe (0.003 in. diameter type K thermocouple microprobe) is inserted into right frontal lobe after making a small hole with the drill. The ICP monitoring transducer (1.4 F ultra miniature pressure transducer) is inserted into the left frontal lobe after making a small hole with the drill (Fig. 3) (see Note 6).

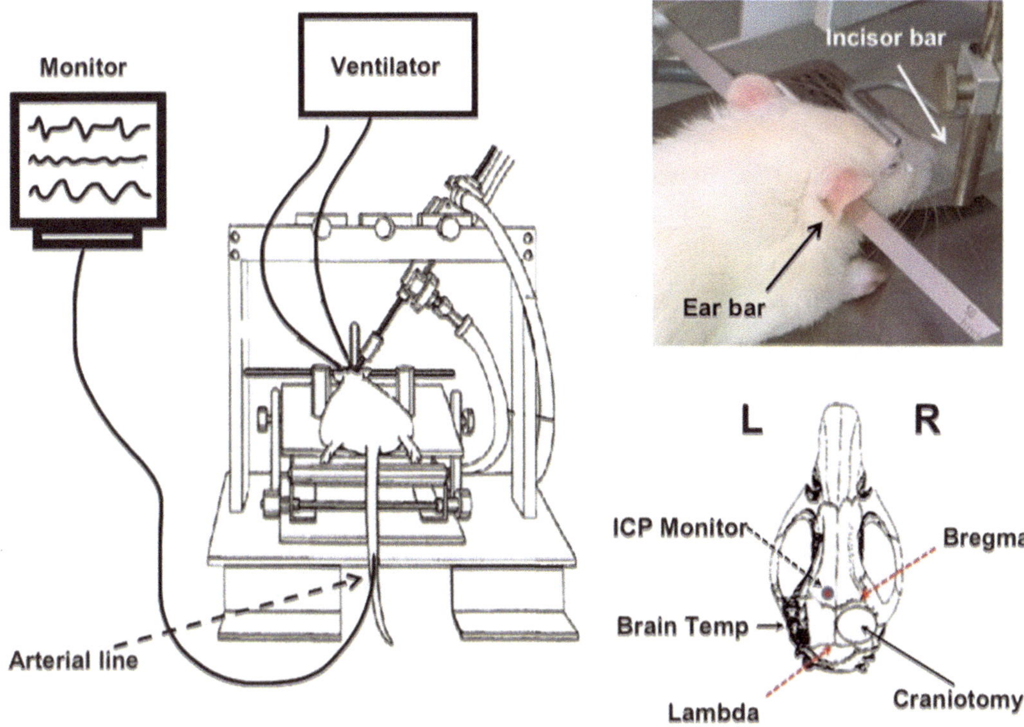

Fig. 3 Position of rat in the stereotactic frame and connections for monitoring various physiological parameters

2. Controlled Cortical Impact: A pneumatic impactor is used for rats and the voltage driven impactor for mice. The pneumatic cylinder (with bore 1.875 and 5 cm stroke) is rigidly mounted on a cross bar so that the position and the angle of impaction can be adjusted over the rats head fixed in the frame. Impactor tip (8 mm) is attached to the lower end of the cylinder. The upper end is attached to the transducer core of the LVDT (Linear variable differential transducer). The impact velocity can be controlled by adjusting the pressure from the gas cylinder. The velocity and duration can be measured from the output of LVDT and is displayed on the monitor (LabVIEW soft ware). Pressures of 150, 35, and 15 psi applied to the impactor usually result in impact velocities of approximately 5 m/s, 3 m/s, and 1 m/s, respectively. The depth of deformation is adjusted by advancing the impactor tip. The parameters depend on the type of study (mild vs. severe TBI) (Table 1). After adjusting the parameters on the pneumatic impactor, a trial impact is made to ensure the values as displayed on the monitor. The animal is fixed on the frame (three point fixation—bilateral

ear bars and incisor bar) and its position is adjusted so that the impactor tip just touches the dural surface (zeroing of the tip). The impactor is withdrawn and then the tip is advanced according to the depth of deformation required. The knob is set to the arm position and the impact is delivered by pressing the button. In mice the velocity and duration of impact can be adjusted by turning the respective knobs on the machine. After the mouse is fixed the impactor is attached and its position is adjusted so that the tip is at 45° to the dural surface. The impactor tip is zeroed at the level of the surface of dura (there is a sensor which alarms when the impactor touches the dural surface). The impactor is withdrawn and the tip is advanced by desired length by using the attached Vernier scale. The impact is then delivered by pressing the button. After securing hemostasis the wound is suture closed using a monofilament 4-0/5-0 prolene or ethilon suture. The parameters used for rats and mice are given in Table 1 (see Notes 6 and 7).

3. Monitoring: The rectal temperature, mean arterial pressure (MAP), intracranial pressure (ICP), and brain temperature are monitored continuously (readings recorder at 30 s intervals). The rectal thermistor is connected to the monitor or a thermocouple thermometer. The brain temperature probe, arterial pressure transducers, ICP transducer are connected to the monitor. Microcapnograph requires zero calibration before each operation. This would require CO_2 (10%) gas cylinder and N_2 (100%) gas cylinders. After calibration both the respiratory rate and $ETCO_2$ are displayed on the monitor. The tip of the sensor is placed into a metal cannula provided at the junction of the "Y" connector or a small hole that is drilled into the plastic "Y" connector that is used for the mice. A noninvasive laser Doppler scan (Periscan, PIM3 system, Perimed AB Stockholm, Sweden) which permits evaluation of microvascular perfusion in real time is used to measure cerebral blood flow on both sides. Measurements are expressed as perfusion units. The values are obtained every 15–30 min during the monitoring, are compared to the baseline values and are expressed as percentage of baseline values (see Note 8).
4. Extubation: After the injury and closure of the skin incision, the animal is transferred along with the heat pad on to the table surface. The isoflurane is stopped and the rebreathing bag is emptied. The animal is maintained on 100% oxygen alone and the reflexes (corneal, pinna, paw, and righting and escape reflexes) are monitored every minute for 20 min till the animal is recumbant. Once the animal gains the righting reflex it is extubated and then oxygen is delivered through a mask (see Notes 9 and 10).
5. Infusion rates: In rats the venous infusion is given at a rate of 1 ml/h to keep it patent and the arterial infusion rate is

Table 2
Phases in the "Mild head injury complicated by hemorrhagic shock" model

	Phase of study		
	Injury + hemorrhage	**Pre-hospital phase**	**Definitive care**
Ventilation	Room air	Room air	100% O_2
Injury	Mild injury (3 m/s, 2.5 mm deformation)	–	–
Hemorrhage	Withdraw blood slowly till MAP < 40 mmHg	–	–
Resuscitation	–	Initial resuscitation with saline to bring MAP > 50 mmHg	Reinfuse shed blood and additional saline to keep MAP > 60 mmHg

0.3 ml/h. It is important to expel the air bubbles in the infusion tubing.

6. Antibiotics and Analgesics: Enrofloxacin (available as 100 mg/ml) is administered in the dosage of 0.05 ml/kg for both rats and mice by subcutaneous route once daily starting on the day of surgery for 3 days. Buprenorphine analgesic (available as 0.3 mg/ml) is given in the dosage of 0.1 mg/kg (rats) and 1 mg/kg (mice) by subcutaneous route twice daily for 3 days starting on the day of surgery.
7. The study drugs (e.g., Epo, 5,000 U/kg) are administered by intraperitoneal route or subcutaneous routes. The rats are restrained in a polythene bag and the drug is injected into the peritoneum. The mouse is held with one hand and stabilized between the palm and fingers and then the drug is injected into the peritoneum with the other hand.
8. TBI model complicated by hemorrhagic shock: After mild TBI, blood is withdrawn into a syringe with citrate phosphate dextrose to slowly drop the mean arterial pressure (MAP) to 40 mmHg. The blood volume withdrawn is usually 2 ml/100 g weight of rat. Blood is slowly withdrawn (50% over first 5 min, 25% over next 5 min and 25% over the next 5 min) to mimic the situation of traumatic blood loss in the battle field. The shed blood is stored at 4°C. The physiological parameters are monitored for 60 min during which anesthesia is maintained with isoflurane and room air only. Following this hypotensive period, lactated Ringer solution is infused at the rate of 1 ml/min to maintain a MAP of 50 mm of Hg. This period mimics the pre-hospital phase where resuscitation is with intravenous fluids. The stored blood is rewarmed to 37°C, reperfused and 100% oxygen is restarted. The stages in this mild TBI with hemorrhagic hypotension model are shown in Table 2.

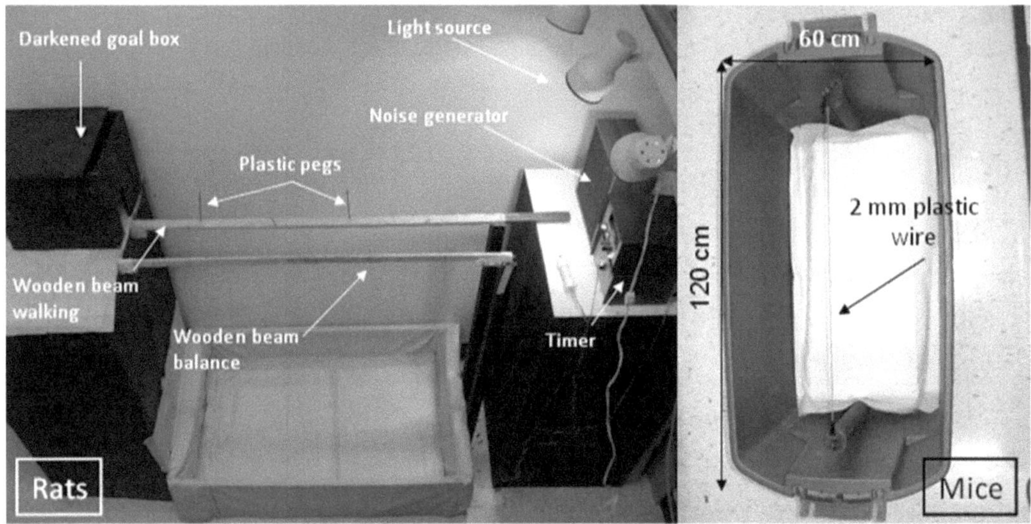

Fig. 4 Beam walking and beam balancing equipment used for rats and mice

3.3 Behavioral Studies

1. Beam Walking Task: The rat is made to walk on a 1-m length wooden beam that is set at 1 m of height above the ground (Fig. 4). It is made to walk from one end of the beam that is close to the noise source, towards the goal box. A wooden beam with appropriate width (depends on weight of the rat; typical width for a 300–350 g rat is 2.5 cm) is selected. The beam is cleaned with 70% alcohol and allowed to dry between the trials. A light source that points towards the starting point is turned on and the room lights are turned off. A typical trial consists of: (a) turning the timer and resetting it to zero; (b) holding the timer switch in the nondominant hand with the thumb on the button; (c) holding the animal by tail with the preferred (dominant) hand and placing it onto the beam at progressively increasing lengths from the goal box (fifths of the way; usually marked on the beam) and simultaneously starting the timer and noise; (d) once the animal learns to walk the whole length, trials are repeated till it is able to cross the whole length for three consecutive trials within 5 s; (e) then the plastic pegs are placed into the holes in the beam at regular intervals alternating from side to side (fifths of total length), within 5 mm along the edge, and the trials are repeated till the animal is able to cross the entire length within 10 s for three consecutive trials. If both trials are met within a total of 30 trials (with pegs and without pegs) the animal qualifies for the next day's trial (the presurgery trial on the day of surgery). The animal is given rest for 30 s between the trials. On the day of surgery the trials are repeated and if the animal is able to cross the beam with pegs within 5 s for three consecutive trials it qualifies for

the surgery (it is disqualified if not able to cross for 3 consecutive trials in a total of 15 trials). From the day after surgery trials are repeated for 5 days, with the pegs in place, and the time taken to cross the beam is recorded. It is taken as a successful trial if the animal crosses the beam within 30 s. Unsuccessful trials and the distance traveled from the starting point towards the goal box in fifths of total length are recorded. Mice: Mice are made to walk on a 2 mm thick wire that is held taut between two poles (50 cm tall) 50 cm apart on a wooden platform that is padded with soft material. This entire assembly can be placed into a large plastic container (Fig. 4) (see Notes 11 and 12).

2. Beam Balancing Task: The rat is made to balance at the center of a 1 m wooden beam (1.5 cm width) that is set 1 m above the ground (Fig. 4). The beam is cleaned with alcohol between the trials. The trial is carried out in the following steps: (a) timer is set to zero and the noise generator is turned off; (b) the light source used for the beam walking is turned off and the room lights are turned on; (c) with the timer button in the non-preferred hand the animal is lowered on to the center of beam length wise and allowed to balance; (d) timer is set on; (e) hand is held 8 in. below the beam so that the animal will not fall to the floor; (f) if the animal balances itself for 60 s for three trials on the day of surgery (before surgery) it qualifies for the surgery. Three trials are repeated every day, on days 1–5 after the surgery. The duration in seconds for which the animal balances on the beam is recorded. Unsuccessful attempts (if it balancing for less than 60 s) are recorded.

3. Morris water maze task: The maze tank is filled with water up to ten inches and white soluble nontoxic, tempura paint is added to make the platform invisible (Fig. 5). The water level is maintained at 2 cm above the plexiform platform. The water is maintained at room temperature. The east, north and west of the room have a cross, triangle and circle signs and there are lights towards the south (these signs and lights serve as visual clues for the rats). The trials are conducted on days (11–15 after surgery). The rat is placed in the water facing the north, south, east or west (4 trials every day) and the order of trials every day and across the days (11–15) are randomized. The same random order is followed every day. If the rat does not find the platform for 120 s the rat is picked up and placed on the platform for 30 s. Once the rat finds the platform by itself it is left over on the platform for 30 s. The rat is given a rest of 4 min between the trials. The rat is kept warm with a heating lamp. On the last day the rat is given two extra trials. One trial consists of placing the rat in the south direction and without the placement of platform and allowed to swim for 120 s. This allows us to find the time spent in the quadrant with platform

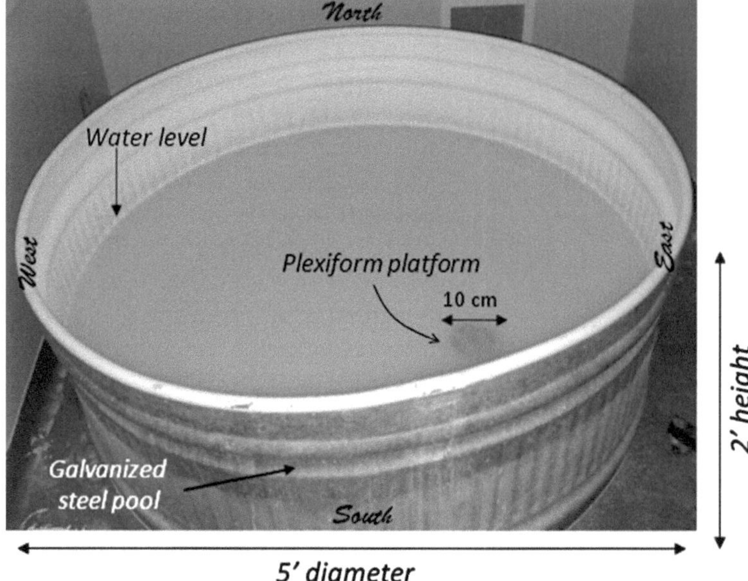

Fig. 5 Morris water maze tank

(indication of memory function). The last trial consists of placing the rat in the east direction, with a visible platform (plexiglass circle with white surface and darker sides is placed above the platform so as to bring it above the surface of water). This trial helps to find the time the rat takes to find the platform (assesses the visual function). The tank is cleaned once in a week. The water level is checked every day and kept constant throughout the trials. The water surface is kept clean with minnow nets after each trial (see Note 13).

3.4 Tissue Preservation

1. Animal sacrifice: The animals are deeply anesthetized and intubated as described above. After intubation the rest of the procedure is carried out inside a chemical fume hood. A midline incision is made over the abdomen and extended upwards and towards both sides of the chest in shape of "V." Once the inferior surface of diaphragm is seen it is cut close towards the chest wall to enter into the thoracic cavity. The beating heart is visualized. The left ventricle is then pierced with an 18 G cannula (rats) or a 21 G cannula (mice) and 0.9% saline is perfused. The right atrium is opened with a scissors and the animal is exsanguinated. When the returns from the right atrium are clear, 4% paraformaldehyde is perfused (about 200–250 ml for rats and 50–75 ml for mice). If the brain vessels are well perfused then the animal will have tonic posturing with diffuse muscle twitching. Then the animal is decapitated using a guillotine. Skull is exposed. Working from the foramen magnum upwards, using a small bone rongeurs, the cranial bones are slowly removed in progressive bites. Once the cranial vault is removed

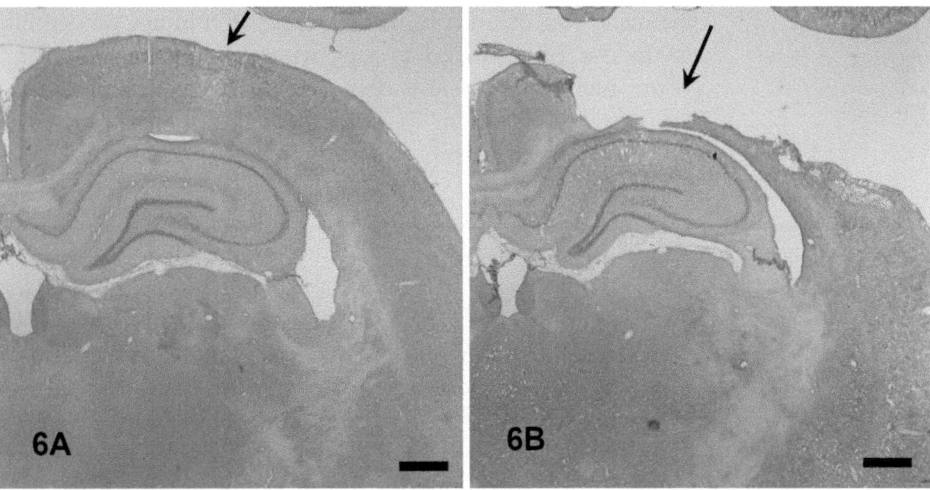

Fig. 6 The impact site in the right parietal cortex 2 weeks after mild TBI, showing typical small contusion in the Epo treated animals (**a**) compared to the much more extensive cortical contusion in animals in the saline group (**b**)

the brain is elevated in front with a spatula, disconnected from the cranial nerves at the base and collected into a bottle with 4% paraformaldehyde. The specimen is later transferred into 30% sucrose solution (after 24–48 h) and preserved at 4°C. The specimens are further processed to study the contusion volumes, assess the cell counts in hippocampus and for immunohistochemical studies (see Note 14).

2. Preparation of 100 ml of 4% Paraformaldehyde (procedure performed in the hood): (a) Heat 50 ml distilled water in a beaker on a hot stir plate to 60–65°C; (b) Measure 4 g of paraformaldehyde and add it to the water; (c) Add few drops (1–4, one by one) of 1 N NaOH till the solution is clear. A pH strip (not pH meter) can be used to check the pH, (d) Cool the solution and then add 50 ml of 0.2 M buffer, and (e) Store the solution at 4°C (see Notes 15 and 16).
3. Preparation of 0.2 M buffer: To prepare 2 l of 0.2 M buffer, add 12.5 g of Sodium Phosphate Monobasic/Monohydrate and 44.2 g of Sodium Phosphate Dibasic anhydrous to 2 l of ddH_2O (see Note 16).
4. Preparation of 30% sucrose: To prepare 100 ml of 30% sucrose, add 30 g of sucrose to PBS and stir it. Store the solution at 4°C (see Note 16).

3.5 Typical Study

A study (3) was conducted in Long Evans rats to evaluate the time window for neuroprotection with Epo after CCI. Recombinant Erythropoietin (rhEpo) was administered (5,000 U/kg, subcutaneous) at various times points after TBI (1, 3, 6, 9, or 12 h). The rats were randomly assigned to one of the Epo groups or to the saline group. Typical examples of the impact site in Epo treated and saline treated animals are shown in Figs. 6a, b, respectively.

The mean contusion volume at 2 weeks after injury was significantly less in the Epo treated groups when compared with the saline group, but this effect was dependent on the administration time. The best improvement in outcome was obtained when the Epo was administered within 6 h after injury.

4 Notes

1. It is important to check that the gas cylinders are full (O_2, air) and the vaporizer has adequate isoflurane. It is important to check (before every set of experiments) that the tubings and their connections are secure. It is prudent to connect the cut end of a glove finger (to serve as a balloon) to the "Y" connector that goes into the ET tube and turn on the ventilator to make sure that it bellows with each cycle of respiration. The infant laryngoscope blade can be cut along its sloping side and the tip so that it will fit into the mouth of the rat. The cannula of the 14/16 G angiocatheter is used as the ET tube in rats. The sharp tip of the catheter is cut and made smooth (to avoid injury to the trachea). The total length of the cannula is kept at 2.5–3 cm to avoid tracheal perforation (if it is too long). An elastic tie (usually the elastic band from face masks can be used) is tied across the hard proximal part of the catheter. The ends can be later used to tie the tube to the lower jaw of the rat, once it is intubated. The position of the cannula is confirmed by giving manual compression and looking for expansion of the chest, by looking for symmetrical movement of the chest, looking for respiratory rate that should corresponds with the rate set on the ventilator, and by looking for progressive distention of abdomen as the stomach will distend if the tube is wrongly placed into the esophagus.
2. It is important to make sure that the mouse is breathing during the procedure. It is better to start the incision from 5 to 8 mm below the mouth in the midline. Stay in the midline while dissecting the pre-tracheal muscles to avoid injury to neck vessels. In mice that are old, the thyroid may be large and the fat may be abundant. This can be usually dissected with a little bit of patience. Slide the tip of cannula along the curve of the tongue till there is a minimal resistance from structures of larynx (tip can usually seen through muscles in the proximal part of the incision). Its tip is then lowered little bit (<1 mm) and advanced into the trachea (this step is visualized through the translucent tracheal wall). Do not use any force while maneuvering the tip into the trachea as it may injure the trachea and cause subcutaneous emphysema. The proximal and outer most PE tube should be long enough to allow its connection with the "Y" junction of the tubing that come from the ventilator. This is helpful especially when the animal is fixed in the stereotactic frame.

3. The tip of the $ETCO_2$ probe is placed into the metal cannula provided at the "Y" junction (for rats) or into a small hole drilled into the "Y" junction (in mice). During the pilot phase of the study it is important to check the arterial blood gases and adjust the tidal volumes and respiratory rates to keep pCO_2 between 30 and 35 mmHg and pO_2 above 100%. It is important to check the correlation between the $ETCO_2$ from the capnometer with the pCO_2 from the arterial blood gases. In rats 0.5 ml of blood can be safely withdrawn 2–3 times during a 6–8 h monitoring period. In mice only 50 μl can be safely withdrawn for arterial blood gas analysis. In a mouse that weighs 20 g the total blood volume is only 1–2.4 ml and the safe bleeding volume is only 0.1–0.2 ml (as removal of >10% of blood volume will result in hypotension and this can affect the results of the study).
4. Double secure the arterial and venous catheters as they can dislodge while turning the animal to prone position on the stereotactic frame.
5. It is important to maintain the temperature of the animal as it can affect the body metabolism, blood flow and contusion volumes.
6. The craniectomy size should be optimal (around 8–10 mm for a 8 mm impactor tip and 4 mm for a 3 mm impactor tip). If the craniectomy is too wide it can result in excessive bleeding, dural tear, or brain herniation and if it is too small (than the tip of impactor) then the bone at the edges can fracture and result in more severe injuries. The drilling is usually done under magnification (loops for rats and microscope for rats). Drilling should be away from the inter-parietal suture in the midline (as shown in Fig. 2), otherwise it can result in fatal venous sinus bleed. During drilling sterile saline is used to irrigate the bone to decrease the amount of heat that is generated during the drilling. If there is bleeding a piece of gel foam of the size of the craniectomy defect is placed over the bleeding site and irrigated with for few minutes. Dura should be kept wet till the impaction because the desiccated dura is more prone for tears. The tip of the impactor is also kept wet before the impact to prevent the dural tear. It is important to apply artificial tears to the eyes especially in chronic (survival) experiments as corneal ulcers can influence the postoperative behavioral studies.
7. It is important to check the cylinder pressure and the impaction parameters during the trial impact. The impactor should be advanced precisely. It is useful to keep a mark (with a permanent marker) on the impactor tip in the pneumatic impactor and calculate the number of rotations advanced (one rotation is equal to 1 mm). This should be calibrated regularly.

8. It is important to record the baseline values (blood pressure, arterial blood gases, and intracranial pressures) before the impact.
9. Animals that do not awake within 1 h after the impact are euthanized. It is important to empty the rebreathing bag after skin closure and to stop the isoflurane as the gas in rebreathing bag can delay the recovery from anesthesia.
10. If performing surgeries in batches it is advisable to sterilize the equipment between two procedures using a hot glass bead sterilizer.
11. Behavior tests should be conducted in a separate room with no visual/noise disturbances. During the days for which the behavior tests are scheduled, the drugs (study drug, analgesic, and antibiotic) are given after the trials to reduce the stress on the rats which can affect the behavior tests. If the rats are above 350 g to start with and chronic studies of longer duration are planned, they are likely to grow further (>400 g) and find it difficult to perform well in the behavior tests.
12. The pegs in the beam walking test are not bent too much otherwise the task will be too easy for the rats. There should be soft bedding on the floor below the beam walking and beam balancing rods, to prevent injury if the rat falls.
13. During the trials it is better for the researcher to stand in a fixed position so as to serve as a fixed visual clue for the rats. Do not keep the heating lamp too close to the rat. Clean up the water maze tank once in a week and follow the sanitizing methods recommended by the institutional protocols.
14. The bone around the craniectomy defect is removed with gentle force with outwards movement of the rongeurs. Magnifying loops can be used to avoid damage to the normal brain.
15. Titration is done using 1 N NaOH as the pH meter will be damaged by paraformaldehyde. Paraformaldehyde powder and solutions should be handled in the hood and after donning the personal protective equipment.
16. Prepared solutions (Buffer, 4% paraformaldehyde, 30% sucrose) are stored at 4°C.

References

1. Langlois JA, Rutland-Brown W, Wald MM (2006) The epidemiology and impact of traumatic brain injury: a brief overview. J Head Trauma Rehabil 21:375–378
2. McAllister TW et al (2001) Differential working memory load effects after mild traumatic brain injury. Neuroimage 14:1004–1012
3. Cherian L, Goodman CJ, Robertson CS (2007) Neuroprotection with erythropoietin administration following controlled cortical impact injury in rats. J Pharmacol Exp Ther 322:789–794
4. Cherian L, Goodman JC, Robertson C (2011) Improved cerebrovascular function and reduced histological damage with darbepoietin alfa administration after cortical impact injury in rats. J Pharmacol Exp Ther 337:451–456
5. Velly L et al (2010) Erythropoietin 2nd cerebral protection after acute injuries: a double-edged sword? Pharmacol Ther 128:445–459

6. Talving P et al (2010) Erythropoiesis stimulating agent administration improves survival after severe traumatic brain injury: a matched case control study. Ann Surg 251:1–4
7. Corwin HL et al (2002) Efficacy of recombinant human erythropoietin in critically ill patients: a randomized controlled trial. JAMA 288:2827–2835
8. Corwin HL et al (2007) Efficacy and safety of epoetin alpha in critically ill patients. NEJM 357:965–976
9. Lin FK et al (1985) Cloning and expression of the human erythropoietin gene. Proc Natl Acad Sci U S A 82:7580–7584
10. Brines M (2010) The therapeutic potential of erythropoiesis-stimulating agents for tissue protection: a tale of two receptors. Blood Purif 29:86–92
11. Sola A et al (2005) Erythropoietin after focal cerebral ischemia activates the Janus kinase-signal transducer and activator of transcription signaling pathway and improves brain injury in postnatal day 7 rats. Pediatr Res 57:481–487
12. Kawakami M et al (2000) Erythropoietin inhibits calcium-induced neurotransmitter release from clonal neuronal cells. Biochem Biophys Res Commun 279:293–297
13. Kawakami M et al (2001) Erythropoietin receptor-mediated inhibition of exocytotic glutamate release confers neuroprotection during chemical ischemia. J Biol Chem 276: 39469–39475
14. Solaroglu I et al (2003) Erythropoietin prevents ischemia-reperfusion from inducing oxidative damage in fetal rat brain. Childs Nerv Syst 19:19–22
15. Wu Y et al (2010) Increased expression of erythropoietin receptor in 1-methyl-4-phenyl-1, 2, 3, 6-tetrahydropyridine-induced Parkinsonian model. Physiol Res 59:281–287
16. Wang L et al (2004) Treatment of stroke with erythropoietin enhances neurogenesis and angiogenesis and improves neurological function in rats. Stroke 35:1732–1737
17. Brines ML et al (2000) Erythropoietin crosses the blood-brain barrier to protect against experimental brain injury. Proc Natl Acad Sci U S A 97:10526–10531
18. Meng Y et al (2011) Dose-dependent neurorestorative effects of delayed treatment of traumatic brain injury with recombinant human erythropoietin in rats. J Neurosurg 115: 550–560
19. Xiong Y et al (2007) Role of gender in outcome after traumatic brain injury and therapeutic effect of erythropoietin in mice. Brain Res 1185:301–312
20. Yatsiv I et al (2005) Erythropoietin is neuroprotective, improves functional recovery, and reduces neuronal apoptosis and inflammation in a rodent model of experimental closed head injury. FASEB J 19:1701–1703
21. Zhu L et al (2009) Erythropoietin prevents zinc accumulation and neuronal death after traumatic brain injury in rat hippocampus: in vitro and in vivo studies. Brain Res 1289:96–105
22. Lu D et al (2005) Erythropoietin enhances neurogenesis and restores spatial memory in rats after traumatic brain injury. J Neurotrauma 22:1011–1017
23. Mahmood A et al (2007) Treatment of traumatic brain injury in rats with erythropoietin and carbamylated erythropoietin. J Neurosurg 107:392–397
24. Bohlius J et al (2009) Recombinant human erythropoiesis-stimulating agents and mortality in patients with cancer: a meta-analysis of randomised trials. Lancet 373:1532–1542
25. Bohlius JF et al (2005) Effectiveness of erythropoietin in the treatment of patients with malignancies: methods and preliminary results of a Cochrane review. Best Pract Res Clin Haematol 18:449–454
26. Xiong Y et al (2010) Delayed administration of erythropoietin reducing hippocampal cell loss, enhancing angiogenesis and neurogenesis, and improving functional outcome following traumatic brain injury in rats: comparison of treatment with single and triple dose. J Neurosurg 113:598–608
27. Leist M et al (2004) Derivatives of erythropoietin that are tissue protective but not erythropoietic. Science 305:239–242
28. Brines M et al (2008) Nonerythropoietic, tissue-protective peptides derived from the tertiary structure of erythropoietin. Proc Natl Acad Sci U S A 105:10925–10930
29. Pankratova S et al (2010) Neuroprotective properties of a novel, non-haematopoietic agonist of the erythropoietin receptor. Brain 133:2281–2294
30. Brines M et al (2004) Erythropoietin mediates tissue protection through an erythropoietin and common beta-subunit heteroreceptor. Proc Natl Acad Sci U S A 101:14907–14912
31. Kanellakis P et al (2010) Darbepoetin-mediated cardioprotection after myocardial infarction involves multiple mechanisms independent of erythropoietin receptor-common beta-chain heteroreceptor. Br J Pharmacol 160: 2085–2096
32. Satake R et al (1990) Chemical modification of erythropoietin: an increase in in vitro activity by guanidination. Biochim Biophys Acta 1038:125–129
33. Erbayraktar S et al (2003) Asialoerythropoietin is a nonerythropoietic cytokine with broad neuroprotective activity in vivo. Proc Natl Acad Sci U S A 100:6741–6746

34. Rodriguez CY et al (2010) Treatment with nasal neuro-EPO improves the neurological, cognitive, and histological state in a gerbil model of focal ischemia. ScientificWorldJournal 10:2288–2300
35. Gao Y et al (2011) Different expression patterns of Ngb and EPOR in the cerebral cortex and hippocampus revealed distinctive therapeutic effects of intranasal delivery of Neuro-EPO for ischemic insults to the gerbil brain. J Histochem Cytochem 59:214–227
36. Adembri C et al (2008) Carbamylated erythropoietin is neuroprotective in an experimental model of traumatic brain injury. Crit Care Med 36:975–978
37. Bouzat P et al (2011) Reduced brain edema and functional deficits after treatment of diffuse traumatic brain injury by carbamylated erythropoietin derivative. Crit Care Med 39:2099–2105
38. Xiong Y et al (2011) Effects of posttraumatic carbamylated erythropoietin therapy on reducing lesion volume and hippocampal cell loss, enhancing angiogenesis and neurogenesis, and improving functional outcome in rats following traumatic brain injury. J Neurosurg 114:549–559
39. Ahmet I et al (2011) A small nonerythropoietic helix B surface peptide based upon erythropoietin structure is cardioprotective against ischemic myocardial damage. Mol Med 17:194–200
40. Ueba H et al (2010) Cardioprotection by a nonerythropoietic, tissue-protective peptide mimicking the 3D structure of erythropoietin. Proc Natl Acad Sci U S A 107:14357–14362
41. Robertson CS et al (2012) Neuroprotection with an erythropoietin mimetic peptide (pHBSP) in a model of mild traumatic brain injury complicated by hemorrhagic shock. J Neurotrauma 2012;29(6):1156–1166
42. Cernak I (2005) Animal models of head trauma. NeuroRx 2:410–422
43. O'Connor WT, Smyth A, Gilchrist MD (2011) Animal models of traumatic brain injury: a critical evaluation. Pharmacol Ther 130:106–113
44. Cherian L et al (1994) Lateral cortical impact injury in rats: cerebrovascular effects of varying depth of cortical deformation and impact velocity. J Neurotrauma 11:573–585
45. Cherian L et al (2003) Neuroprotective effects of L-arginine administration after cortical impact injury in rats: dose response and time window. J Pharmacol Exp Ther 304:617–623
46. Goodman JC et al (1994) Lateral cortical impact injury in rats: pathologic effects of varying cortical compression and impact velocity. J Neurotrauma 11:587–598
47. Liu H, Goodman JC, Robertson CS (2002) L-arginine reduces neuronal damage after traumatic brain injury in the mouse. J Neurotrauma 19:327–334
48. Dixon CE et al (1991) A controlled cortical impact model of traumatic brain injury in the rat. J Neurosci Methods 39:253–262
49. Michael LH et al (1995) Myocardial ischemia and reperfusion: a murine model. Am J Physiol 38:H2147-H2154
50. Michael LH et al (1995) Myocardial ischemia and reperfusion: a murine model. Am J Physiol 269:H2147–H2154

Chapter 10

Therapeutic Efficacy of Erythropoietin in Experimental Autoimmune Encephalomyelitis in Mice, a Model of Multiple Sclerosis

Ilaria Cervellini, Pietro Ghezzi, and Manuela Mengozzi

Abstract

Erythropoietin (EPO) has neuroprotective effects in many models of damage and disease of the nervous system where neuroinflammation plays a substantial role, including experimental autoimmune encephalomyelitis (EAE), the animal model of multiple sclerosis (MS). Since the first pioneering studies, in which EPO was shown to protect rats with acute EAE mainly by inhibiting inflammation, many other studies have pointed out other mechanisms of protection, including oligodendrogenesis and inhibition of axonal damage.

Here we review the preclinical studies in which EPO has shown therapeutic efficacy in several models of EAE in mice and rats. Moreover, we report in detail the protocol to administer EPO to mice with myelin oligodendrocyte glycoprotein (MOG)-induced chronic progressive EAE, and a representative result. In this model, EPO inhibits the clinical score of the disease when administered according to a preventive but also to a therapeutic schedule, and therefore at disease onset, suggesting that it might not only inhibit inflammation but also actively stimulate repair.

Key words Neuroinflammation, Demyelination, Neuroprotection, Animal model, Myelin oligodendrocyte glycoprotein

1 Introduction

Several therapies can slow down disease progression, but to date there is no cure for multiple sclerosis (MS), and patients suffer from progressive neurological disability. Most treatment options target the immune system, inhibiting the autoimmune response and the associated inflammation, but are not able to reverse axonal damage, the main cause of permanent disability (reviewed in (1)). Classical neuroprotective and neurotrophic factors, like brain derived neurotrophic factor (BDNF), have been difficult to administer because they hardly penetrate the blood–brain barrier (2). A neuroprotective neuroregenerative strategy is still needed.

Pietro Ghezzi and Anthony Cerami (eds.), *Tissue-Protective Cytokines: Methods and Protocols*, Methods in Molecular Biology, vol. 982, DOI 10.1007/978-1-62703-308-4_10,

Table 1
Studies describing the use of EPO and its derivatives in EAE models in rat and mouse

MBP-EAE (rat)	MOG-EAE (rat)	MOG-EAE (mouse)	PLP-EAE (mouse)
Erythropoietin (EPO)			
(3)	(8)	(10)	(16)
(7)	(9)	(21)	
(18)		(22)	
		(23)	
Carbamylated EPO (CEPO)			
		(20)	
		(21)	
AsialoEPO			
		(21)	
Darbepoetin alfa			
		(24)	

The observation that erythropoietin (EPO) is neuroprotective and, more importantly, when administered systemically reaches the brain in high enough amounts to protect from brain injury opened new therapeutic possibilities for neurodegenerative diseases. Interestingly, EPO had a striking blocking effect on the neuroinflammation associated with brain injury, suggesting that it might be effective in diseases with a substantial neuroinflammatory component, like MS (3).

Since then, preclinical studies have reported efficacy of systemically administered EPO, and EPO derivatives devoid of erythropoietic activity, in several models of experimental autoimmune encephalomyelitis (EAE), the experimental model of MS (reviewed in (4, 5)). Overall, these studies have shown that EPO ameliorates the clinical course of EAE targeting not only inflammation but also demyelination and axonal damage, and inducing neurogenesis and oligodendrogenesis, suggesting that it might be an effective neuroregenerative therapy. Accordingly, in a pilot clinical trial EPO improved cognitive functions in patients with chronic progressive MS (6).

Results obtained with several schedules of EPO treatment in different models of EAE are reviewed here (Table 1), together with a specific example of efficacy of EPO and of a non-erythropoietic EPO derivative, carbamylated EPO (CEPO), in myelin oligodendrocyte glycoprotein (MOG)-induced EAE, a widely used model of chronic progressive EAE in mice.

1.1 Efficacy of EPO in Several Models of EAE

In the first study in Lewis rats with myelin basic protein (MBP)-induced EAE, EPO delayed disease onset and decreased clinical symptoms (3). In a later study in the same model, EPO was shown to inhibit the production of inflammatory cytokines, in particular TNF and IL6, in spinal cord and brain (7).

MBP-induced EAE in Lewis rat is a monophasic disease with onset around 10 days after immunization, after which the animals recover completely, and it is characterized by a strong inflammatory component but limited demyelination, that is the primary feature of MS. EPO (50 μg/kg) was administered daily, intraperitoneally (i.p.), starting 3 days after immunization, well before the onset of the disease. It was clear from these studies that EPO inhibited inflammation, but not if it could also target the neurodegenerative aspect of EAE.

In a model of MOG-induced EAE in Brown Normay rats, characterized by optic neuritis with early axonal damage of the optic nerve, similar to the one observed in MS patients, EPO treatment (50 μg/kg, i.p., daily) protected retinal ganglion cells (RGC) from cell death (8). In a later study from the same group, a combination of EPO with methylprednisolone (MP), the classical anti-inflammatory therapy for the treatment of relapses in MS, inhibited also axonal degeneration, whereas MP monotherapy, although effective at inhibiting inflammation, slightly promoted neurodegeneration (9).

In MOG-induced chronic progressive EAE in C57BL/6 mice, EPO at 50 μg/kg significantly decreased the clinical score of the disease, inflammation, blood–brain barrier breakdown, and also demyelination and axonal injury (10). EPO was administered 1–2 days after the disease onset (onset at day 14, EPO administered at day 15–16), and the treatment was repeated daily for 14 days. Although given at a high dose (50 μg/kg), in a dose response experiment the lowest effective dose was 500 ng/kg.

The pathological features and clinical relevance of the different animal models of EAE have been reviewed elsewhere (11–13). MOG-induced EAE in C57BL/6 mice shares with MS extensive axonal injury, marked demyelination and a mononuclear cell infiltrate. Axonal injury occurs early in this model. Inflammation peaks at the maximum of disease, around day 15 post-immunization (p.i.), and then decreases to about 20% at the later disease phase. At day 9 p.i., at the beginning of the disease, there is already a massive decrease of axonal densities in the lesion, which later decrease further, decreasing also in the normal appearing white matter. The prevalence of axonal degeneration in chronic MOG-EAE has been recently confirmed with transmission electron microscopy (14). Axonal loss seems to be dissociated from acute inflammatory infiltration and correlates with clinical impairment (15). In this model, EPO could block most of the neurological disease progression when administered after disease onset (10), when inflammation and axonal injury were established already, suggesting that it might also actively stimulate repair.

PLP-induced relapsing-remitting EAE in SJL mice resembles relapsing-remitting MS, the most common form of MS. It is characterized by an onset around day 11–14, in which the mice develop the first wave of paralysis, after which they recover, going towards a remitting phase, but then one or more relapses may occur. Mononuclear cell infiltrates and extensive primary demyelination and remyelination account for relapses and remissions. Therapeutic EPO (5,000 IU/kg) administered for 7 days starting from disease onset reduced clinical score, inflammation and demyelination. Importantly, it also increased oligodendrocyte progenitor cell proliferation and BDNF positive cell density, indicating also in this model induction of an active repair process (16).

Interestingly, in both MOG-induced chronic progressive and PLP-induced relapsing-remitting models of EAE, we found an increased stable expression of endogenous EPO in the spinal cord (17). In a subsequent study Kang et al. also found elevated levels of EPO in spinal cord of rats with MBP-induced acute EAE at the peak of the disease, thus confirming our observation (18). It is therefore likely that EPO is part of an endogenous mechanism of repair induced by the disease itself.

It should be noted that EPO is safe when administered to anemic patients, but when administered to non anemic patients has side effects, and it may be prothrombotic due to its erythropoietic activity (19). This may be a problem when treating patients with chronic diseases, like MS, who may obviously need long treatments. Interestingly, Leist et al. found that carbamylated EPO (CEPO), a derivative of EPO in which all lysines are transformed to homocitrulline by carbamylation, modification already known to abolish EPO's erythropoietic activity, retained EPO's neuroprotective activity and also protected mice with MOG-induced EAE (20). In this study, CEPO was administered at day 3 after MOG immunization, according to the preventive schedule, and at week 10, according to the late therapeutic schedule. The fact that CEPO protected the mice also when administered at week 10, after 4 weeks of stable chronic disease, reinforces previous observations suggesting that EPO and derivatives may act not only inhibiting inflammation but also at some level as neuroregenerative agents.

In a systematic study in the same model of MOG-induced chronic progressive EAE in C57BL/6 mice, in which EPO and CEPO were administered in parallel at different doses according to preventive, therapeutic, or late therapeutic schedules, both EPO and CEPO were found to reduce clinical symptoms and inhibit inflammatory cytokines in spinal cord and peripheral lymphocytes (21). Although the preventive schedule was obviously more efficacious, both EPO and CEPO worked at comparable level even when administered according to a therapeutic schedule at the lowest dosage tested (0.5 μg/kg, three times a week). Also asialoEPO, that since desialylated has a very short half-life and therefore is not erythropoietic, ameliorated the clinical score of the disease. However, in this study it was administered at the highest dose according to the preventive schedule, and therefore, more studies would be needed to compare its efficacy to that of EPO or CEPO.

More recently, Yuan et al. have showed that EPO acts also as an immunomodulator in MOG-induced EAE in C57BL/6 mice. In this study, EPO (5,000 IU/kg) was administered 3 days after immunization, according to a preventive schedule, or 7 days after immunization, still before onset, and was shown to reduce disease severity and to have immunomodulatory effects in the periphery and spinal cord, inhibiting the proliferation of pathogenic T-cells and favoring the expansion of protective T-cell subsets (22). Using a very similar treatment schedule (5,000 IU, preventive), Chen et al. confirmed EPO immunomodulatory effects on pathogenic and protective T-cell subsets, and also found increased expression of heme oxygenase-1 (HO-1), an enzyme with antioxidant potential, in central nervous system and spleen of mice with MOG-induced EAE (23).

In the same model darbepoetin alfa, a hyperglycosylated EPO mutant, administered on day 13 and 17 after MOG immunization (2.5 μg/kg, i.p., therapeutic schedule) reduced clinical symptoms, apparently through induction of tissue inhibitor of metalloprotease (TIMP-1), since it increased the number of astrocytes expressing TIMP-1 in brain and spinal cord and did not have any protective effect in Timp-1 null mice. Of note, at this dose darbepoetin alfa did not increase the hematocrit (24).

Importantly, these results have been translated to the clinic in a small proof of concept clinical trial in patients with chronic progressive MS, in which treatment with a high dose of EPO improved motor and cognitive functions. The efficacy of EPO in MS patients has been reviewed by Erenreich (4). Although more data corroborate EPO's anti-inflammatory activity, preclinical data obtained in EAE models suggesting that EPO also inhibits axonal damage and induces oligodendrogenesis and repair have been confirmed in this pilot clinical trial, in which it has been clearly shown that EPO can improve cognitive functions.

More studies are needed to elucidate the mechanism by which EPO may have neuroprotective and neurogenerative effects in EAE and MS, focusing on neurogenic/neurotrophic effects and induction of remyelination. Importantly, studies with asialoEPO and CEPO have shown that derivatives of EPO devoid of erythropoietic activity have therapeutic efficacy in EAE.

1.2 Conclusions

EPO and derivatives of EPO devoid of erythropoietic activity (e.g., CEPO, asialoEPO) ameliorate the clinical course of EAE. Efficacy in several models has been reported. Other than inhibiting inflammation, EPO and derivatives inhibit demyelination and promote neuronal and axonal protection, suggesting that they might actively induce nervous system repair. In a pilot clinical trial, EPO has shown efficacy in patients with primary progressive MS, improving motor and cognitive functions.

In the following section, we report an example of effectiveness of EPO and CEPO in ameliorating the clinical score of EAE (21). We describe the protocol to induce EAE in C57BL/6 by immunization with MOG peptide 35–55 and the schedule and doses of EPO and CEPO administration. Importantly, in this model both EPO and CEPO work not only when administered before onset but also when given according to a therapeutic schedule. Detailed protocols to induce EAE in C57BL/6 and SJL mouse strains by immunization with several neuroantigens, including MOG peptide 35–55, have been previously reported (13, 25).

2 Materials

1. Female C57BL/6 mice, 6–8 weeks old.
2. Recombinant human erythropoietin (rhEPO).
3. Carbamylated EPO (CEPO) and asialoEPO, prepared as described (20, 26).
4. MOG peptide 35–55 (MEVGWYRSPFSRVVHLYRNGK).
5. Incomplete Freund's adjuvant.
6. Heat-inactivated *Mycobacterium tuberculosis*, strain H37Ra.
7. Pertussis toxin from *Bordetella pertussis.*
8. Pyrogen-free phosphate buffered saline (PBS).
9. Pyrogen-free saline solution.
10. Two 5 ml glass syringes connected by a 2-way luer lock adapter.
11. 1 ml syringes with removable needle.
12. 21-gauge needles.
13. Two small glass beakers.
14. Magnetic stirrer.
15. Infrared lamp.
16. Mouse restrainer for intravenous injections.
17. Ice.

3 Methods

3.1 Preparation of EPO or CEPO

Concentrated solutions of EPO or CEPO (about 1 mg/ml in PBS) are stored in small aliquots at −80 °C (10 μl each). Each aliquot can be thawed a maximum of 2 times. EPO or CEPO are administered by intraperitoneal injection (i.p.) at the dose of 50 μg/kg, in a final volume of 200 μl/mouse. For instance, if the mice weight 20 g, each mouse is injected with 1 μg of EPO or CEPO in 200 μl of saline solution (1 μl stock solution up to 200 μl saline).

3.2 Treatments

EPO or CEPO diluted in saline (see above) are injected i.p. three times a week. For preventive schedule, they are administered starting 3 days after immunization; for therapeutic schedule, starting at the onset; for late therapeutic schedule, treatment is initiated 27 days after immunization.

3.3 Preparation of Emulsion of Antigen and Adjuvant for Immunization

1. *Emulsion volume.* Each mouse receives 300 μl of emulsion. To calculate the volume, consider 7–8 mice in excess (2–3 ml in excess) because there is a waste during the emulsification process (e.g., for 20 mice, instead of 6 ml prepare 8.4 ml of emulsion, calculated considering 28 mice).
2. *Emulsion composition.* Each mouse receives 200 μg of MOG 35–55 in complete Freund's adjuvant (CFA), that is incomplete Freund's adjuvant (IFA) supplemented with 8 mg/ml of *Mycobacterium tuberculosis.* The solutions of MOG in PBS (1.33 mg/ml) and *Mycobacterium tuberculosis* in IFA (8 mg/ml) are prepared separately, in two different glass beakers, and then mixed in a 1:1 volume ratio (e.g., for 20 mice, prepare 8.4 ml of emulsion: 4.2 ml of 1.33 mg/ml MOG in PBS and 4.2 ml of 8 mg/ml *Mycobacterium tuberculosis* in IFA).
3. *Preparation of the two solutions.* After calculating the volume needed as above, prepare the two solutions: (a) in a small glass beaker dissolve MOG 35–55 in PBS (1.33 mg/ml); (b) in a second glass beaker prepare a homogeneous suspension of *Mycobacterium tuberculosis* in IFA at 8 mg/ml, using a magnetic stirrer.
4. *Preparation of the emulsion.* Emulsify the two solutions in a 1:1 volume ratio with two glass syringes connected with a 2-way luer lock adapter for 4–5 min, until the solution is firm enough (see Note 1). Work on ice. Fill the insulin syringes with the emulsion using the 21-gauge needle (see Note 2). Leave the syringes on ice for 5 min to settle. Eliminate all the air in the syringes and refill them again. When the syringes are ready and contain 900 μl of emulsion (enough for three mice) without air, change the needle again with the 25-gauge one. Keep them always on ice.

Table 2
Scale used for clinical evaluation

0	Healthy
1	Flaccid tail. The mouse is held by the base of the tail. If it is healthy the tail is always straight, otherwise it hangs down
1.5	If the mouse stumbles at least twice on the grid cage (unsteady gait, or ataxia)
2	Ataxia and/or hind-limbs paresis, or slow righting reflex (see Note 6)
3	Paralysis of hind limb and/or paresis of forelimbs
4	Paraparesis of fore limb
5	Moribund or death

3.4 Induction of EAE

1. Hold the mice from the neck and the tail.
2. Inject the mice subcutaneously (s.c.), slowly, with 100 μl of emulsion in each flank and 100 μl at the tail base (see Note 3).
3. After immunization (4–5 h), inject the mice intravenously (i.v., see Note 4) with 500 ng of pertussis toxin resuspended in 100 μl of PBS. After 48 h, inject the mice again with pertussin toxin, i.v., 500 ng/100 μl each.
4. Administer EPO or CEPO i.p., 200 μl/mouse in saline solution, three times per week, starting at different times after immunization depending on the schedule (preventive, therapeutic or late therapeutic, see above).

3.5 Clinical Evaluation

The onset of the disease is usually 10–18 days after immunization. Mice need to be checked daily for body weight and clinical score (see Note 5). They will usually lose 1–2 g of weight in the days before the onset. For clinical evaluation a scale with five steps is used (Table 2).

3.6 Efficacy of Preventive or Therapeutic EPO or CEPO in Chronic-Progressive EAE

C57BL/6 mice immunized with MOG suffer chronic progressive EAE with onset around day 10. EPO inhibits the clinical score of the disease when administered according to a preventive schedule, and also according to a therapeutic schedule, starting at disease onset (Fig. 1 (21)). Also with a late therapeutic schedule of treatment some improvement in the status of the animals was obtained, as detected by comparing the curves reporting the body weight, although no significant difference was observed in the clinical score; interestingly, a late therapeutic schedule of CEPO treatment significantly improved also the clinical score (21).

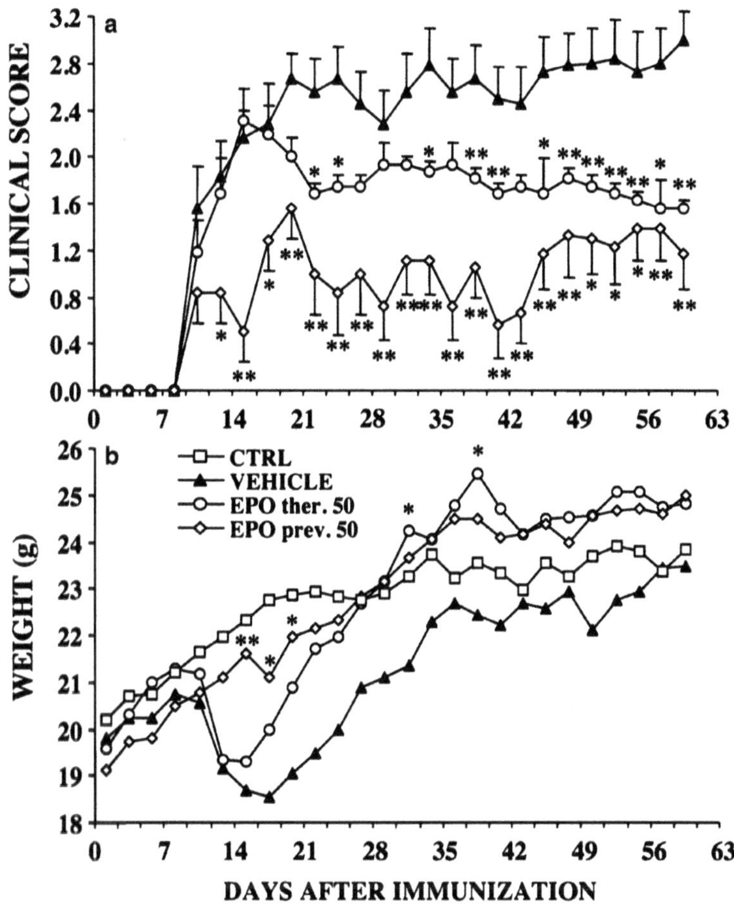

Fig. 1 Preventive and therapeutic EPO treatment improves the clinical course of EAE. Mice were immunized with MOG and treated with vehicle or EPO (50 μg/kg), i.p., 3 days a week, starting at day 3 after immunization (preventive schedule) or at disease onset (therapeutic schedule). The clinical score (*panel A*) and the body weight (*panel B*) are reported. Data are the mean ± S.E.M. Vehicle, $n=9$; EPO preventive, $n=9$; EPO therapeutic, $n=8$; healthy mice $n=5$. *$P<0.05$, **$P<0.01$ by Mann-Whitney test (reproduced with permission from Savino et al. J Neuroimmunol. 2006;172:27–37)

4 Notes

1. The emulsion is ready when a small drop placed on the surface of cold distilled water does not disperse instantly but maintains its integrity.
2. It is better to fill a syringe completely to avoid waste of the emulsion.
3. If the subcutaneous injection is well done it should be possible to see a lump in the injection area. The emulsion is injected subcutaneously in three different sites to allow a better adsorption.

4. The tail can be warmed up with hot water or leaving the cage under infrared lamp for about 5 min to ease the detection of the lateral tail vein. Mice are held in the restrainer for i.v. injection.
5. During the course of the disease mice are not able to move properly. The food pellet and the water need to be placed in a petri dish into the cage to enable sick mice to eat and drink.
6. It is almost impossible to put a healthy mouse on its back, because it will turn over immediately to regain the upright position. When a mouse is sick, it will lie on its back at least for a few seconds, or it will be unable to regain the upright position. Any impairment of the righting reflex is scored as 2.

Acknowledgments

This work was realized as part of the TC2N "Trans Channel Neuroscience Network" Interreg IV A 2 Mers Seas Zeeëns program, "Investing in your future" crossborder cooperation programme 2007–2013 part financed by the European Union (European Regional Development Fund).

References

1. Trapp BD, Nave KA (2008) Multiple sclerosis: an immune or neurodegenerative disorder? Annu Rev Neurosci 31:247–269
2. Makar TK, Bever CT, Singh IS et al (2009) Brain-derived neurotrophic factor gene delivery in an animal model of multiple sclerosis using bone marrow stem cells as a vehicle. J Neuroimmunol 210:40–51
3. Brines ML, Ghezzi P, Keenan S et al (2000) Erythropoietin crosses the blood–brain barrier to protect against experimental brain injury. Proc Natl Acad Sci U S A 97:10526–10531
4. Bartels C, Spate K, Krampe H, Ehrenreich H (2008) Recombinant human erythropoietin: novel strategies for neuroprotective/neuro-regenerative treatment of multiple sclerosis. Ther Adv Neurol Disord 1:193–206
5. Sargin D, Friedrichs H, El-Kordi A et al (2010) Erythropoietin as neuroprotective and neuroregenerative treatment strategy: comprehensive overview of 12 years of preclinical and clinical research. Best Pract Res Clin Anaesthesiol 24:573–594
6. Ehrenreich H, Fischer B, Norra C et al (2007) Exploring recombinant human erythropoietin in chronic progressive multiple sclerosis. Brain 130:2577–2588
7. Agnello D, Bigini P, Villa P et al (2002) Erythropoietin exerts an anti-inflammatory effect on the CNS in a model of experimental autoimmune encephalomyelitis. Brain Res 952:128–134
8. Sattler MB, Merkler D, Maier K et al (2004) Neuroprotective effects and intracellular signaling pathways of erythropoietin in a rat model of multiple sclerosis. Cell Death Differ 11(Suppl 2):S181–S192
9. Diem R, Sattler MB, Merkler D et al (2005) Combined therapy with methylprednisolone and erythropoietin in a model of multiple sclerosis. Brain 128:375–385
10. Li W, Maeda Y, Yuan RR et al (2004) Beneficial effect of erythropoietin on experimental allergic encephalomyelitis. Ann Neurol 56: 767–777
11. Gold R, Linington C, Lassmann H (2006) Understanding pathogenesis and therapy of multiple sclerosis via animal models: 70 years of merits and culprits in experimental autoimmune encephalomyelitis research. Brain 129:1953–1971
12. Steinman L, Zamvil SS (2006) How to successfully apply animal studies in experimental allergic encephalomyelitis to research on multiple sclerosis. Ann Neurol 60:12–21
13. Furlan R, Cuomo C, Martino G (2009) Animal models of multiple sclerosis. Methods Mol Biol 549:157–173

14. Aharoni R, Vainshtein A, Stock A et al (2011) Distinct pathological patterns in relapsing-remitting and chronic models of experimental autoimmune enchephalomyelitis and the neuroprotective effect of glatiramer acetate. J Autoimmun 37:228–241
15. Herrero-Herranz E, Pardo LA, Gold R et al (2008) Pattern of axonal injury in murine myelin oligodendrocyte glycoprotein induced experimental autoimmune encephalomyelitis: implications for multiple sclerosis. Neurobiol Dis 30:162–173
16. Zhang J, Li Y, Cui Y et al (2005) Erythropoietin treatment improves neurological functional recovery in EAE mice. Brain Res 1034:34–39
17. Mengozzi M, Cervellini I, Bigini P et al (2008) Endogenous erythropoietin as part of the cytokine network in the pathogenesis of experimental autoimmune encephalomyelitis. Mol Med 14:682–688
18. Kang SY, Kang JH, Choi JC et al (2009) Expression of erythropoietin in the spinal cord of lewis rats with experimental autoimmune encephalomyelitis. J Clin Neurol 5:39–45
19. Bennett CL, Silver SM, Djulbegovic B et al (2008) Venous thromboembolism and mortality associated with recombinant erythropoietin and darbepoetin administration for the treatment of cancer-associated anemia. JAMA 299: 914–924
20. Leist M, Ghezzi P, Grasso G et al (2004) Derivatives of erythropoietin that are tissue protective but not erythropoietic. Science 305:239–242
21. Savino C, Pedotti R, Baggi F et al (2006) Delayed administration of erythropoietin and its non-erythropoietic derivatives ameliorates chronic murine autoimmune encephalomyelitis. J Neuroimmunol 172:27–37
22. Yuan R, Maeda Y, Li W et al (2008) Erythropoietin: a potent inducer of peripheral immuno/inflammatory modulation in autoimmune EAE. PLoS One 3:e1924
23. Chen SJ, Wang YL, Lo WT et al (2010) Erythropoietin enhances endogenous haem oxygenase-1 and represses immune responses to ameliorate experimental autoimmune encephalomyelitis. Clin Exp Immunol 162:210–223
24. Thorne M, Moore CS, Robertson GS (2009) Lack of TIMP-1 increases severity of experimental autoimmune encephalomyelitis: effects of darbepoetin alfa on TIMP-1 null and wild-type mice. J Neuroimmunol 211:92–100
25. Miller SD, Karpus WJ (2007) Experimental autoimmune encephalomyelitis in the mouse. Curr Protoc Immunol Chapter 15, Unit 15 11.
26. Erbayraktar S, Yilmaz O, Gokmen N et al (2003) Erythropoietin is a multifunctional tissue-protective cytokine. Curr Hematol Rep 2:465–470

Chapter 11

Deciphering the Intracellular Signaling of Erythropoietin in Neuronal Cells

Murat Digicaylioglu

Abstract

The search for potential drugs to treat neurodegenerative diseases has been intense in the last two decades. Among many candidates, erythropoietin (EPO) was identified as a potent protectant of neurons suffering from various adverse conditions. A wide array of literature indicates that endogenous or exogenous recombinant human erythropoietin and its variants activate cell signaling that initiates survival-promoting events in neurons and neuronal cells. This chapter gives an overview of the pro-survival signaling induced by endogenous and exogenous erythropoietin in vitro and in vivo and provides methods to further investigate the intracellular signaling. It is important to know *that* EPO is neuroprotective, but it will greatly enhance our chances to establish EPO as a new drug candidate if we know *how* EPO protects neurons.

The descriptions below summarize our current knowledge in non-neuronal and neuronal signaling pathways induced by EPO. The signaling pathways involved in EPO are multiple; some are well known whereas others are still under intense investigation and few are observed in very specific cell types. It is important to note that neuronal signaling events triggered by EPO are still incomplete and require further research. Therefore, excellent review articles that explore specific EPO-signaling events are referenced.

Key words JAK2, Primary neuronal cultures, Akt kinase assay, Real time PCR, Western blotting, Immunoprecipitation

1 Introduction

Erythropoietin is a class I cytokine with a molecular mass of 30.4 kDa (1). In humans, circulating EPO is produced by the hepatocytes in fetal liver and the peritubular fibroblast-like type-1 interstitial cells in the renal cortex and outer medulla of the adult kidney (2). In addition, EPO is also produced in spleen, lung, and testis. Particularly important for neuroprotection is the production of EPO in neuronal cells, such as neurons and astrocytes (3). Brain derived EPO can be detected in the cerebrospinal fluid (4) of human neonates and at significantly lower levels in adults. Circulating EPO and brain derived EPO are likely to have different functions. Under physiological conditions the blood brain barrier

Pietro Ghezzi and Anthony Cerami (eds.), *Tissue-Protective Cytokines: Methods and Protocols*, Methods in Molecular Biology, vol. 982, DOI 10.1007/978-1-62703-308-4_11, © Springer Science+Business Media, LLC 2013

(5) separates the blood from the brain, strictly limiting the access of blood born EPO into the brain. In the same manner BBB also restricts the access of brain derived EPO into the circulating blood. Therefore, under physiological conditions blood born EPO is unlikely to exert any neurological effects. Due to its very limited access and significantly lower concentration than circulating EPO levels, brain derived EPO is more likely to act in an autocrine or paracrine manner and is not likely to modulate hematopoiesis. Similar to EPO regulation for hematopoiesis, intrinsic levels of EPO in the brain are also regulated by HIF. As in the fetal liver and adult kidney, hypoxia-induced regulation of EPO production in the central nervous system depends mainly of HIF1. This is evidenced by the fact that in animal models with deficiency in HIF1 expression, HIF2 expression cannot substitute for absence of the former (6).

The identification of circulating EPO as the activator for erythropoiesis initiated the search for its cellular target, the EPO-R. The EPO-R is a transmembrane receptor consisting of 484 amino acids. This glycoprotein has an extracellular domain with WSXWS-motif, a single hydrophobic transmembrane section, and a cytoplasmic domain (7–9). Homodimerization of two EPO-R's allows the binding of a single EPO molecule. The latter is variable and changes its confirmation upon activation by EPO, enabling the activation of EPO-R-associated Janus Kinase-2 (JAK-2) (10) by autophosphorylation (11). Eight tyrosine residues within the cytoplasmic section of the EPO-R are phosphorylated by the activation of JAK-2 and subsequently attract signaling proteins containing Src homology-2 (SH2) domains to phosphotyrosines. The variety of recruited proteins containing SH2 domains determines the signaling pathway activated and indicates the diversity of cellular responses to EPO/EPO-R interaction. Among those are signal transducers and activators of transcription (STAT-1, STAT-3, STAT-5), phosphatidyl-inositol 3-kinase (PI-3K), MAP-kinase, and p38 (12).

The "classical" EPO-induced anti-apoptotic pathway in erythroid cells involves the activation of signal transducer and activator of transcription 5 (STAT-5), which results in enhanced production of Bcl-X_L (13). In non-neuronal cells the upregulation of this anti-apoptotic factor prevents programmed cell death of the immature erythroid cells and is therefore essential for the production of erythrocytes (14).

The other well-known main signaling pathway activated by EPO/EPO-R and JAK-2 is the PI-3K pathway. This pathway is also necessary to prevent apoptotic cell death of proerythrocytes. PI-3K can bind to the EPO-R either directly at the tyrosine Y479 residue or indirectly via adaptors proteins, such as Grb2 and RAS. Direct binding of PI-3K to EPO-R results in phosphorylation and subsequent activation of AKT, a key element in pro-survival signaling in many cell types. Activated AKT phosphorylates transcription factors GATA-1 (15) and forkhead O3A (FoxO3A) (16, 17). Phosphorylation and activation of GATA-1 enables this transcription factor to translocate into the nucleus and induce the transcription of its target genes such as Bcl-X_L. In contrast, phosphorylation of FoxO3A inhibits its upregulation of pro-apoptotic genes encoding for TNF-related apoptosis-inducing ligand (TRAIL), members of the Bcl-2 family and cyclin dependent kinase (CDK) inhibitor. Indirect association of PI-3K with EPO-R results in inhibition of caspase 3. Caspases are endopeptidases that cleave cysteine residues, resulting in proteolysis and subsequent apoptosis (18).

It is necessary to point out that the function of neuronal EPO is likely to be non-hematopoietic. Therefore, intracellular signaling involved in erythrocyte survival and maturation might not occur in neurons. In a similar manner, neuron-specific EPO-dependent events might be unique to this cell type and not observed in non-neuronal cells. However, in proerythrocytes and in neurons, EPO induces pro-survival and anti-apoptotic mechanisms, which do involve the same signaling elements.

In the adult mammalian brain EPO mRNA is expressed in astrocytes, neurons and endothelial cells (19–21). Numerous studies have revealed that endogenous expression of EPO in the brain starts in the embryonic stage but is strongly reduced at birth (22). It continues in adult CNS, albeit at significantly lower levels. EPO plays a key role in embryonic neuronal development. In animals deficient in either EPO or EPO-R expression neurogenetic development is disrupted, resulting in incomplete closure of the neural tube and significantly reduced proliferation of neuronal progenitor cells (NPC) in the subventricular zone (23). Moreover, *Epo-* or *Epor-null* mice display normal vasculogenesis but abnormal angiogenesis (24). The developmental deficiencies and severe anemia due to failed erythropoiesis caused by the absence of EPO or EPO-R expression results in in utero death. To circumvent the lethal effect of EPO and EPO-R depletion at the embryonic stage, Tsai et al. developed an animal model with conditional and brain-specific deletion of EPO-R expression (23). In this model lack of EPO-R expression resulted in significantly reduced proliferation of neuroblasts and NPCs in the subventricular zone, impairing neurogenesis in brain injuries.

The first confirmation that EPO-R's are expressed in neuron-like cell lines was reported by Masuda et al. (21). This report was supported later by detection of binding sites for EPO in situ in the mouse brain (22, 25).

EPO production in the brain was first suggested by studies in dogs and rats (26). In adults, astrocytes and neurons are the main sources for EPO production in brain (20, 27). Furthermore, microglial cells, endothelial cells and oligodendrocytes are capable of expressing EPO in vitro (21, 28). As in the renal system, EPO production in the brain is also oxygen-dependent and responds to hypoxia with increased expression, albeit response in the brain is magnitudes smaller than in the kidney (29). Targeted deletion of the *HIF1* α and *HIF2* α in astrocytes resulted in over 50% reduction of hypoxia-dependent upregulation of EPO production in the brain (29). However, this did not cause systemic anemia, further underlining the distinct non-hematopoietic function of EPO in the CNS.

2 Materials

2.1 Detection of EPO and EPO-Receptor mRNA in the Brain Tissue

1. TRIzol® (Life Technologies).
2. High Capacity cDNA Reverse Transcription Kit (4368814, Life Technologies).
3. Primers: primers: EPO: 5′-TCTGCGACAGTCGAGTTCT-3′ (sense) and 5′-GTATCCACTGTGAGTGTTCG-3′ (antisense), located in exon 2 and exon 5 of the murine *EPO gene*, and 5′-GGACACCTACTTGGTATTGG-3′ (sense) and 5′-GACGTTGTAGGCTGGAGTCC-3′ (antisense), located in exon 8 and the 3′ untranslated region of the murine *EPO-R gene*.
4. SYBR Green PCR Master Mix (4367659, Life Technologies).

2.2 Detection of EPO-Receptor Protein in the Brain Tissue (see Note 1)

1. Cell lysis buffer (50 mM Tris–Cl buffer at pH 8.0, 150 mM NaCl, 100 mg/ml phenylmethyl sulphonyl fluoride, 1 mg/ml aprotinin, 1% Triton X-100).
2. BCA-Protein assay kit (Thermo Scientific, # 23225).
3. 4× LDS sample buffer (NuPAGE® LDS Sample Buffer, Invitrogen, NP0007).
4. 10% Bis–Tris SDS-PAGE gels (Invitrogen, # NP0301BOX) and Nitrocellulose Membrane (Invitrogen, #LC2009).
5. ChemiBlocker (Millipore, #2170).
6. Antibodies raised against EPO-R (R&D Systems), p-JAK2 (Santa Cruz), JAK2 (Santa Cruz), p-STAT-5 (Cell Signaling) or STAT-5 (Cell Signaling).
7. Horseradish peroxidase (HRP) (1:20,000, Vector Laboratories).
8. LumiGLO® Substrate (Cell Signaling, #7003).
9. X-Ray films.

2.3 Immunoprecipitation

1. Protein A agarose beads (Cell signaling, #9863).
2. 4× LDS sample buffer (NuPAGE® LDS Sample Buffer, Invitrogen, NP0007).

2.4 Akt Kinase Activity Assay

1. AKT Kinase Assay Kit (Nonradioactive; Cell Signaling, #9840).
2. Assay Lysis buffer (20 mM Tris–HCl (pH 7.5), 150 mM NaCl, 1 mM Na2EDTA, 1 mM EGTA, 1% Triton X, 2.5 mM sodium pyrophosphate, 1 mM b-glycerophosphate, 1 mM Na_3VO_4, 1 μg/ml leupeptin, 1 mM phenylmethylsulfonyl fluoride).
3. Kinase buffer (25 mM Tris–HCl (pH 7.5), 5 mM b-glycerophosphate, 2 mM dithiothreitol (DTT), 0.1 mM Na_3VO_4, 10 mM $MgCl_2$).
4. ATP (included in the Akt Kinase assay kit).
5. GSK-3 Fusion Protein (included in the Akt Kinase assay kit).
6. LumiGLO® Substrate (Cell Signaling, #7003).

2.5 Primary Mixed Cultures

Materials for primary rat neuronal cultures:

1. Poly-L-lysine.
2. Borate Buffer (pH 8.0).
3. Sprague Dawley Rat, pregnant (Charles Rivers).
4. Culture plates.
5. Glass coverslips (Fisher Scientific, Fisherbrand Coverglasses, Circles No. 2—0.17–0.25 mm thick; Size: 22 mm; Catalog no. 12-546-1) (see Note 2).
6. 70% nitric acid.
7. 100% ethanol.
8. Two #5 tweezers with sharp and pointy tips.
9. 1 retina scissor.
10. Hank's balanced salt solution (pH 7.2).
11. 0.25% Trypsin-EDTA.
12. Dulbecco's Modified Eagle Medium (DMEM).
13. F-12 Nutrient Mixture.
14. Heat inactivated bovine serum.
15. L-Glutamine.
16. HEPES buffer.
17. Penicillin.
18. Streptomycin.
19. Pasteur pipets with fire polished tips (final opening diameter at tip >300 μm).
20. Cell strainer, 40 μm (BD Falcon, 352340).

3 Methods

3.1 Detection of EPO and EPO-Receptor mRNA in the Brain Tissue

1. Isolate total RNA with TRIzol® (Life Technologies).
2. Total RNA (1 μg) is reverse-transcribed with the commercially available High Capacity cDNA Reverse Transcription Kit.
3. Use cDNA as a template for PCR amplification of either EPO or EPO-R with the following primers: 5′-TCTGCGA-CAGTCGAGTTCT-3′ (sense) and 5′-GTATCCACTGTGAG-TGTTCG-3′ (antisense), located in exon 2 and exon 5 of the murine EPO gene, and 5'-GGACACCTACTTGGTATTGG-3′ (sense) and 5′-GACGTTGTAGGCTGGAGTCC-3′ (antisense), located in exon 8 and the 3′ untranslated region of the murine EPO-R gene.
4. Real time PCR is performed using a commercially available SYBR Green PCR Master Mix. For each sample, a 20 μl reaction prepare a mixture containing 10 μl of SYBR Green PCR Master Mix, 200 nM of the reverse primer, 200 nM of the forward primer and 10–100 ng of cDNA template.
5. Program a thermal cycler to hold for 10 min at 95°C, followed by 40 cycles of 95°C for 15 s and 60°C for 60 s. Real time PCR products will be 395 bp for EPO and 452 bp for EPO-R.

3.2 Detection of EPO and EPO-Receptor Protein in the Brain Tissue

3.2.1 Western Blotting

1. Load equivalent amounts of protein (15–30 μg) on 10% Bis–Tris SDS-PAGE gels under reducing conditions at (80 V, 3 h, RT).
2. Transfer separated proteins onto nitrocellulose membranes. After nonspecific binding is blocked with ChemiBlocker for 1 h, the blots are subsequently incubated overnight at 4°C with EPO-R (1:200), p-JAK2 (1:200), JAK2 (1:200), p-STAT-5 (1:1,000), or STAT-5 (1:1,000). After washing, membranes are incubated with secondary antibodies conjugated to horseradish peroxidase (HRP) (1:20,000) for 1 h and developed with an enhanced chemiluminescence kit and exposed to X-ray film.

3.3 Methods

3.3.1 Lysates

To detect JAK2 and STAT-5 signaling in neurons, primary cerebrocortical cultures are lysed in ice-cold cell lysis buffer (50 mM Tris–Cl buffer at pH 8.0, 150 mM NaCl, 100 mg/ml phenylmethyl sulphonyl fluoride, 1 mg/ml aprotinin, 1% Triton X-100). The lysates are vortexed for 15 s and centrifuged at 10,000 × *g* for 20 m. The supernatants are used to determine protein concentrations using a BCA-Protein assay kit (Pierce).

3.3.2 Western Blotting

Equivalent amounts of protein (15–30 μg) is resolved on 10% Bis–Tris SDS-PAGE gels under reducing conditions, and transferred onto nitrocellulose membranes. After nonspecific binding is blocked with ChemiBlocker (Chemicon) for 1 h, the blots are subsequently incubated overnight at 4°C with EPO-R (1:200, R&D Systems), p-JAK2 (1:200, Santa Cruz), JAK2 (1:200, Santa Cruz), p-STAT-5 (1:1,000, Cell Signaling), or STAT-5 (1:1,000, Cell Signaling). After washing, membranes are incubated with secondary antibodies conjugated to horseradish peroxidase (HRP) (1:20,000, Vector Laboratories) for 1 h and developed with an enhanced chemiluminescence kit and exposed to X-ray film.

3.3.3 Immunoprecipitation of STAT5 to Detect JAK2 Activation

1. Preclear 200 μl of cell lysate by adding Protein A agarose beads (20 μl).
2. Incubate at 4°C for 60 min, spin for 10 min, and transfer the supernatant to a fresh tube.
3. Add STAT5 antibody (1:50, Cell Signaling) and incubate overnight at 4°C on an orbital rocker.
4. Then add protein A agarose beads (20 μl) and incubate for 3 h at 4°C.
5. Microcentrifuge for 30 s and wash the bead pellet five times with 500 μl of cell lysis buffer.
6. Resuspend the pellet with 30 μl of 4× LDS sample buffer and heat to 70°C for 10 min.
7. Microcentrifuge for 1 min at 10,000 × *g*.
8. Load the samples (15–30 μl) on a 10% Bis–Tris SDS-PAGE gel under reducing conditions, and transfer onto nitrocellulose membranes.
9. After blocking for 1 h, incubate the blots overnight at 4°C with p-JAK2 (1:200, Santa Cruz), JAK2 (1:200, Santa Cruz), p-STAT-5 (1:1,000, Cell Signaling), or STAT-5 (1:1,000, Cell Signaling). After washing, membranes are incubated with secondary antibodies and developed with an enhanced chemiluminescence kit and exposed to X-ray film.

3.3.4 Akt Kinase Activity Assay

Activation of the PI-3K/AKT signaling pathway also provides another mechanism for EPO induced signaling. Akt activity can be measured using a commercial nonradioactive AKT Kinase Assay (Cell Signaling).

1. To measure Akt activity, lyse cells using ice-cold lysis buffer (20 mM Tris–HCl (pH 7.5), 150 mM NaCl, 1 mM Na2EDTA, 1 mM EGTA, 1% Triton X, 2.5 mM sodium pyrophosphate, 1 mM b-glycerophosphate, 1 mM Na_3VO_4, 1 μg/ml leupeptin, 1 mM phenylmethylsulfonyl fluoride), centrifuge for 10 min at 10,000 × *g*, and transfer supernatant to a new tube.
2. Add 20 μl of Immobilized Phospho-Akt (Ser473) antibody bead slurry to 200 μl of cell lysate. Incubate overnight at 4°C.

3. Centrifuge the cell lysate/immobilized antibody at 10,000 × *g* for 30 s.
4. Wash the pellets twice with 500 μl of lysis buffer, and then with 500 μl of kinase buffer (25 mM Tris–HCl (pH 7.5), 5 mM b-glycerophosphate, 2 mM dithiothreitol (DTT), 0.1 mM Na_3VO_4, 10 mM $MgCl_2$).
5. Add 50 μl of kinase buffer, 1 μl of 10 mM ATP, and 1 μl of GSK-3 Fusion Protein (Cell Signaling) to the pellets. Incubate for 30 min at 30°C.
6. Terminate the reaction with 25 μl of 4× LDS Sample Buffer (Life Technologies).
7. Heat the sample to 70°C for 10 min.
8. Load 5–15 μl of sample per well on a 10% Bis–Tris SDS-PAGE gel and transfer to a nitrocellulose membrane. After blocking, incubate the membrane with Phospho-GSK-3a/b (Ser21/9) Antibody (1:1,000, Cell Signaling) overnight at 4°C. After washing, incubate the membrane with HRP-conjugated secondary antibody (1:2,000) and HRP-conjugated anti-biotin antibody (1:1,000) to detect biotinylated protein markers. After washing, incubate the membrane with 10 mL LumiGLO® Substrate (Cell Signaling) for 1 min and expose to X-ray film.

3.3.5 Primary Neuronal Cultures

1. Treated glass coverslips with a wash in 70% nitric acid under strong agitation for 2–3 days, followed by a rinse with water for 4–5 h and a wash in 100% ethanol for 20–30 min. Autoclave the coverslips on a dry cycle for 15–20 min.
2. Coat tissue culture plates or glass coverslips with poly-L-lysine (100 μg/ml in borate buffer) overnight at room temperature. Then wash the plates three times in water and allow to dry overnight.
3. Euthanize apregnant Sprague-Dawley rat at embryonic day 17 (E17) with an isoflurane overdose.
4. Sterilize the lower abdomen is with ethanol and open with scissors. Expose the uterus and remove each embryo and place them in ice-cold Hank's balanced salt solution (HBSS).
5. Remove the heads from each embryo with scissors and place in fresh ice-cold HBSS. To remove the brains from the skull, cut up the midline of the skull with scissors, peel away the top layers of tissue (skin + skull) to expose the cortical hemispheres, then pinch out the brain from the spinal cord using curved forceps, and transfer the brain to ice-cold HBSS.
6. Under a dissecting microscope, turn the brain to the dorsal view. Dissect one hemisphere at a time by cutting the proximal and distal ends of the cortex with small scissors (about 1 mm).

7. Unroll the cortex and remove the hippocampus and striatum. Pinch off the cortex from the rest of the brain and remove the meninges (see Note 3). Then transfer the cortex to ice-cold HBSS.
8. When all the cortices have been collected, transfer the tissue to a 15 ml tube and incubate tissue in 0.125% Trypsin-EDTA at 37°C for 25–30 min.
9. Add 1 ml of bovine calf serum to stop the reaction and centrifuge at 2,500 × *g* for 2 min.
10. Aspirate out the trypsin and FBS and replace with 2 ml of medium (Dulbecco's Modified Eagle Medium: Nutrient Mixture F-12 (DMEM/F12), 10% heat-inactivated bovine calf serum (see Note 4), 2 mM L-glutamine, 25 mM HEPES, 25 IU/mL penicillin, 25 μg/ml streptomycin).
11. Using a fire polished glass Pasteur pipet, *gently* triturate up and down 20 (max. 30) times avoiding bubbles (see Note 5).
12. Filter the cells through a sterile 40 μm cell strainer into a 50 ml tube and further dilute the cell suspension with 20–30 ml of media.
13. Count the number of cells using Trypan Blue exclusion.
14. Plate cells at approximately 100,000 cells/cm^2 and keep in a humidified incubator with 5% CO_2 at 37°C.
15. Four to 5 days after plating, replace 50% of the media with fresh media. Then add 25% fresh media to the cells every 4–5 days.

To test the neuronal viability in mixed cultures we first developed a method to distinguish between neurons and astrocytes. For this purpose we have chosen the cell specific markers for each cell type. NeuN is a neuronal protein that is almost exclusively expressed in neuronal nuclei (30). Astrocytes abundantly express glial fibrillary acidic protein (GFAP), which is widely used as a specific marker for these glial cells. By using two different fluorochromes for NeuN and GFAP, respectively, we were able to determine the percentage of neurons in normal cultures and distinguish between these two cell types. Furthermore, we added TUNEL staining as a third "color" to visualize DNA fragmentation, a hallmark of apoptosis, in either cell type. Nuclei of cells in normal conditions are negative for TUNEL, whereas DNA fragmentation and associated TUNEL staining is abundant in apoptotic cells (31). Cells under pathological conditions undergoing apoptosis display pycnotic nuclei, that are condensed or fragmented, which are displayed by Hoechst 33342 staining as dense dots with significantly smaller nuclei from normal cells (32). The Hoechst dye is excited at 350 nm (ultraviolet light) and emits a blue/cyan fluorescent light at 461 nm (Hoechst AG, Germany, manufacturer's specification).

Using this differential staining of the cells and nuclei and a confocal microscope capable of exciting and capturing fluorophores at three different wavelengths, we can determine the number of apoptotic cells in control conditions and in presence of a neurotoxic insult (33). We have used two chemical inducers of neurodegeneration: *N*-methyl-D-aspartate (NMDA) and SNOC (s-nitrosocystein). Both result in oxidative stress in neurons and induce neuronal apoptosis by free oxygen radicals, similar to events in ischemic brain (34–36). The determination of neuronal apoptosis requires that at least several criteria for neuronal degeneration are simultaneously detectable in the exposed cells (37). Meeting these requirements helps us to distinguish apoptosis from necrosis, which is an irreversible event and will not respond to neuroprotective treatment. In a third model of neuronal injury cerebrocortical cells were incubated with HIV/gp120, an envelope protein of the HIV virus known to induce HIV-associated neurocognitive disorders in transgenic mice and humans (38). Cerebrocortical cultures 17–21 days in culture are incubated with HIV/gp120 (50 mM) for 24 h and apoptotic neurons displaying NeuN and TUNEL staining counted.

4 Notes

1. It is important to notice that expression of a functional EPO-R in brain tissue was disputed recently and the specificity of antibodies used questioned by Sinclair and colleagues (39). However, in a response focusing on neuroprotective effect of EPO (10) we challenged that conclusion.
2. Glass coverslips are critical. We found that this type of glass used to produce the coverslips results in superior adherence of neuronal cells.
3. This step is determining for the quality of the primary neuronal cultures. All meninges have to be removed diligently (and rapidly) to minimize overgrowth of cultures with fibroblasts. Furthermore, it is crucial that this procedure is carried out in petri dishes filled with Hanks Medium that is equilibrated at RT and completely cover the brains. Successful isolation will yield in square cortex sections measuring 1×1 mm.
4. Quality and suitability of the fetal calf serum varies from batch to batch. It is highly recommended to test several batches from one or more manufacturers and to determine the one that results in the best outcome. Once identified, our laboratories buy a large quantity of the most suitable batch of fetal calf serum and store at −80°C.
5. Proper trituration will determine the final number and viability of neurons in primary cultures. Triturating with not perfectly

polished glass pipettes (check under microscope for sharp edges and opening of >300 μm), too many times, too forceful to create bubbles will significantly decrease the number of neurons. Moreover, remaining neurons might not be viable and become necrotic within few days in culture.

References

1. Davis JM, Arakawa T, Strickland TW, Yphantis DA (1987) Characterization of recombinant human erythropoietin produced in Chinese hamster ovary cells. Biochemistry 26(9): 2633–2638
2. Yoon D, Ponka P, Prchal JT (2011) Hypoxia. 5. Hypoxia and hematopoiesis. Am J Physiol Cell Physiol 300(6):C1215–C1222. doi: 10.1152/ajpcell.00044.2011
3. Arcasoy MO (2008) The non-haematopoietic biological effects of erythropoietin. Br J Haematol 141(1):14–31. doi:10.1111/j.1365-2141.2008.07014.x
4. Juul SE, Stallings SA, Christensen RD (1999) Erythropoietin in the cerebrospinal fluid of neonates who sustained CNS injury. Pediatr Res 46(5):543–547
5. Fu A, Hui EK, Lu JZ, Boado RJ, Pardridge WM (2010) Neuroprotection in stroke in the mouse with intravenous erythropoietin-Trojan horse fusion protein. Brain Res 1369:203–207. doi:S0006-8993(10)02440-6 (pii) 10.1016/j.brainres.2010.10.097
6. Kapitsinou PP, Liu Q, Unger TL, Rha J, Davidoff O, Keith B, Epstein JA, Moores SL, Erickson-Miller CL, Haase VH (2010) Hepatic HIF-2 regulates erythropoietic responses to hypoxia in renal anemia. Blood 116(16):3039–3048. doi:10.1182/blood-2010-02-270322
7. Youssoufian H, Longmore G, Neumann D, Yoshimura A, Lodish HF (1993) Structure, function, and activation of the erythropoietin receptor. Blood 81(9):2223–2236
8. Watowich SS (2011) The erythropoietin receptor: molecular structure and hematopoietic signaling pathways. J Investig Med. doi:10.231/JIM.0b013e31820fb28c
9. Jelkmann W (2004) Molecular biology of erythropoietin. Intern Med 43(8):649–659
10. Ghezzi P, Bernaudin M, Bianchi R, Blomgren K, Brines M, Campana W, Cavaletti G, Cerami A, Chopp M, Coleman T, Digicaylioglu M, Ehrenreich H, Erbayraktar S, Erbayraktar Z, Gassmann M, Genc S, Gokmen N, Grasso G, Juul S, Lipton SA, Hand CC, Latini R, Lauria G, Leist M, Newton SS, Petit E, Probert L, Sfacteria A, Siren AL, Talan M, Thiemermann C, Westenbrink D, Yaqoob M, Zhu C (2010) Erythropoietin: not just about erythropoiesis. Lancet 375(9732):2142. doi:10.1016/S0140-6736(10)60992-0
11. Witthuhn BA, Quelle FW, Silvennoinen O, Yi T, Tang B, Miura O, Ihle JN (1993) JAK2 associates with the erythropoietin receptor and is tyrosine phosphorylated and activated following stimulation with erythropoietin. Cell 74(2):227–236
12. Chateauvieux S, Grigorakaki C, Morceau F, Dicato M, Diederich M (2011) Erythropoietin, erythropoiesis and beyond. Biochem Pharmacol. doi:10.1016/j.bcp. 2011.06.045
13. Socolovsky M, Nam H, Fleming MD, Haase VH, Brugnara C, Lodish HF (2001) Ineffective erythropoiesis in Stat5a(-/-)5b(-/-) mice due to decreased survival of early erythroblasts. Blood 98(12):3261–3273
14. Koury MJ, Bondurant MC (1991) The mechanism of erythropoietin action. Am J Kidney Dis 18(4 Suppl 1):20–23
15. Simon MC, Pevny L, Wiles MV, Keller G, Costantini F, Orkin SH (1992) Rescue of erythroid development in gene targeted GATA-1- mouse embryonic stem cells. Nat Genet 1(2):92–98. doi:10.1038/ng0592-92
16. Lei H, Quelle FW (2009) FOXO transcription factors enforce cell cycle checkpoints and promote survival of hematopoietic cells after DNA damage. Mol Cancer Res 7(8):1294–1303. doi:10.1158/1541-7786.MCR-08-0531
17. Kashii Y, Uchida M, Kirito K, Tanaka M, Nishijima K, Toshima M, Ando T, Koizumi K, Endoh T, Sawada K, Momoi M, Miura Y, Ozawa K, Komatsu N (2000) A member of Forkhead family transcription factor, FKHRL1, is one of the downstream molecules of phosphatidylinositol 3-kinase-Akt activation pathway in erythropoietin signal transduction. Blood 96(3):941–949
18. Thornberry NA, Lazebnik Y (1998) Caspases: enemies within. Science 281(5381):1312–1316
19. Marti HH, Wenger RH, Rivas LA, Straumann U, Digicaylioglu M, Henn V, Yonekawa Y, Bauer C, Gassmann M (1996) Erythropoietin gene expression in human, monkey and murine brain. Eur J Neurosci 8(4):666–676
20. Bernaudin M, Marti HH, Roussel S, Divoux D, Nouvelot A, MacKenzie ET, Petit E (1999)

A potential role for erythropoietin in focal permanent cerebral ischemia in mice. J Cereb Blood Flow Metab 19(6):643–651. doi: 10.1097/00004647-199906000-00007
21. Masuda S, Okano M, Yamagishi K, Nagao M, Ueda M, Sasaki R (1994) A novel site of erythropoietin production. Oxygen-dependent production in cultured rat astrocytes. J Biol Chem 269(30):19488–19493
22. Liu C, Shen K, Liu Z, Noguchi CT (1997) Regulated human erythropoietin receptor expression in mouse brain. J Biol Chem 272(51):32395–32400
23. Tsai PT, Ohab JJ, Kertesz N, Groszer M, Matter C, Gao J, Liu X, Wu H, Carmichael ST (2006) A critical role of erythropoietin receptor in neurogenesis and post-stroke recovery. J Neurosci 26(4):1269–1274. doi:10.1523/JNEUROSCI.4480-05.2006
24. Wu H, Lee SH, Gao J, Liu X, Iruela-Arispe ML (1999) Inactivation of erythropoietin leads to defects in cardiac morphogenesis. Development 126(16):3597–3605
25. Digicaylioglu M, Bichet S, Marti HH, Wenger RH, Rivas LA, Bauer C, Gassmann M (1995) Localization of specific erythropoietin binding sites in defined areas of the mouse brain. Proc Natl Acad Sci U S A 92(9):3717–3720
26. Baciu I, Oprisiu C, Derevenco P, Vasile V, Muresan A, Hriscu M, Chis I (2000) The brain and other sites of erythropoietin production. Rom J Physiol 37(1–4):3–14
27. Siren AL, Knerlich F, Poser W, Gleiter CH, Bruck W, Ehrenreich H (2001) Erythropoietin and erythropoietin receptor in human ischemic/hypoxic brain. Acta Neuropathol 101(3):271–276
28. Sugawa M, Sakurai Y, Ishikawa-Ieda Y, Suzuki H, Asou H (2002) Effects of erythropoietin on glial cell development; oligodendrocyte maturation and astrocyte proliferation. Neurosci Res 44(4):391–403
29. Weidemann A, Kerdiles YM, Knaup KX, Rafie CA, Boutin AT, Stockmann C, Takeda N, Scadeng M, Shih AY, Haase VH, Simon MC, Kleinfeld D, Johnson RS (2009) The glial cell response is an essential component of hypoxia-induced erythropoiesis in mice. J Clin Invest 119(11):3373–3383. doi: 10.1172/JCI39378
30. Mullen RJ, Buck CR, Smith AM (1992) NeuN, a neuronal specific nuclear protein in vertebrates. Development 116(1):201–211
31. Hensey C, Gautier J (1998) Programmed cell death during Xenopus development: a spatiotemporal analysis. Dev Biol 203(1):36–48. doi:10.1006/dbio.1998.9028
32. Digicaylioglu M, Lipton SA (2001) Erythropoietin-mediated neuroprotection involves cross-talk between Jak2 and NF-kappaB signalling cascades. Nature 412(6847):641–647. doi:10.1038/35088074, 35088074 (pii)
33. Kaul M, Lipton SA (1999) Chemokines and activated macrophages in HIV gp120-induced neuronal apoptosis. Proc Natl Acad Sci U S A 96(14):8212–8216
34. Budd SL, Lipton SA (1999) Signaling events in NMDA receptor-induced apoptosis in cerebrocortical cultures. Ann N Y Acad Sci 893:261–264
35. Bonfoco E, Krainc D, Ankarcrona M, Nicotera P, Lipton SA (1995) Apoptosis and necrosis: two distinct events induced, respectively, by mild and intense insults with N-methyl-D-aspartate or nitric oxide/superoxide in cortical cell cultures. Proc Natl Acad Sci U S A 92(16):7162–7166
36. D'Emilia DM, Lipton SA (1999) Ratio of S-nitrosohomocyst(e)ine to homocyst(e)ine or other thiols determines neurotoxicity in rat cerebrocortical cultures. Neurosci Lett 265(2):103–106
37. Nicotera P, Ankarcrona M, Bonfoco E, Orrenius S, Lipton SA (1997) Neuronal necrosis and apoptosis: two distinct events induced by exposure to glutamate or oxidative stress. Adv Neurol 72:95–101
38. Garden GA, Budd SL, Tsai E, Hanson L, Kaul M, D'Emilia DM, Friedlander RM, Yuan J, Masliah E, Lipton SA (2002) Caspase cascades in human immunodeficiency virus-associated neurodegeneration. J Neurosci 22(10):4015–4024. doi:20026351
39. Sinclair AM, Coxon A, McCaffery I, Kaufman S, Paweletz K, Liu L, Busse L, Swift S, Elliott S, Begley CG (2010) Functional erythropoietin receptor is undetectable in endothelial, cardiac, neuronal, and renal cells. Blood 115(21):4264–4272. doi:10.1182/blood-2009-10-248666

Chapter 12

Assessment of Allodynia Relief by Tissue-Protective Molecules in a Rat Model of Nerve Injury-Induced Neuropathic Pain

Maarten Swartjes, Marieke Niesters, and Albert Dahan

Abstract

Neuropathic pain following nerve injury is a chronic disease characterized by allodynia and hyperalgesia of either mechanical or thermal origin. The mechanism underlying this disease is poorly understood leading to pharmacologic and physiotherapeutic control that is often insufficient. In this chapter, we describe a method to induce nerve injury in rats to create a robust animal model for studying neuropathic pain. Additionally we describe a method to follow up on animals in the process of testing treatments for efficacy in alleviating allodynia by testing for both mechanical and thermal allodynia with reproducible results.

Key words Neuropathic pain, Allodynia, Spared nerve injury, ARA290, Erythropoietin, Erythropoietin derivative

1 Introduction

Neuropathic pain is a chronic disease with a mechanism that is diverse and not yet completely understood. It is characterized by allodynia (increased sensitivity to a non-painful stimulus) and hyperalgesia (increased sensitivity to a painful stimulus) of either mechanical (touch or pressure) or thermal (cold or heat) origin (1). These pain states can become invalidating to patients resulting in reduced social participation and inability to maintain a job (2). Up until now, pharmacological (i.e., treatment with opioids, NSAIDS, antidepressants) or non-pharmacological treatment (spinal cord stimulation, physiotherapy) of neuropathic pain with has shown limited efficacy. The mechanism leading to neuropathic pain includes central and peripheral sensitization, neuronal plasticity, and neurogenic inflammation. These elements share intrinsic pathways that ultimately lead to altered nociception (3, 4). In animal experiments EPO has shown to cross the blood-brain barrier and to be neuroprotective (5, 6). Additionally it has shown to be able to alleviate neuropathic pain following nerve injury presumably

Pietro Ghezzi and Anthony Cerami (eds.), *Tissue-Protective Cytokines: Methods and Protocols*, Methods in Molecular Biology, vol. 982, DOI 10.1007/978-1-62703-308-4_12, © Springer Science+Business Media, LLC 2013

due to the tissue protective effects of EPO, resulting in increased survival of neuronal cells and reduced inflammation of the nervous system. In 2003, Campana and Meyers showed that treating rats with rhEpo following L5 spinal nerve crush (SNC) alleviated allodynia and decreased the time to recover from SNC, whereas animals in the vehicle treatment group showed a higher degree of allodynia and a longer time to reach recovery. This effect was supported by the observation that rhEPO prevented apoptosis of dorsal root ganglion (DRG) cells and induction of phosphorylated Jak-2, a molecule when phosphorylated induces apoptosis (7). In addition to the peripheral effects observed, EPO showed a central effect by protecting neurons in the spinal cord in a rat model of neuropathic pain. Following L5 proximal nerve root crush, rhEPO treated animals showed less allodynia when compared to vehicle treated animals which was accompanied by less apoptosis of neurons in both the ventral and dorsal horns of the spinal cord and identification of the EPO receptor (EPOR) and lower levels of TNF-α in spinal cord neurons (8). A study performed by Keswani et al. (9) assessed the role of the EPOR and showed neuroprotective effects in both in vitro and in vivo models. They showed in vitro that EPO is being produced by neurons and Schwann cells and that the EPOR is being expressed predominantly by neurons and was not restricted to the soma of the neuron. Additionally they showed beneficial effects of EPO in neurotoxicity. In an animal model of nerve damage they showed that EPO mRNA was increased in dorsal root ganglia (DRG) as well as in the sciatic nerve, while the EPOR mRNA was increased solely in the DRG. Additionally, in acrylamide induced neuropathy, EPO protected denervation of the skin, improved motor function in the grip strength test and prevented hyperalgesia in the paw withdrawal test. The role of EPO and TNF-α in neuropathic pain states was again explored by Campana et al. (10) in a chronic constriction injury model (CCI). They showed that rhEPO was able to reduce pain behavior in animals with nerve injury and that TNF-α was increased in injured nerves proximal to the injury. The induction of TNF-α was counteracted by rhEPO resulting in lower levels of the cytokine. Additionally, Jia et al. (11) showed that treating animals that had received an L5 spinal nerve transection with rhEPO showed decreased mechanical and thermal hyperalgesia with respect to control animals. This coincided with less microglia activation, decreased pro-inflammatory cytokine production (IL-1, 6 and TNF-α), increased anti-inflammatory cytokine production (IL-10) and decreased the expression of NF-κB, a signaling molecule important in pain processing. Both the expression of the cytokines and NF-κB was shown to be dose dependent (12). Also EPO derivatives devoid of erythropoietic properties show these effects. In a model of neuropathic pain where nucleus pulposus was applied to the DRG of animals EPO and asialo-EPO, an EPO derivative without erythropoietic properties, decreased mechanical allodynia and decreased levels of phospho-P-38, a signaling molecule important in pain processing and inflammation, and TNF-α (13). The EPO-derivative ARA290, an 11-amino-acid peptide mimicking the 3-dimensional structure of the B helix of EPO (14) has shown to be able to prevent the onset of allodynia in animals with nerve injury. In a rat model of neuropathic pain where animals received a spared nerve injury (SNI), a short treatment paradigm resulted in a delay of onset of allodynia, while the same paradigm complemented with a once-per-week maintenance treatment prevented the onset of allodynia for the duration of 15 weeks. It was shown that ARA290 works through the EPOR-β-common-receptor (EPOR-BCR) complex. Mice devoid of the BCR showed no response to ARA290, whereas wild type mice showed reduced levels of allodynia (15). EPO and its derivatives show efficacy in neuropathic pain making these molecules promising agents as treatment modalities.

2 Materials

2.1 Induction of the Neuropathic Pain Model: Spared Nerve Injury

1. Female Sprague-Dawley rats, 8 weeks old.
2. Ethanol 70% and wipes.
3. Absorbing under pad.
4. Syringe equipped with a 25 G needle containing buprenorphin.
5. Vapor anesthetics (Sevoflurane, isoflurane: see Note 1).
6. (Animal) shaver.
7. Tape.
8. Disinfectant.
9. Gauzes.
10. Small cotton swabs.
11. 5-0 silk sutures.
12. 4-0 nylon sutures.
13. Standard pattern forceps, straight (Fine Science Tools, Heidelberg, Germany).
14. Metzenbaum scissors, straight 14.5 cm (Fine Science Tools, Heidelberg, Germany).
15. Bonn micro forceps, smooth 7 cm (Fine Science Tools, Heidelberg, Germany).
16. Vannas spring scissors, straight 4 mm blade (Fine Science Tools, Heidelberg, Germany).

17. Student iris scissors, straight 11.5 cm (Fine Science Tools, Heidelberg, Germany).
18. Halsey needle holder, smooth (Fine Science Tools, Heidelberg, Germany).

2.2 Assessment of Pain: Tactile Allodynia

1. Plateau with grid (UGO Basile, Varese, Italy).
2. Perspex cages with lid (UGO Basile, Varese, Italy).
3. Semmes–Weinstein monofilaments (North Coast Medical Inc., San Jose, CA, USA).

2.3 Assessment of Pain: Cold Allodynia

1. Syringe (1 ml).
2. Needle 25 G, bent 90°.
3. Acetone, analytical grade.

2.4 Treatment with ARA290

1. PBS.
2. ARA290 (Araim Pharmaceuticals, Ossining, NY, USA).
3. Syringe, 1 ml equipped with 25 G needle.

3 Methods

3.1 Induction of the Neuropathic Pain Model: Spared Nerve Injury

1. Sterilize the instruments, for instance with a tabletop sterilizer.
2. Disinfect the surgical area of the table with 70% ethanol.
3. Place an absorbing under pad on the surgical area and place the surgical tools.
4. Fifteen minutes prior to surgery, administer a single dose of 0.01–0.05 mg/kg buprenorphin subcutaneously in the scruff of the neck for the relief of acute post operative pain.
5. To start surgery, induce and maintain anesthesia (6% induction, 3% maintenance in medicinal air mixture).
6. Place animal on the stomach and shave the leg that is going to be operated on.
7. Disinfect the shaved hind leg and direct and fixate it with a piece of tape towards yourself.
8. Draw an imaginary line between the patella and the crest of the ilium and locate the center of the line. This is approximately where the trifurcation of the nerve is situated (Fig. 1a).
9. Lift the skin of the hind leg with the standard pattern forceps.
10. Make a small incision with the Metzenbaum scissors perpendicular to the imaginary line 1 cm distally from where the trifurcation is supposed to be.

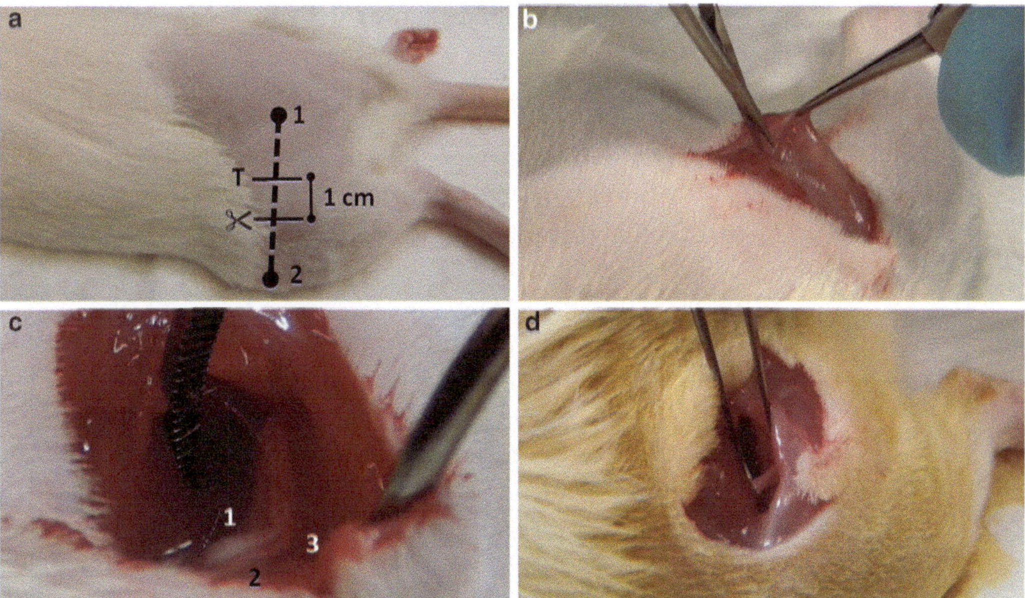

Fig. 1 Surgery for induction of the spared nerve injury. (**a**) Superficial landmarks for orientation: (1) Crest of ilium, (2) Patella, (T) Site of trifurcation, (✂): Location of first incision. (**b**) Making the incision in between the two heads of the biceps femoris muscle with a micro scissor to enter the site of the location where the trifurcation is being situated. (**c**) Sciating nerve and trifurcation: (1) Common peroneal nerve, (2) Tibial nerve, (3) Sural nerve. (**d**) Ligation and transection of the common peroneal nerve. Lifting the nerve produces a bridge that allows safe transection of the nerve

11. Insert the Metzenbaum scissors horizontally and closed into the small incision between the skin and the muscle layer and detach the skin from the underlying tissue by opening the scissors and carefully withdrawing it. Repeat this procedure until the skin is sufficiently detached.
12. Make an incision to proximal with a total length of 3–4 cm following the femoral bone.
13. Retract the skin to expose the underlying muscles.
14. Locate the margins of the two heads of the biceps femoris muscle, which is characterized by a white line of adjoining fascia.
15. Carefully lift the medial part of the muscle with the Bonn micro forceps to create a small indentation (Fig. 1b).
16. Carefully cut the fascia with the Vannas spring scissors to detach the muscles. This allows the exposure of the space where the nerves and vessels are situated.
17. Expose the sciatic nerve and its trifurcation carefully by blunt preparation with the standard pattern curved forceps. Insert the forceps in a closed manner and allow it to open in order to make space (see Note 2). Be careful not to touch or stretch the

sciatic nerve, its branches, or the vessels that are situated in that area.

18. Identify tibial, common peroneal and caudal cutaneous sural nerve (Fig. 1c). The tibial and common peroneal will be the nerves that are going to be transected. The cutaneous sural nerve will be spared.
19. Carefully free tibial and common peroneal nerve from their surroundings with a cotton swab.
20. Place the curved Moria iris forceps under the tibial nerve and use it to guide a 5-0 suture to pass under the nerve. Ligate the nerve at approximately 1 cm distal from the trifurcation.
21. Repeat the previous step for the common peroneal nerve and ligate with 5-0 suture at approximately 1 cm distal from the trifurcation.
22. Lift the tibial nerve with the curved Moria forceps closed and allow the forceps to open to have the nerve form a bridge of about 4 mm between the two legs of the forceps.
23. Cut the nerve approximately 4 mm from the ligature (see Note 3).
24. Lift the tibial nerve and cut away approximately 3 mm of nerve distal from the suture.
25. Lift the common peroneal nerve with the curved Moria forceps closed and allow the forceps to open to have the nerve form a bridge of about 4 mm between the two legs of the forceps.
26. Cut the nerve approximately 4 mm from the ligature (see Note 4).
27. Lift the common peroneal nerve and cut away approximately 3 mm of nerve distal from the suture.
28. Carefully displace the proximal nerve stumps with a cotton swab.
29. Restore muscle integrity and suture the fascia with 5-0 silk suture
30. Close skin with four 4-0 nylon sutures.
31. Allow animal to awake and monitor for 30–60 min under a heating source maintained at 38°C.
32. Transfer the animal to a cage with fresh sawdust, food and water available. Animals can be housed 2 per cage.

3.2 Assessment of Pain: Tactile Allodynia

1. Place the animal in the Perspex cage on the grid and allow to acclimatize for 10–20 min.
2. Stimulate the hind paw with the Semmes–Weinstein monofilaments just lateral from the midline. Maintain the filament perpendicular to the paw. Start with the filament that applies the lowest amount of force (1,65). Apply at a rate of 1 Hz to a total of ten stimuli.

Table 1
Scoring table for the assessment of cold allodynia

Response	Score
No response	0
Startle response lasting less than 1 s	1
Clear withdrawal lasting between 1 and 5 s	2
Clear withdrawal lasting between 5 and 30 s (with or without licking)	3
Clear withdrawal lasting over 30 s (with or without licking and repeated shaking)	4

3. When a response is observed in the form of an acute withdrawal upon stimulation at any point during stimulation, this is noted and the paw will no further be stimulated with the same filament or with a filament of a higher force.
4. When no response is observed, continue with the next filament. Continue increasing the filament and repeat until the animal responds. This response is noted.

Repeat the entire testing sequence to obtain results in duplex.

3.3 Assessment of Pain: Cold Allodynia

1. Spray 20 μl of acetone in one fluent application on the plantar surface by using the 1 ml syringe with bent needle.
2. Observe the response of the animal and score according to the scoring table (Table 1).
3. After 2 min rest, repeat the sequence to obtain results in duplex.

3.4 Treatment with ARA290

1. Make stock solution of ARA290 of 1 mg/ml in PBS and store at 4°C.
2. Administer 30 μg/kg ARA290 or vehicle (PBS) in a total volume of 200 μl intra peritoneally (i.p.) with a 1 ml syringe mounted with a 25 G needle (see Note 4).

3.5 Typical Results

Sixteen animals were given the spared nerve injury as previously described and were randomly allocated to a treatment group. Eight animals received a sham operation. In short, animals were anesthetized with sevoflurane (6% induction, 3% maintenance) and the trifurcation of the nerve was exposed. No ligation and transection was performed and the wound was closed in two layers. Twenty-four hours post injury animals received treatment with ARA290 or vehicle 5 times at 2 day intervals followed by once a week maintenance therapy. Within the first 2 weeks following nerve injury, vehicle-treated animals showed rapid development of tactile allodynia to the lowest applicable force of 0.004 g. In contrast, i.p. injections with ARA290 produced long-term relief of tactile allodynia lasting

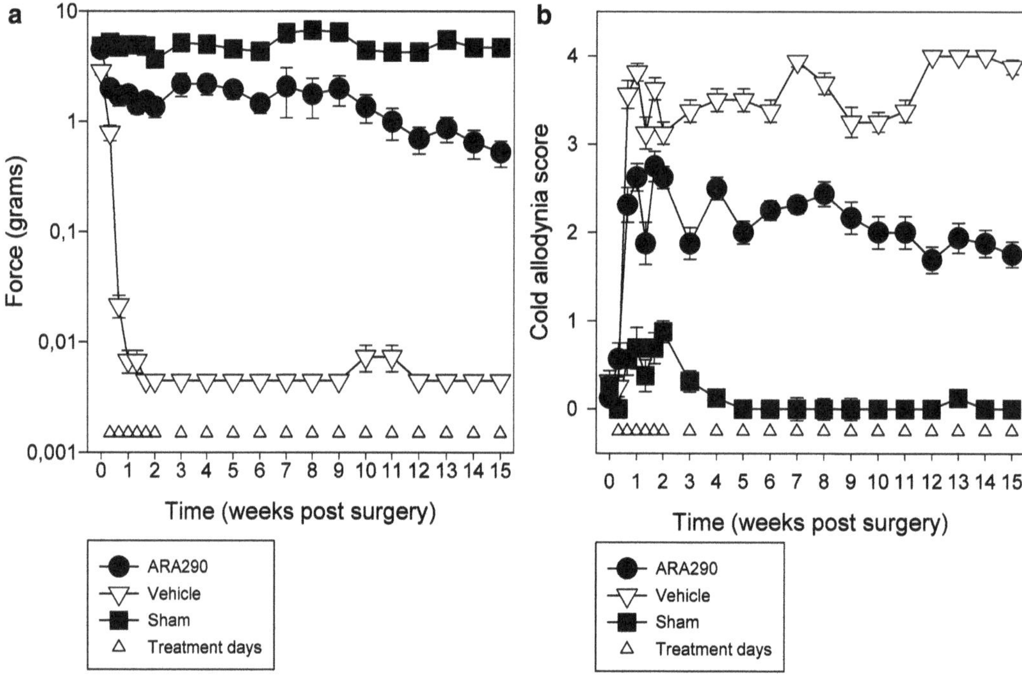

Fig. 2 Effect of ARA290 on the development of neuropathic pain. Treatment with ARA290 results in the prevention of developing: (**a**) tactile allodynia and (**b**) allodynia for cold for a period of 15 weeks

at least 15 weeks (Fig. 2a). The allodynic responses differed significantly between treatment groups (repeated measures ANOVA, post hoc Holm-Sidak: $P < 0.001$ vs. vehicle-treated animals).

Similarly, cold allodynia developed in animals treated with vehicle following nerve lesion with mean scores between 3 and 4 (4 being the maximum score) during the 15 week study period (Fig. 2b). Treatment with ARA290 was associated with significantly less cold allodynia with mean scores between 1.8 and 2.9 (Kruskal-Wallis, post hoc Tukey test: $P < 0.001$ vs. vehicle-treated animals).

4 Notes

1. Anesthesia is induced and maintained with vaporized anesthetic agents (i.e., sevoflurane, isoflurane) rather than ketamine, for ketamine and other NMDA receptor antagonists, the class of drugs ketamine belongs to, have shown to reduce neuropathic pain in both humans and animals (16, 17).
2. Literature describes this procedure to be done by making an incision through the muscle (18). This induces collateral damage and may cause blood loss. The method described in this chapter has been developed to perform the procedure without any to minimal blood loss.

3. The most important thing to remember while performing the surgery is to maintain a visual on every action in order not to cause additional damage.
4. Treatment is being given after the behavioral tests to minimize influence from stress on behavioral tests due to handling the animals during i.p. administration.

References

1. Baron R, Binder A, Wasner G (2010) Neuropathic pain. Diagnosis, pathophysiological mechanisms, and treatment. Lancet Neurol 9:807–819
2. Breivik H, Collett B, Ventafridda V, Cohen R, Gallacher D (2006) Survey of chronic pain in Europe: Prevalence, impact on daily life, and treatment. Eur J Pain 10:287–333
3. Kuner R (2010) Central mechanisms of pathological pain. Nat Med 16:1258–1266
4. Tanga FY, Raghavendra V, DeLeo JA (2007) Quantitative real-time RT-PCR assessment of spinal microglial and astrocytic activation markers in a rat model of neuropathic pain. Neurochem Int 45:397–407
5. Brines M, Ghezzi P, Keenan S, Agnello D, De Lanorelle N, Cerami A (2000) Erythropoietin crosses the blood-brain barrier to protect against experimental brain injury. Proc Natl Acad Sci U S A 97:10531–10536
6. Liao ZB, Zhi XG, Shi QH, He ZH (2008) Recombinant human erythropoietin administration protects cortical neurons from traumatic brain injury in rats. Eur J Neurol 15:140–149
7. Campana MW, Myers RR (2003) Exogenous erythropoietin protects against dorsal root ganglion apoptosis and pain following peripheral nerve injury. Eur J Neurosci 18:1497–1506
8. Sekiguchi Y, Kikuchi S, Myers RR, Campana WM (2003) Erythropoietin inhibits spinal neuronal apoptosis and pain following nerve root crush. Spine 28:2577–2584
9. Keswani S, Buldanlioglu U, Fischer A, Reed N, Polley M, Liang H, Zhou C, Jack C, Leitz G, Hoke A (2004) A novel endogenous erythropoietin mediated pathway prevents axonal degeneration. Ann Neurol 56:815–826
10. Campana WM, Li X, Shubayev VI, Angert M, Cai K, Myers RR (2006) Erythropoietin reduces Schwann cell TNF-a, Wallerian degeneration and pain-related behaviors after peripheral nerve injury. Eur J Neurosci 23:617–626
11. Jia H, Feng X, Li W, Hu Y, Zeng Q, Liu J, Xu J (2009) Recombinant human erythropoietin attenuates spinal neuroimmune activation of neuropathic pain in rats. Ann Clin Lab Sci 39:84–91
12. Jia H, Jin Y, Ji Q, Hu Y, Zhou Z, Xu J, Yang J (2009) Effects of recombinant human erythropoietin on neuropathic pain and cerebral expressions of cytokines and nuclear factor-kappa B. Can J Aneasth 56:597–603
13. Sasaki N, Sekiguchi M, Kikuchi S, Konno S (2011) Effects of asialo-erythropoietin on pain-related behavior and expression of phosphorylated-P38 MAP kinase and tumor necrosis factor-alpha induced by application of autologous nucleus pulposus on nerve root in rat. Spine 36:E86–E94
14. Brines M, Patel NSA, Villa P, Brines C, Mennini T, De Paola M, Erbayraktar Z, Erbayraktar S, Sepodes B, Thiemermann C, Ghezzi P, Yamin M, Hand CC, Xie QW, Coleman T, Cerami A (2008) Nonerythropoietic, tissue-protective peptides derived from the tertiary structure of erythropoietin. Proc Natl Acad Sci U S A 105:10925–10930
15. Swartjes M, Morariu A, Niesters M, Brines M, Cerami A, Aarts L, Dahan A (2011) ARA290, a peptide derived from the tertiary structure of erythropoietin, produces long-term relief of neuropathic pain. An experimental study in rats and β-common receptor knockout mice. Anesthesiology 115:1084–1092
16. Sigtermans M, van Hilten JJ, Bauer M, Arbous M, Marinus J, Sarton E, Dahan A (2009) Ketamine produces effective and long-term pain relief in patients with complex regional pain syndrome type 1. Pain 145:304–311
17. Swartjes M, Morariu A, Niesters M, Aarts L, Dahan A (2011) Nonselective and NR2B-selective N-methyl-d-aspartic acid receptor antagonists produce antinociception and long-term relief of allodynia in acute and neuropathic pain. Anesthesiology 115:165–174
18. Decosterd I, Woolf CJ (2000) Spared nerve injury: an animal model of persistent peripheral neuropathic pain. Pain 87:149–158

Chapter 13

Intra-epidermal Nerve Fibers Density and Nociception in EPO-Treated Type 1 Diabetic Rats with Peripheral Neuropathy

Bianchi Roberto, Lombardi Raffaella, Porretta-Serapiglia Carla, and Lauria Giuseppe

Abstract

Small-diameter nerve fibers, which subserve nociception, can be affected early in peripheral neuropathies, although their injury may not be detectable by routine neurophysiologic tests. On the other hand, skin biopsy has proved to be a reliable tool to examine nonmyelinated nerve fibers, as assessed by the quantification of intra-epidermal nerve fiber (IENF) density not only along with the degenerative process but, noteworthy, IENF density could be very helpful in evaluating drug efficacy such as erythropoietin (EPO) treatment.

Key words Diabetes, Erythropoietin, Diabetic neuropathy, Epidermal nerve fibers, PGP 9.5, Skin

1 Introduction

The global prevalence of type 2 diabetes appears to be increasing dramatically, possibly as a consequence of a more sedentary lifestyle and the adoption of Western diets. The World Health Organization estimates the number of people with diabetes to be approximately 180 million and this figure is projected to more than double by the year 2030. Furthermore, there is convincing evidence that the risk of developing long-term complications, such as retinopathy, nephropathy, and neuropathy, is related to the degree of hyperglycemia (1). Strict glycemic control by intensive insulin administration, pancreas or islets transplantation, can significantly reduce, but nor abolish, the incidence of diabetic complications such as neuropathy, nephropathy and retinopathy, which are the major cause of morbidity and mortality in type 1 insulin-dependent diabetic patients (2).

1.1 Diabetic Peripheral Neuropathy

Diabetic Peripheral Neuropathy (DPN) that occurs in about 34% of patients who have been hyperglycemic for more than 15 years, is the most common complication of diabetes mellitus, and

Pietro Ghezzi and Anthony Cerami (eds.), *Tissue-Protective Cytokines: Methods and Protocols*, Methods in Molecular Biology, vol. 982, DOI 10.1007/978-1-62703-308-4_13,

contributes the greatest morbidity and mortality and severely impairs the quality of life, (3, 4). Early disorders of nerve function include slowing in nerve conduction velocity (NCV) followed by axonal degeneration, paranodal demyelination, and loss of myelinated fibers (5). DPN are painful in more than 15% of patients, a substantial proportion of who complain of chronic pain and poor response to conventional analgesics (6, 7). Patients often develop a compromised ability to perceive tactile sensation, exaggerated sensitivity to nociceptive stimuli (hyperalgesia or hypoalgesia) or may perceive normal stimuli as painful (allodynia) (8–11).

1.2 Small-Fiber Neuropathy

Small-diameter nerve fibers innervating the skin subserve thermal sensation and nociception. Skin nerve fibers firing can be triggered by epidermal cells which may have a role in the pathogenesis of neuropathic pain (12). Small-diameter nerve fibers can be affected early in peripheral neuropathies, although their injury may not be detectable by routine neurophysiologic tests (13). On the contrary, skin biopsy has proved to be a reliable tool to examine unmyelinated nerve fibers, as assessed by the quantification of intra-epidermal nerve fiber (IENF) density and to estimate their degeneration over time in human neuropathies (14). IENFs are unmyelinated axons with exclusively somatic function, as demonstrated by their degeneration after axotomy or gangliotomy and their sparing after rhizotomy or sympathectomy (15). Their density can be quantified in experimental models. Studies have focused on cutaneous innervation in mouse, but data are scanty on control and disease bearing rats (16–18).

1.3 Streptozotocin-Diabetic Rat Model

The chemically induced diabetes such as the STZ model, however, has several advantages (19). It shares a number of features with human diabetic neuropathy at the functional and biochemical levels (20). Decreased NCV, together with reductions in Na^+, K^+-ATPase activity, is the hallmark of diabetic neuropathy, but these rats present various types of early neurological dysfunction (21), including altered pain sensation suggesting early involvement of small nociceptive sensory neurons (14, 22).

1.4 Erythropoietin and Neuroprotection

EPO is a cytokine originally used for its effect on erythropoiesis, since it supports the survival, proliferation and differentiation of erythroid progenitor cells. However, in the past few years it has become clear that EPO is a multifunctional trophic factor, with potent neurotrophic activity on a variety of neural cells in the central and peripheral nervous systems (23–25). EPO confers protection by preventing the development of or treating peripheral neuropathies associated with HIV (26), cisplatin (27, 28), and taxane (29).

EPO acts by binding with its receptors (EPOR), which belong to the cytokine receptor type I superfamily (30, 31). EPO and EPOR are expressed in cerebral and spinal cord neurons, dorsal root ganglia, nerve axons and Schwann cells (32). They are over-expressed after nerve injury, and there appears to be an endogenous neuroprotective pathway, this sets the basis for therapeutic use of exogenous EPO (26, 33). Exogenous EPO can be beneficial when locally or virus-mediated delivered by promoting the regeneration of peripheral nervous system axons and preventing the progression of diabetic neuropathy (34, 35). In vivo experimental models have shown that EPO and its carbamylated non-hematopoietic derivative CEPO can both prevent and treat diabetic peripheral neuropathy in rats and mice (21).

The procedure to investigate the innervation of the skin by bright-field immunohistochemistry in rats with STZ-induced DPN is reported. The technique of IENF density count is described with the aim of evaluate the number of IENF that clearly cross the epidermal–dermal junction (extending off dermal fibers) and are located in the epidermis and the length (in mm) of the epidermis in order to determine the mean number of IENF/mm/skin biopsy (length density).

2 Materials

2.1 Solutions

Prepare all solutions using ultrapure water (prepared by purifying deionized water to attain a sensitivity of 18 MΩ cm at 25°C) and analytical grade reagents.

1. 0.4 M Sorenson's phosphate buffer (PB; use at 0.1 M): Dissolve 7.176 g sodium phosphate monobasic monohydrate in 100 mL water. Dissolve 49.4 g sodium phosphate dibasic anhydrate in 750 mL water and mix solutions the two solutions in a 1-L graduated cylinder adding water up to 900 mL. Adjust the pH to 7.6; with base or acid. Make up to 1 L with water. Store at 4°C up to 1 month.
2. 0.1 M Phosphate saline buffer (PBS). Dissolve 2.759 g sodium phosphate monobasic monohydrate in 200 ml water (MONO). Dissolve 8.898 g sodium phosphate dibasic dihydrate in 500 mL water (DIBA). Dissolve 4.5 g NaCl in 100 mL DIBA; add to this solution DIBA and MONO to obtain 500 mL PBS 0.1 M pH 7.4 (see Note 1). Store at 4°C up to 1 month.
3. 8% paraformaldehyde stock (see Note 2): Add 40 g of "Prill Grade" Paraformaldehyde to 500 mL PB and stir continuously heating up to 55°C (see Note 3) on stir plate until diluted. Do not allow powder to settle on bottom of beaker. After 2–3 h the solution will be transparent. Stir until solution is clear. A small amount of flocculent material will remain. Vacuum filter (paper filter) through a 3 mm Whatman filter in a Buchner funnel on a filter flask. Store at 4°C up to 1 month.

4. 0.1 M Lysine stock solution for Paraformaldehyde-lysine-perio date (PLP) mix. Dissolve 1.64 g of L-Lysine in 30 mL water, in a 150 mL graduated cylinder. Adjust pH to 7.4 with 0.1 M sodium phosphate dibasic dihydrate (DIBA). Add water up to 45 ml. Add 0.1 M PB up to 90 ml. Store at 4°C for up to 3 weeks.
5. 2% PLP fix for skin biopsies (see Note 4). To 30 mL of lysine stock solution add 0.085 g of sodium periodate powder. Shake until powder dissolves (5–10 min). Add 10 ml of 8% Paraformaldehyde stock solution. Keep at 4°C.
6. Cryoprotective solution. Mix 20% glycerol, 20% PB 0.1 M, and 60% water. Store at 4°C up to 1 month.
7. Glycerol-based antifreeze solution. Mix 300 mL glycerol, 300 mL ethylene glycol, 200 mL PBS. Make up to 1 L with water. Store at 4°C up to 1 month.
8. 0.25% potassium permanganate solution. Dissolve 25 mg of potassium permanganate in 10 mL water. A slightly purple colored solution is obtained.
9. Oxalic acid (OX) solution: Add 0.45 mg of oxalic acid to 9 ml water and stir for 2 min.
10. Tris Buffered Saline (TBS). Dissolve 12.11 g Tris base and 9.0 g sodium chloride in about 900 mL of water, stir and adjust the pH to 7.4 with HCl. Make up to 1 L with water. Store at 4°C up to 1 month.
11. Block solution: 4 mL normal goat serum and 100 μL Triton X-100; make up to 100 mL with TBS.
12. Quenching solution: 30 mL of methanol and 1 mL of hydrogen peroxide; make up to 100 mL with PBS.
13. 0.09 M sodium citrate Buffer: 0.53 g of sodium citrate and .0.38 g of citric acid in about 15 mL of water, stir and eventually adjust the pH to 4.2 with NaOH. Make up to 20 mL. Store at 4°C up to 1 month.
14. 10 mM Tris-sucrose buffer solution (see Note 5): 0.303 g of Tris base, 21.4 g sucrose, and 0.12 g of EGTA adjust to pH 7 with HCl and make up to 250 mL with water.
15. Buffer A: 15.86 g Tris base, 0.399 g EGT, 7.4 g NaCl, 0.976 g KCl, 0.341 g sodium azide in 1 L water. pH to 7.4 at 37°C and adjust to final volume. Store at 4°C up to 1 month.
16. Solution A1: Dissolve 27 mg of phosphoenolpyruvate and 18.44 mg nicotinamide adenine dinucleotide (NADH) in 50 mL.
17. Ouabain solution: Dissolve 54.66 mg ouabain in 25 mL with solution A1.

18. ATP solution: in 20 mL tube, to 10 mL water add 0.4 g $MgCl_2$ and 1.1 g adenosine 5′-triphosphate (ATP).
19. Xylazine–ketamine solution. Solution A: Dissolve 1 g of Xylar in 10 mL of saline (0.9% NaCl). Solution B: dissolve 1 g of ketamine in 10 mL of saline as above. Take 1.0 mL of solution A and 1.5 mL of solution B and adjust to 10 mL with saline. Administer 10 mL/kg body weight to animals.

2.2 Animals

The experimental examples shown were done according to national legislation. Statement of Compliance (Assurance) with Standards for Humane Care and Use of Laboratory Animals has been recently reviewed (10/28/2008) and will expire on 10/31/2013 and has been accepted (id. #A5023-01) by the NIH-Office for Protection from Research Risks.

Animals are housed in groups of two rats (male Sprague Dawley, 180–200 g). Animal room temperature and relative humidity are set at 22 ± 2°C and 55 ± 10%. Artificial lighting provides a 12 h light–12 h dark cycle (7 a.m. to 7 p.m.). The animals have free access to diet and water.

3 Methods

The methods described below outline (1) the use of animals, (2) the induction of diabetes, (3) the experimental design and drug treatments, (4) the characterization of the model in terms of basic changes and EPO effects, (5) and skin biopsy and immunohistochemistry technologies and determination of IENF density.

1. Diabetes is induced in rats fasted overnight by a single intraperitoneal (i.p.) injection of 60 mg/kg of STZ dissolved in sodium citrate buffer (pH 4.5). The other rats are injected with vehicle and serve as nondiabetic controls. Hyperglycemia is confirmed by measuring glycosuria 72 h after STZ injection, using commercially-available strips. Only animals with glycosuria >5% are classified as diabetic and included in the study.
2. Two studies were performed with rats in which were tested the efficacy of EPO in protecting (preventive schedule) or reversing (therapeutic schedule) nerve dysfunction in STZ-induced diabetes and their correlation with IENF density. The flowchart of the studies is reported in Fig. 1. Study 1 investigated the efficacy of EPO in protecting the development of DPN in the STZ-induced diabetes rat model. Study 2 examined the in vivo efficacy of EPO chronically administered in treating established DPN in STZ-diabetic rats. For study 1, diabetic and control rats were treated with rhEPO (Dragon Pharmaceuticals, Vancouver) 40 μg/kg of b.w., i.p., three times per week for 5 weeks starting on week 2.

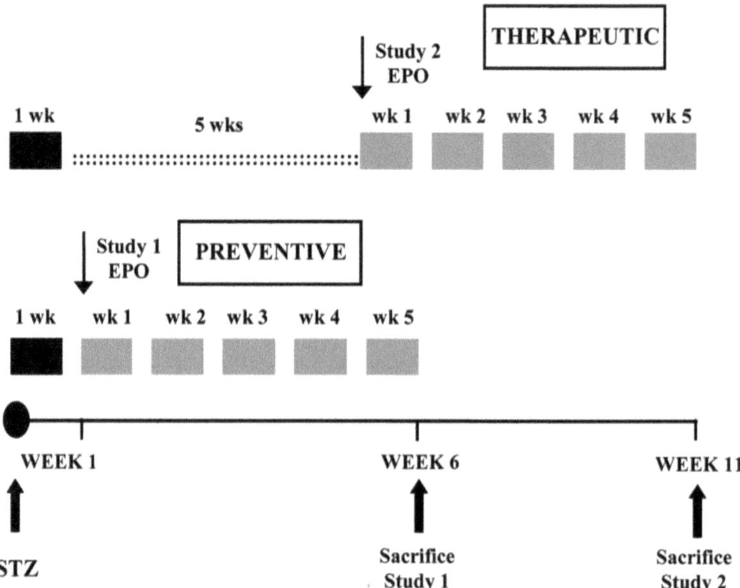

Fig. 1 Flow chart of the studies. Two studies were performed with rats in which were tested the efficacy of EPO in protecting (protective schedule) or reversing (therapeutic schedule) nerve dysfunction in STZ-induced diabetes and their correlation with IENF density. For study 1 (protection), diabetic and control rats were treated with rhEPO 40 μg/kg of b.w., i.p., three times per week for 5 weeks starting on week 2 following STZ. For therapeutic schedule (study 2) the same EPO protocol was applied, starting 6 weeks after diabetes induction and lasting an additional 5 weeks

In the therapeutic schedule, treatment with rhEPO was started 6 weeks after diabetes induction and lasted an additional 5 weeks with the same EPO treatment protocol.

3. In addition to general observations, growth and glycemia, specific measurements were performed. Among them we evaluate nociceptive thresholds, as markers of painful neuropathy (which correlate with small-fiber neuropathy), NCV and Na^+/K^+-ATPase-activity, largely considered the hallmarks of diabetic neuropathy. We have previously shown that in normal rats overall quantification of IENF density correlate with changes in NCV (see Note 6).

3.1 Nociceptive Thresholds

The thermal nociceptive threshold response utilizes a polysynaptic pathway involving higher centers, whereas the mechanical nociceptive threshold is a monosynaptic response. Rats were accustomed to the devices 3 days before performing the tests.

The nociceptive threshold to radiant heat was quantified using the hot plate paw withdrawal test, as previously described (29). Briefly, a 40 cm high Plexiglas cylinder was suspended over the hot plate (Ugo Basile, Comerio, Italy), and the temperature was maintained at

50°C to give a latency of about 10 s for control rats. Withdrawal latency was defined as the time between placing the rat on the hot plate and the time of withdrawal, or licking the hind paw. Each animal was tested twice, separated by a 30-min rest interval.

The mechanical nociceptive threshold was quantified by using the Randal–Selitto paw withdrawal test with an analgesy meter (Ugo Basile, Comerio, Italy), which generates a linearly increasing mechanical force. The paw withdrawal reflex is automatically recorded using the latency until withdrawal, in seconds, and the force at which paw was withdrawn, in grams.

In agreement with others, in these experiments, diabetes was associated with an increase in the threshold of thermal withdrawal and a decrease in threshold for mechanical stimulation (Fig. 2). As expected, after STZ treatment hind paw thermal withdrawal latency significantly changed from week 2 until week 6, being 1.8 times higher in STZ untreated group as compared with the control group (Fig. 2a). EPO treatment progressively and significantly did prevent the increases in latency. The mechanical paw force withdrawal threshold in diabetic rats was lower at all times than for controls (by about 30%) and EPO significantly increase: the mechanical threshold in diabetic rats (Fig. 2b). EPO did not affect the response latencies in nondiabetic rats at any time (Fig. 2).

Figure 3a shows the hind paw thermal withdrawal threshold measured over the 5 weeks after beginning EPO treatment according to the Therapeutic schedule. Five weeks after STZ injection the thermal response latency was significantly changed, with hypoalgesia in the diabetic untreated group. EPO did not affect the latencies in control treated rats at any time. Noteworthy, EPO was able to significantly ameliorate thermal nociceptive threshold from week 9 until week 11.

Figure 3b presents the hind paw force withdrawal thresholds according to the Therapeutic protocol. Diabetic rats showed a decrease in the mechanical thresholds at all times (by 30–45%). Starting from a comparable situation, EPO only partially restore the diabetic mechanical hyperalgesia.

3.2 Nerve Conduction Velocity

Antidromic tail NCV was measured for each animal by a method already used in neuroprotection experiments (36). Briefly, the antidromic NCV in the tail nerve was assessed by placing recording ring electrodes distally in the tail, and stimulating electrodes (stimulus duration 100 ms, filter 1 Hz to 5 MHz) were placed 5 and 10 cm proximally from the recording point. The latencies of the potentials recorded at the two sites after nerve stimulation were determined (peak-to-peak) and NCV was calculated. All the neurophysiological determinations were done under standard conditions in room and animal controlled temperature.

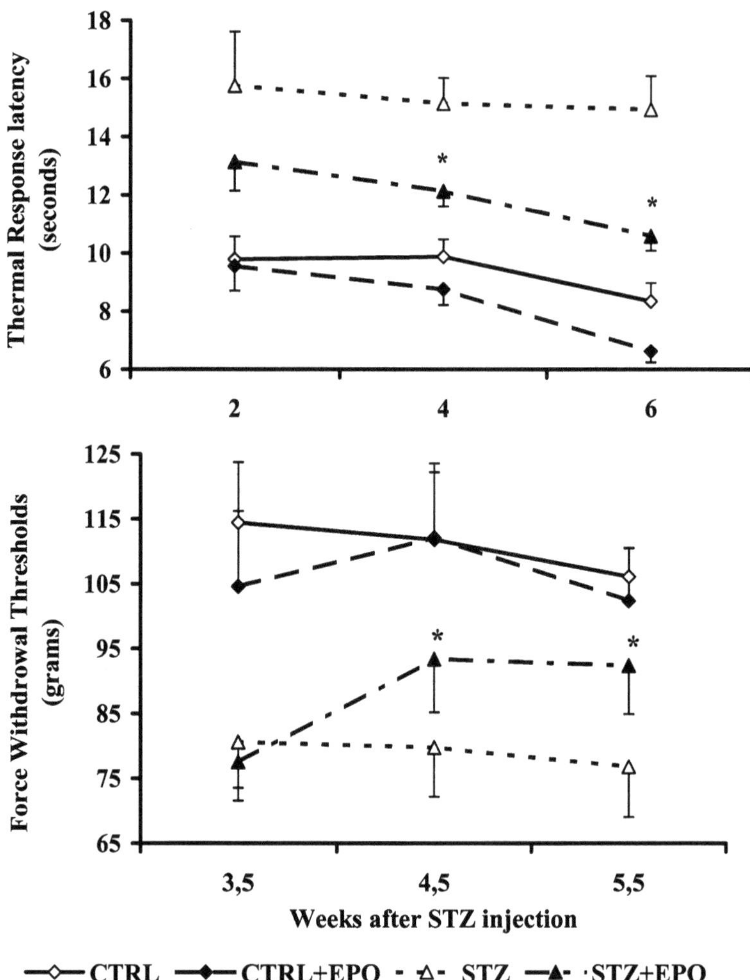

Fig. 2 EPO prevent changes in thermal and mechanical thresholds in diabetic rats. Control or STZ-diabetic rats were treated according to the preventive schedule (study 1). (**a**) Thermal sensitivity threshold is expressed as thermal response latency in seconds. (**b**) Mechanical threshold is expressed as force withdrawal latency in grams. CTRL: untreated control rats; CTRL+EPO: Healthy rats treated with EPO; STZ: untreated diabetic rats; STZ+EPO: diabetic rats treated with EPO. Data are mean ± SEM. **P* < 0.05 vs. STZ

Five weeks after STZ injection the reduction in NCV in the diabetic group (−21%) of CTRL group was partially but significantly counteracted (by 14%) by EPO in rats treated according to the protection schedule (Fig. 4). Immediately before the beginning of treatment, at randomization, NCV was significantly reduced ($P < 0.001$) in STZ compared to control groups. EPO according to the therapeutic schedule counteracted this decrease in the diabetic groups (from −42% of the diabetic rats to −23% of EPO-treated rats), while it did not have any effect in control rats (Fig. 4).

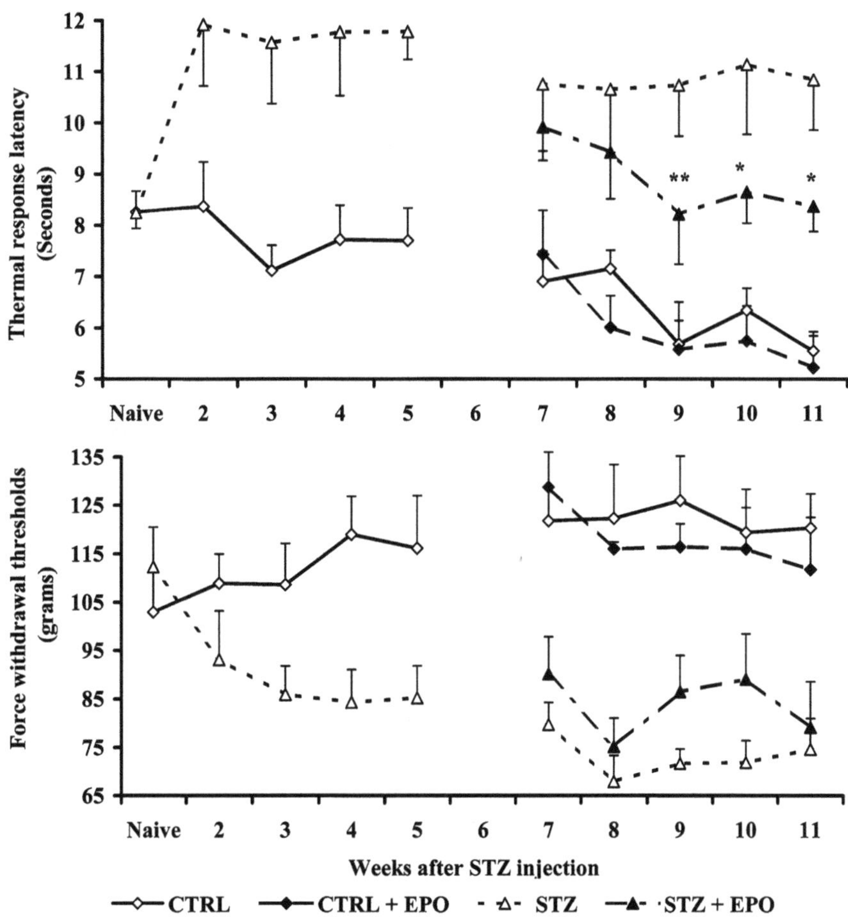

Fig. 3 EPO restores changes in thermal and mechanical thresholds in diabetic rats. Control or STZ-diabetic rats were treated according to the therapeutic schedule (study 2). (**a**)Thermal sensitivity threshold is expressed as thermal response latency in seconds. (**b**) Mechanical threshold is expressed as force withdrawal latency in grams. Groups as in Fig. 2. Data are mean ± SEM. $^{**}P<0.01$ vs. STZ, $^{*}P<0.05$ vs. STZ

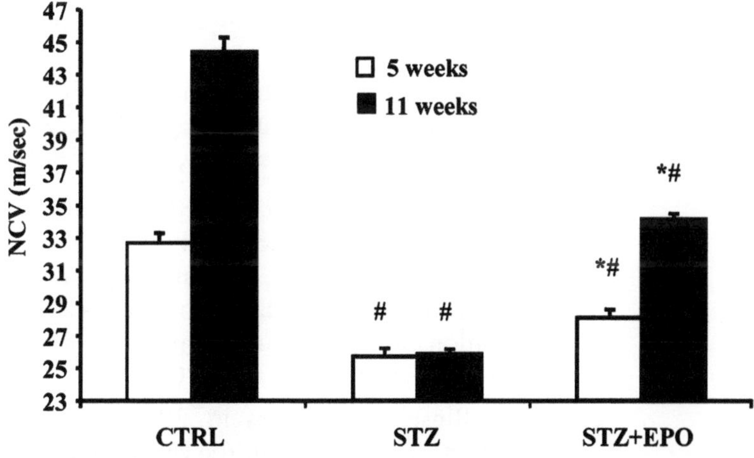

Fig. 4 Tail nerve conduction velocity in rats at 5 and 11 weeks, after EPO according to the protection and the therapeutic protocols. Groups as in Fig. 2. Data are mean ± SEM. $^{*}P<0.01$ vs. STZ, # $P<0.001$ vs. CTRL

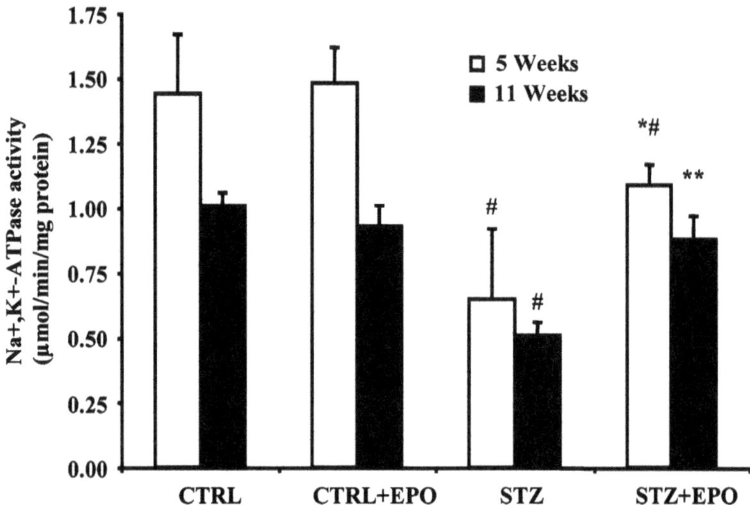

Fig. 5 EPO prevents and restores changes in Na^+, K^+-ATPase activity in diabetic rats. Groups names as in Fig. 2. Data are mean ± SEM. *$P < 0.05$ vs. STZ, # $P < 0.01$ vs. CTRL and CTRL + EPO

3.3 Na^+, K^+-ATPase Activity

For Na^+, K^+-ATPase activity tibial stumps (from the two sciatic nerves) were de-sheathed at death and homogenized in chilled solution of 10 mL Tris-sucrose buffer at 1:20 (w/v) in a glass-glass Potter homogenizer and stored at −80°C for ATPase determinations. Na^+, K^+-ATPase activity was determined spectrophotometrically as previously described (37).

Diabetes greatly reduced in sciatic nerve Na^+, K^+-ATPase activity by about 60%. The Na^+, K^+-ATPase activity in diabetic rats treated with EPO prevention schedule for 5 weeks was only 22% different from control rats (Fig. 5). The change in diabetic rats treated with EPO according to the therapeutic schedules for 5 weeks was only 16% reduced in respect to control rats.

3.4 Skin Biopsy and IENF Quantification

This protocol starts at indicated times, namely 6 and 11 weeks following diabetes induction in rats (see Note 7) for preventive and therapeutic schedule, respectively, and lasts 5 days.

1. Day 1. Hind paws were collected at sacrifice. Animals were anesthetized using xylazine–ketamine and sacrificed by means of exsanguination. After separating plantar glabrous skin underlines metatarsal bone, punch biopsies (3 mm) under sterile conditions were taken and immediately fixed by immersion in 1 mL of 2% PLP for 24 h at 4°C.
2. Day 2. Rinse the biopsies in PB twice for 10 min each (2 × 10 min). Put biopsies in the cryoprotective solution overnight at 4°C.

3. Day 3. Twenty-micron-thick sections perpendicular to the epidermis were cut serially with a cryostat, sequentially labeled and stored in antifreez solution at –20°C.
4. Day 4. Three sections from a single footpad were randomly chosen, rinsed in 0.1 M TBS at room temperature on shaker table, placed in 0.25% potassium permanganate for 15 min at room temperature on shaker table, rinsed in TBS at room temperature on shaker table, placed in 5% oxalic acid for 2 min, rinsed twice in TBS, placed in block solution (4% normal goat serum, 0.1% Triton X-100, in TBS) for 1 h on shaker table at room temperature and then incubated overnight with rabbit polyclonal anti-protein gene product 9.5 (PGP 9.5, Biogenesis Ltd.) antibody AbD serotec (1:1,000 diluted in 2% normal goat serum, 0.05% Triton X-100, TBS), using modified free-floating protocol (38).
5. Day 5: After rinsing in TBS at room temperature, sections were incubated with biotinylated goat anti-rabbit IgG for 1 h (1:100 diluted in TBS; Vector), quenched in 30% methanol/1% hydrogen peroxide (in PBS) for 30 min, rinced in PBS and then placed in avidin-biotin complex (Vector) for 1 h. The reaction product was demonstrated by the blue chromogen/peroxidase substrate (Vector SG substrate kit) according to manufacturer indications.

Two expert measurers, blinded to the healthy or neuropathic status of the rats, independently count the total number of PGP 9.5-positive IENF in each section under a light microscope (40×), with the assistance of a microscope-mounted video camera. Individual fibers were counted as they crossed the dermal–epidermal junction, while secondary branching within the epidermis was excluded from the quantification. The length of the epidermis was measured using a computerized system (Microscience Inc.), and the linear density of IENF (IENF/mm) was obtained. Difference between groups was assessed by two-tailed Student's *t* test. Pearson's test was used to assess the interobserver agreement in IENF counts (see Note 8).

IENFs were intensely immunostained by PGP 9.5 (see Note 9) as shown in a representative microphotograph of in the footpad of control (Fig. 6a), diabetic (Fig. 6b), and EPO-treated diabetic rats (Fig. 6c). It is clearly seen that diabetic rats have a marked decrease in cutaneous innervation density; this is completely reversed by EPO administration (Fig. 6).The density of footpad IENF was quantified in healthy and diabetic Sprague Dawley rats and in rats treated with EPO (Fig. 6). In diabetic rats, the IENF density is only slightly changed at 5 weeks of diabetes (data not shown), but it was significantly reduced in diabetic rats at week 11 at the end of therapeutic schedule: A quantification of IENF density, indicating a statistically significant protective effect of EPO, is shown in Fig. 6d.

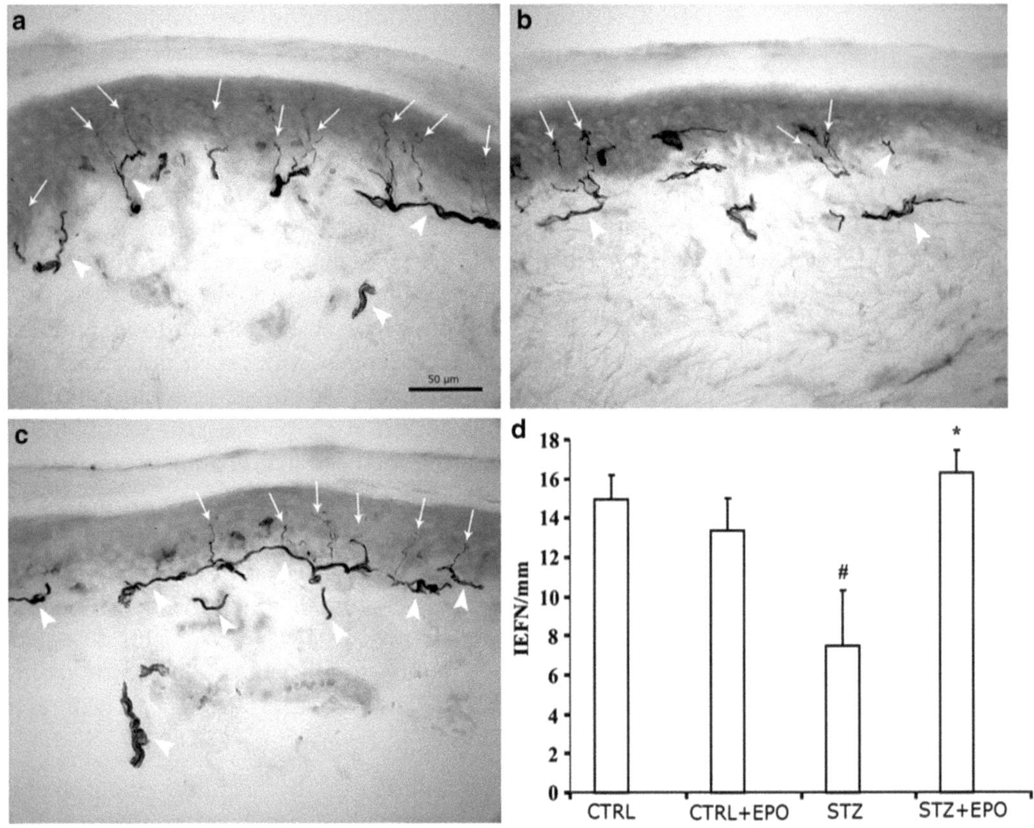

Fig. 6 EPO restores intra-epidermal nerve fiber (IENF) density in diabetic rats. At the end of therapeutic schedule STZ significantly reduces IENF density in diabetic group and EPO completely restore the skin innervation. (**a**) Microphotographs are PGP 9.5 immunostaining in 20 μm-thick sections (bar = 50 μm). (**b**) Quantification of IENF density. Data are expressed as the number of linear density of IENF and are mean ± SEM. Groups names as in Fig. 2. ***P*< 0.01 vs. STZ, **P*< 0.05 vs. STZ, #*P*< 0.01 vs. CTRL and CTRL + EPO

4 Notes

1. pH of 7.4 extremely important. pH too low: add DIBA—pH too high: add MONO.
2. Neurotoxic: prepare solutions under hood, behind protective glass.
3. Do not allow temperature to exceed 60°C or you will make formic acid.
4. To make PLP 2% fix for skin biopsies: make fresh on the same day as biopsy.
5. This solution has to be done fresh for each time needed.
6. Although NCV and cutaneous fibers affect different fiber populations (i.e., predominately large-diameter vs. small-diameter myelinated fibers), the decrease in IENF density

showed a significant correlation with the decrease in tail NCV (18). The present findings are in keeping with previous studies after sciatic nerve injury in rats (12, 17) and with experimental neuropathy in mice (39). The linear correlation between neurophysiologic examination, which represents a primary outcome measure in experimental neuropathies, and IENF quantification strengthens the potential usefulness of skin biopsy as a tool in experimental trials.

7. We chose to examine cutaneous innervation of rat instead of mouse hind paw because of the availability of a well-characterized model of neuropathy in our laboratory in which (as above described) behavioral and neurophysiologic methods can be easily performed.

8. Interobserver agreement was significant for IENF counts ($r=0.818$, $p<0.01$). This study demonstrated that the protocol used to quantify the density of IENF, which is the same used in human clinical practice (40, 41), is reliable, as shown by the high interobserver agreement obtained by two blinded expert examiners in healthy and neuropathic rats.

9. An exact count of the number of nerve fibers should *not* be expected. The counts cannot provide more than an estimated range of the IENF density at a particular site. It is also important to note that some cases can be much more difficult to quantify than others, due to the sheer number of variables involved. There will be situations in which the determination is ambiguous or involves conflicting information. It is in these cases that the experience or practice of the measurer plays a critical role in making the final decision whether a stained profile is or is not an IENF. Through experience with an already trained measurer, the beginning measurer will be able to determine when to apply which rules and hints most effectively. More importantly, consistency and reliability will be established between the measurer's own style and that of the trained measurer. The measuring process requires skills and acquired experiences.

References

1. Palumbo PJ (2001) Gycemic control, mealtime glucose excursions, and diabetic complications in type 2 diabetes mellitus. Mayo Clin Proc 76:609–618
2. Meloche RM (2007) Transplantation for the treatment of type 1 diabetes. World J Gastroenterol 13:6347–6355
3. Vinik AI, Milicevic Z, Pittenger GL (1995) Beyond glycemia. Diabetes Care 18:1037–1041
4. Galer BS, Gianas A, Jensen MP (2000) Painful diabetic polyneuropathy: epidemiology, pain description, and quality of life. Diabetes Res Clin Pract 47:123–128
5. Vinik AI, Park TS, Stansberry KB, Pittenger GL (2000) Diabetic neuropathies. Diabetologia 43:957–973
6. Hoybergs YM, Meert TF (2007) The effect of low-dose insulin on mechanical sensitivity and allodynia in type I diabetes neuropathy. Neurosci Lett 417:149–154
7. McQuay HJ (2002) Neuropathic pain: evidence matters. Eur J Pain 6(Suppl A):11–18

8. Norrsell U, Eliasson B, Frizell M, Wallin BG, Wesslau C, Olausson H (2001) Tactile directional sensibility and diabetic neuropathy. Muscle Nerve 24:1496–1502
9. Kiguchi S, Imamura T, Ichikawa K, Kojima M (2004) Oxcarbazepine antinociception in animals with inflammatory pain or painful diabetic neuropathy. Clin Exp Pharmacol Physiol 31:57–64
10. Kapur D (2003) Neuropathic pain and diabetes. Diabetes Metab Res Rev 19(Suppl 1):S9–S15
11. Navarro X, Kennedy WR (1991) Evaluation of thermal and pain sensitivity in type I diabetic patients. J Neurol Neurosurg Psychiatry 54:60–64
12. Lindenlaub T, Sommer C (2002) Epidermal innervation density after partial sciatic nerve lesion and pain-related behavior in the rat. Acta Neuropathol 104:137–143
13. Lacomis D (2002) Small-fibres neuropathy. Muscle Nerve 26:173–188
14. Lauria G, Morbin M, Lombardi R, Borgna M, Mazzoleni G, Sghirlanzoni A, Pareyson D (2003) Axonal swellings predict the degeneration of epidermal nerve fibers in painful neuropathies. Neurology 61:631–636
15. Li Y, Hsieh ST, Chien HF, Zhang X, McArthur JC, Griffin JW (1997) Sensory and motor denervation influence epidermal thickness in rat foot glabrous skin. Exp Neurol 147:452–462
16. Stankovic N, Johansson O, Hildebrand C (1999) Increased occurrence of PGP 9.5-immunoreactive epidermal Langerhans cells in rat plantar skin after sciatic nerve injury. Cell Tissue Res 298:255–260
17. Lin YW, Tseng TJ, Lin WM, Hsieh ST (2001) Cutaneous nerve terminal degeneration in painful mononeuropathy. Exp Neurol 170:290–296
18. Lauria G, Lombardi R, Borgna M, Penza P, Bianchi R, Savino C, Canta A, Nicolini G, Marmiroli P, Cavaletti G (2005) Intraepidermal nerve fibres density in rat foot pad: neuropathologic-neurophysiologic correlation. J Peripher Nerv Syst 10:202–208
19. Szkudelski T (2001) The mechanism of alloxan and streptozotocin action in B cells of the rat pancreas. Physiol Res 50:537–546
20. Cameron NE, Cotter MA, Archibald V, Dines KC, Maxfield EK (1994) Anti-oxidant and pro-oxidant effects on nerve conduction velocity, endoneurial blood flow and oxygen tension in non-diabetic and streptozotocin-diabetic rats. Diabetologia 37:449–459
21. Bianchi R, Buyukakilli B, Brines M, Savino C, Cavaletti G, Oggioni N, Lauria G, Borgna M, Lombardi R, Cimen B, Comelekoglu U, Kanik A, Tataroglu C, Cerami A, Ghezzi P (2004) Erythropoietin both protects from and reverses experimental diabetic neuropathy. Proc Natl Acad Sci U S A 101:823–828
22. Stevens MJ, Zhang W, Li F, Sima AA (2004) C-peptide corrects endoneurial blood flow but not oxidative stress in type 1 BB/Wor rats. Am J Physiol Endocrinol Metab 287:E497–E505
23. Brines M, Cerami A (2005) Emerging biological roles for erythropoietin in the nervous system. Nat Rev Neurosci 6:484–494
24. Elfar JC, Jacobson JA, Puzas JE, Rosier RN, Zuscik MJ (2008) Erythropoietin accelerates functional recovery after peripheral nerve injury. J Bone Joint Surg Am 90:1644–1653
25. Siren AL, Fasshauer T, Bartels C, Ehrenreich H (2009) Therapeutic potential of erythropoietin and its structural or functional variants in the nervous system. Neurotherapeutics 6:108–127
26. Keswani SC, Leitz GJ, Hoke A (2004) Erythropoietin is neuroprotective in models of HIV sensory neuropathy. Neurosci Lett 371:102–105
27. Bianchi R, Brines M, Lauria G, Savino C, Gilardini A, Nicolini G, Rodriguez-Menendez V, Oggioni N, Canta A, Penza P, Lombardi R, Minoia C, Ronchi A, Cerami A, Ghezzi P, Cavaletti G (2006) Protective effect of erythropoietin and its carbamylated derivative in experimental Cisplatin peripheral neurotoxicity. Clin Cancer Res 12:2607–2612
28. Bianchi R, Gilardini A, Rodriguez-Menendez V, Oggioni N, Canta A, Colombo T, De Michele G, Martone S, Sfacteria A, Piedemonte G, Grasso G, Beccaglia P, Ghezzi P, D'Incalci M, Lauria G, Cavaletti G (2007) Cisplatin-induced peripheral neuropathy: neuroprotection by erythropoietin without affecting tumour growth. Eur J Cancer 43:710–717
29. Cervellini I, Bello E, Frapolli R, Porretta-Serapiglia C, Oggioni N, Canta A, Lombardi R, Camozzi F, Roglio I, Melcangi RC, D'Incalci M, Lauria G, Ghezzi P, Cavaletti G, Bianchi R (2010) The neuroprotective effect of erythropoietin in docetaxel-induced peripheral neuropathy causes no reduction of antitumor activity in 13762 adenocarcinoma-bearing rats. Neurotox Res 18:151–160
30. Wojchowski DM, Gregory RC, Miller CP, Pandit AK, Pircher TJ (1999) Signal transduction in the erythropoietin receptor system. Exp Cell Res 253:143–156
31. Leist M, Ghezzi P, Grasso G, Bianchi R, Villa P, Fratelli M, Savino C, Bianchi M, Nielsen J, Gerwien J, Kallunki P, Larsen AK, Helboe L, Christensen S, Pedersen LO, Nielsen M, Torup L, Sager T, Sfacteria A, Erbayraktar S, Erbayraktar Z, Gokmen N, Yilmaz O, Cerami-Hand C, Xie QW, Coleman T, Cerami A, Brines M (2004) Derivatives of erythropoietin that are tissue protective but not erythropoietic. Science 305:239–242
32. Li X, Gonias SL, Campana WM (2005) Schwann cells express erythropoietin receptor

and represent a major target for Epo in peripheral nerve injury. Glia 51:254–265

33. Campana WM, Myers RR (2003) Exogenous erythropoietin protects against dorsal root ganglion apoptosis and pain following peripheral nerve injury. Eur J Neurosci 18:1497–1506
34. Toth C, Martinez JA, Liu WQ, Diggle J, Guo GF, Ramji N, Mi R, Hoke A, Zochodne DW (2008) Local erythropoietin signaling enhances regeneration in peripheral axons. Neuroscience 154:767–783
35. Chattopadhyay M, Walter C, Mata M, Fink DJ (2009) Neuroprotective effect of herpes simplex virus-mediated gene transfer of erythropoietin in hyperglycemic dorsal root ganglion neurons. Brain 132:879–888
36. Pisano C, Pratesi G, Laccabue D, Zunino F, Lo Giudice P, Bellucci A, Pacifici L, Camerini B, Vesci L, Castorina M, Cicuzza S, Tredici G, Marmiroli P, Nicolini G, Galbiati S, Calvani M, Carminati P, Cavaletti G (2003) Paclitaxel and Cisplatin-induced neurotoxicity: a protective role of acetyl-L-carnitine. Clin Cancer Res 9:5756–5767
37. Bianchi R, Marini P, Merlini S, Fabris M, Triban C, Mussini E, Fiori MG (1988) ATPase activity defects in alloxan-induced diabetic sciatic nerve recovered by ganglioside treatment. Diabetes 37:1340–1345
38. McCarthy BG, Hsieh ST, Stocks A, Hauer P, Macko C, Cornblath DR, Griffin JW, McArthur JC (1995) Cutaneous innervation in sensory neuropathies: evaluation by skin biopsy. Neurology 45:1848–1855
39. Ko MH, Chen WP, Hsieh ST (2000) Cutaneous nerve degeneration induced by acrylamide in mice. Neurosci Lett 293:195–198
40. Lauria G, Hsieh ST, Johansson O, Kennedy WR, Leger JM, Mellgren SI, Nolano M, Merkies IS, Polydefkis M, Smith AG, Sommer C, Valls-Sole J (2010) European Federation of Neurological Societies/Peripheral Nerve Society Guideline on the use of skin biopsy in the diagnosis of small fiber neuropathy. Report of a joint task force of the European Federation of Neurological Societies and the Peripheral Nerve Society. Eur J Neurol 17(903–12):e44–e49
41. Lauria G, Bakkers M, Schmitz C, Lombardi R, Penza P, Devigili G, Smith AG, Hsieh ST, Mellgren SI, Umapathi T, Ziegler D, Faber CG, Merkies IS (2010) Intraepidermal nerve fiber density at the distal leg: a worldwide normative reference study. J Peripher Nerv Syst 15:202–207

Chapter 14

ARA290 in a Rat Model of Inflammatory Pain

Andrew Dilley

Abstract

Chronic pain affects as many as one in five people. A proportion of patients with symptoms of neuropathic pain do not have clinical signs of any obvious tissue or nerve injury. Such patients include those with diffuse limb pain, back pain, and complex regional pain syndrome type 1. These patients remain a clinical enigma. However, through the development of the neuritis model, it has become apparent that local nerve inflammation in the absence of gross pathology (i.e., axonal degeneration and demyelination) may underlie part of the mechanisms of pain. In this chapter, we describe a method to induce the neuritis model. We also describe in detail a reliable method to test for mechanical allodynia and heat hyperalgesia. Data that demonstrates the potential benefits of the neuroprotective agent ARA290 in reducing pain behavior in the neuritis model are presented.

Key words Sciatic nerve, Nerve inflammation, Neuropathic pain, Von Frey testing, Hargreaves' method, Nociception

1 Introduction

Chronic pain is a debilitating condition that affects as many as one in five people. It has huge economic impact through lost work days and healthcare costs. For some patients with the symptoms of neuropathic pain, routine clinical examination provides evidence of a peripheral nerve injury. Much of our understanding of the physiological mechanisms of pain in these patients has been gained through the use of animal models such as the chronic constriction injury, spinal nerve ligation and spared nerve injury models. In these models, animals exhibit behavioral changes that are consistent with clinical signs of allodynia and hyperalgesia (1–3). Electrophysiological recordings from primary sensory neurons reveal signs of increased excitability, whereby injured axons develop ongoing (spontaneous) activity and become responsive to mechanical stimulation from their regenerating tips (4–7). However, many patients do not have injuries that are comparable to these models. In fact a proportion of patients with chronic musculoskeletal pain lack signs of nerve or tissue injury on routine clinical testing despite

Pietro Ghezzi and Anthony Cerami (eds.), *Tissue-Protective Cytokines: Methods and Protocols*, Methods in Molecular Biology, vol. 982, DOI 10.1007/978-1-62703-308-4_14, © Springer Science+Business Media, LLC 2013

symptoms of neuropathic pain. Examples of such conditions include diffuse limb pain, back pain, and complex regional pain syndrome (8, 9).

Studies using a model of localized nerve inflammation, the neuritis model, suggest that in patients where there is no clinical evidence of a nerve injury, painful symptoms may in part be generated from inflamed peripheral nerves (10–13). In the neuritis model, a small strip of absorbable sponge (i.e., Gelfoam) saturated in an immune-stimulant such as complete Freund's adjuvant (killed bacterial cell wall), is loosely wrapped around the sciatic nerve. Within hours following induction, a robust local inflammatory response occurs in the epineurium (11, 14). Immune cells, which include activated macrophages, neutrophils, T lymphocytes, and granulocytes, accumulate outside of the perineurium (10, 11, 13, 14). The numbers of these cells peak during the first month following neuritis and persist for at least three months without entering the endoneurium (14). The neuritis model contrasts from inflammatory pain models that are induced by subcutaneous injection of immune stimulants into the hind paw (e.g., (15)). Whereas the neuritis model results in a localized mid-axonal inflammatory lesion, injection of immune stimulants causes chronic inflammation and observable edema of the hind paw that predominantly affects the peripheral terminals of primary sensory neurons. The neuritis model is therefore more likely to resemble the pathology in those patients with neuropathic pain who do not show signs of apparent tissue injury.

The most notable histological feature of the neuritis model is the lack of axonal degeneration and demyelination compared to other pain models (10, 12, 13). Despite this lack of gross nerve pathology, animals develop signs of mechanical and cold allodynia as well as heat and mechanical hyperalgesia (10, 13, 16). These behavioral changes occur early. Mechanical allodynia for example develops from 3 h and peaks on days 3–4 following surgery. Such behavioral changes show signs of recovery by 7 days.

Electrophysiological studies on the neuritis model have shown that uninjured primary sensory neurons exhibit signs of hyperexcitability. Intact unmyelinated C- as well as myelinated Aδ-nociceptive axons develop ongoing activity, which peaks within a week of neuritis induction (10, 12, 17–19). Ongoing activity in nociceptor neurons may correspond to the symptoms of spontaneous pain that are often reported by patients. Ongoing activity is also considered to play a role in the development of central neuropathic symptoms such as allodynia (20–22).

The immune mechanisms that underlie pain behavior and ongoing activity in the neuritis model are only starting to be explored. The immune cells that congregate at the neuritis site secrete a host of immune mediators, which include the pro-inflammatory cytokines interleukin (IL) 1β, IL6, and tumor necrosis factor α (TNFα) as well as reactive oxygen species (11, 23). Such immune mediators are considered to be important in the development of pain behavior in the neuritis model, since the administration of antagonists or scavengers against these components can prevent mechanical allodynia (24). In line with their likely role in the mechanisms of pain behavior, both TNFα and IL1β are reported to induce ongoing activity in sensory axons following direct exposure (25–27). More specifically, in the neuritis model, TNFα and the chemokine CCL2 may act to modulate ongoing activity in C-fiber neurons that are already modified by the inflammatory process (19).

Another prominent electrophysiological feature of the neuritis model is the development of axonal mechanical sensitivity at the treatment site, whereby axons respond to direct mechanical stimulation. Similar to ongoing activity, axonal mechanical sensitivity also peaks within a week of neuritis induction (10, 12, 17, 28). Axonal mechanical sensitivity develops along intact, uninjured C- and A-fiber axons, thereby contrasting from the original findings in the nerve injury models where it was the tips of degenerated axons that became sensitive to mechanical stimulation (4, 7). Clinically, axonal mechanical sensitivity corresponds to movement-evoked radiating pain that is reported by patients. Axonal mechanical sensitivity most likely results from the inflammation-induced disruption of axoplasmic transport, whereby mechanosensitive channel components that are transported along axons towards the terminals dam up at the site of the neuritis. This accumulation leads to a "hot spot" of mechanosensitivity (29).

Whilst the majority of studies have examined the role of inflammation in the development of axonal hyperexcitability and pain behaviors, investigations into the role of neuroprotective agents in the neuritis model have yet to be published. Of interest is the role of erythropoietin and its derivative ARA290 in inflammatory pain pathways (reviewed by Swartjes and colleagues in Chapter 12). Previous studies that have focused on nerve injury models (chronic constriction injury, spinal nerve crush, and spared nerve injury) have demonstrated the reversal of pain behavior following the administration of erythropoietin or ARA290 (30–32). Based on these observations, such agents can undoubtedly provide a protective role in pain pathways. The data presented in this chapter demonstrates the potential benefits of ARA290 in the neuritis model, where there is no apparent nerve injury. It provides detailed methodology for inducing the neuritis and testing heat hyperalgesia, as well as an alternative method to that described by Swartjes and colleagues (Chapter 12) for examining mechanical allodynia.

2 Materials

2.1 Preparation of Complete Freund's Adjuvant

1. Complete Freund's adjuvant (Sigma-Aldrich, Dorset, UK).
2. Eppendorf tube (Sterile), 1.5 ml.
3. Saline (Sterile), 0.9%.
4. Sterile filter, 0.22 μm (Millipore, MA, USA).
5. Syringe, 1 ml.
6. Vortexer.

2.2 Induction of Neuritis Model

1. Male Sprague–Dawley rats, 250–300 g (Harlan, UK)—weight on day of surgery.
2. Autoclave bags and tape.
3. Sterile paper sheet (e.g., inside of an autoclave bag).
4. Sterile latex gloves.
5. Ethanol, 70%.
6. Videne antiseptic solution.
7. Betadine Anti-Microbial Dry Powder Spray.
8. Gauze swabs (5 cm × 5 cm).
9. Q tips/wooden applicators.
10. Gelfoam (Spongostan, Ethicon, West Lothian, UK).
11. Gaseous anesthesia (isoflurane).
12. Animal shaver.
13. Heated blanket.
14. Scalpel, size 10 blade.
15. Curved forceps.
16. ToughCut Mayo straight scissors.
17. 2× Dumont angled 45°, medical #5/45 forceps.
18. Curved 8.5 cm Iris scissors.
19. Castroviejo sharply curved scissors.
20. Goldstein Retractor with sharp teeth.
21. Olsen-Hegar needle holders (15 mm cutting, 2 mm jaw, 16 cm).
22. Coated Vicryl suture, 19 mm 3/8 curved reverse cutting needle.
23. Hot bead sterilizer.

2.3 Assessment of Pain: Heat Hyperalgesia

1. Hargreaves's Plantar Test (catalogue number: 37370; Ugo Basile, VA, Italy), which includes raised glass pane and Perspex animal enclosures.
2. Radiometer (Ugo Basile, Italy).

2.4 Assessment of Pain: Mechanical Allodynia

1. Raised platform with wire mesh floor and Perspex animal enclosures with lids.
2. von Frey filaments (catalogue number: 37450; Ugo Basile, VA, Italy).

2.5 Treatment with ARA290

1. Saline, 0.9%.
2. ARA290 (Araim Pharmacueticals, Ossining, NY, USA).
3. Syringe, 1 ml with 25 gauge needle.
4. Eppendorf tube (Sterile), 1.5 ml.
5. Sterile filter, 0.22 μm (Millipore, MA, USA), and syringe, 1 ml.

3 Methods

3.1 Preparation of Complete Freund's Adjuvant

1. Sterile filter 0.5 ml saline, using a 0.22 μm Millipore filter attached to a 1 ml syringe, into a 1.5 ml sterile Eppendorf tube.
2. Vortex the complete Freund's adjuvant and add 0.5 ml of the adjuvant to the saline using a 1 ml syringe.
3. Vortex the final solution.

3.2 Induction of the Neuritis Model (see Note 1)

1. A sterile gown should be worn for all surgical procedures.
2. All instruments, gauze, and cotton-tipped applicators should be autoclaved prior to use.
3. Surgical area should be cleaned with 70% alcohol or disinfectant.
4. Place instruments onto a sterile paper sheet, such as the inside of an autoclave bag.
5. Using a size 10 scalpel blade, cut a small strip of Gelfoam (10 × 3 × 3 mm) in preparation for wrapping around the nerve.
6. Switch heated blanket onto low setting 15 min prior to surgery and place a sterile sheet onto the blanket.
7. Induce and maintain anesthesia in induction chamber (isoflurane in oxygen; 3% induction (4 l/min O_2) and 1.75% (0.8 l/min O_2) maintenance).
8. Once the rat is anesthetized, place prone onto the heated blanket with the nose into the nose cone.
9. Prior to commencing surgery, check depth of anesthesia (see Note 2).
10. Place 2–3 gauze swabs under the hind limb/thigh to be operated on. This will raise the limb and improve visibility for surgery.

11. Shave the lateral aspect of the thigh from the sciatic notch to the knee.
12. Clean surgical site with 70% ethanol, followed by disinfection with Videne antiseptic solution, working out from the center.
13. Change into sterile gloves.
14. Using your fingers, feel for the ends of the femur (i.e., the hip and knee joints).
15. Note a point mid way along the length of the femur.
16. Lift the skin using the curved forceps and at this midpoint, make a small incision through the skin, just posterior to the bone, with the ToughCut Mayo straight scissors.
17. Extend the length of the incision to approximately 10 mm following the long axis of the femur.
18. Separate the skin from the underlying subcutaneous tissue by blunt dissection (see Note 3).
19. Locate the plane of fascia between the two heads of the biceps femoris muscle. This plane can be identified as a white line parallel to the femur.
20. Support the myofascia of the biceps femoris muscle with the Dumont angled forceps and use the curved 8.5 cm Iris scissors to carefully cut through the fascia. Use the Iris scissors to blunt dissect between the muscle heads.
21. Once the sciatic nerve can be seen, use the Dumont angled forceps to blunt dissect around the nerve (see Notes 4 and 5). It may be necessary to use the Castroviejo sharply curved scissors to cut the surrounding connective tissue.
22. Use the Goldstein Retractor or the curved forceps to open the space between the muscle heads.
23. Just proximal to the trifurcation, clear approximately 5 mm of connective tissue from around the nerve.
24. Take the strip of Gelfoam and submerge into the Eppendorf of complete Freund's adjuvant (see Note 6). Place the Gelfoam onto a gauze swab close to the surgery site.
25. Pass the tips of the curved forceps under the nerve. Using the second set of curved forceps, carefully feed the Gelfoam below the nerve. It is important not to stretch or damage any neural tissue. Carefully draw the Gelfoam around the nerve, feeding it through with the second pair of curved forceps (see Note 7). Loosely wrap the ends of the Gelfoam over the top of the nerve (Fig. 1).
26. Remove any excess complete Freund's adjuvant using the cotton tipped applicators.

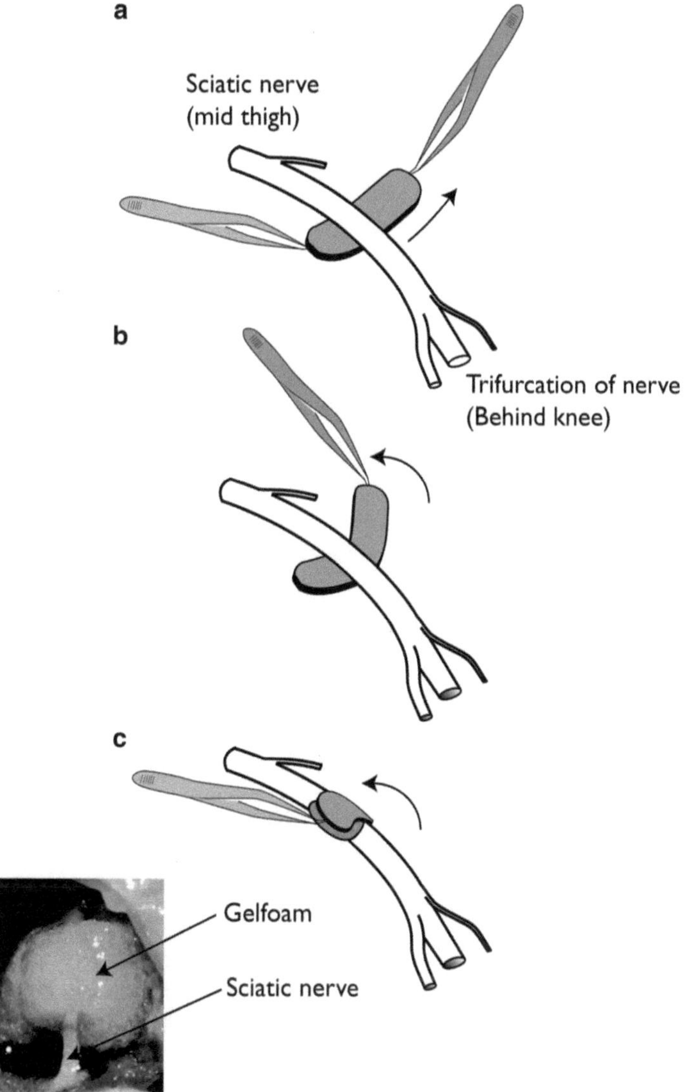

Fig. 1 Illustration of the neuritis induction. (**a**) A strip of Gelfoam saturated in complete Freund's adjuvant is carefully fed under the cleared length of the sciatic nerve in the mid thigh (proximal to the trifurcation of the nerve). (**b**) The Gelfoam is drawn around the nerve and (**c**) the ends are loosely wrapped over the top. The *photo insert* shows the Gelfoam surrounding the sciatic nerve

27. Close the overlying fascia with 4-0 coated Vicryl sutures. Two simple interrupted sutures (surgical knots) are usually required. Use the Dumont angled forceps to support the fascia.
28. Close skin with 4-0 coated Vicryl sutures. Four interrupted sutures are usually required. Use the needle holders and the curved forceps to support the skin (see Note 8).
29. Spray surgical site with Betadine powder.

30. Allow animal to recover either on a heated blanket or in a warm air recovery chamber.
31. Instruments should be sterilized (using ethanol and the hot bead sterilizer) between animals.

3.3 Assessment of Pain: Heat Hyperalgesia

1. Habituate rats for 1 h in their home cage in the testing room on day one, followed by 1 h in the individual Perspex animal enclosures of the Hargreaves rig for at least three consecutive days (see Notes 9 and 10).
2. Calibrate the intensity of the radiant heat source according to manufacturer's instructions using the Radiometer.
3. Set the radiant heat source intensity such that the mean withdrawal latency is 10–15 s for the size and strain of rats being tested.
4. Testing to establish baseline withdrawal latencies should be performed on three separate, preferably consecutive, days prior to surgery (see Note 11).
5. On each test day, allow rats to acclimate in rig for 15 min prior to testing.
6. If rat urinates, remove animal and dry glass floor as well as the paws of the rat. Do not test for at least 2 min after drying (see Note 12).
7. Aim the moveable radiant heat source at the glabrous skin on the mid-plantar surface of the hind limb (see Note 13). Do not start the test when the rat is grooming or sleeping (see Note 14).
8. When the radiant heat source is switched on, it will activate a timer that will automatically stop when the animal withdraws their foot. It is important to set the cutoff of the radiant light source to 20 s to avoid damage to skin.
9. Record the withdrawal latency or 20 s if the cut off for the radiant light source is reached.
10. If the rat moves away from heat source rather than rapidly withdrawing the paw, wait for 10 min and then retest.
11. Repeat the test on the first hind paw of each rat before testing the other paw (or allow at least 10 min between sides).

3.4 Assessment of Pain: Mechanical Allodynia

1. Habituate rats for 1 h in their cage in the testing room on day one, followed by 1 h in the individual Perspex animal enclosures for at least three consecutive days (see Notes 9 and 10).
2. Baseline testing should then be performed on three separate, preferably consecutive, days prior to surgery (see Note 11).
3. On each test day, allow rats to acclimate in the rig for 15 min prior to testing.

4. Before applying the von Frey filaments, rats must be standing and weight bearing, i.e., not lying down.
5. When applying the von Frey filaments, aim for the glabrous skin on the mid-plantar surface of the hind limb (see Note 13). Apply the monofilament perpendicular to the skin so that it bends slightly. Hold it in this position for 5 s. Avoid the hairy skin as well as the foot pads. If the monofilament slips or the skin is brushed, wait 1 min before repeating the application. A positive response is a rapid withdrawal, or a lift, followed by a hold or lick of the paw. If the rat steps away, this should not be considered a positive response.
6. Both hind limbs should be tested. Randomize which side to test first and leave at least five min between sides.
7. Begin testing with the 4 g von Frey filament. Apply the monofilament up to five times (or six if the rats' first response is to the fifth repeat). If the rat ambulates during testing or the response is ambiguous, dismiss the trial and apply the monofilament up to a maximum of eight times. Leave 10 s between repeat applications with the same monofilament.
8. If the rat responds twice in a row to the 4 g monofilament, stop testing and repeat with the 0.4 g monofilament. If the rat does not respond twice to the 4 g monofilament, repeat with the 6 g monofilament. Work your way up through monofilaments of ascending thicknesses (from either the 0.4 g or 6 g monofilament) until a repeat response is achieved. Leave 1 min interval between different monofilaments.
9. The threshold is the lowest thickness monofilament that produces two withdrawals.
10. Test animals as pairs (i.e., Test the first hind paw of two animals before repeating on the second paw). Allow at least 5 min between sides.

3.5 Treatment with ARA290

1. Administer 120 μg/kg sterile filtered (0.22 μm Millipore filter) ARA290 or vehicle (0.9% Saline) in a total volume of 200 μl i.p. using a 1 ml syringe mounted with a 25 G needle (see Note 15).
2. Animals are injected immediately after behavioral testing at 1, 2, 3, and 4 days following surgery (see Note 16).

3.6 Typical Results

A study was carried out in 18 adult male Sprague–Dawley rats (250–300 g) to assess the effects of ARA290 on neuritis-induced mechanical allodynia and heat hyperalgesia. Prior to induction of the neuritis, baseline behavioral values were obtained on three consecutive days. Behavioral testing was further performed on days 1, 2, 3, 4, and 7 post-neuritis. Animals were assigned into two groups ($n=9$ per group). One group was administered 120 μg/kg ARA290, whereas the other group received only the vehicle. All testing was performed blind with respect to the drug.

Following neuritis, there were signs of mechanical allodynia on the ipsilateral side that peaked on day 4, which was indicated by a decrease in monofilament threshold (Fig. 2a, vehicle group). By day 7 mechanical allodynia had partly recovered. Treatment with 120 μg/kg ARA290 significantly reduced the development of mechanical allodynia on day 4 (Fig. 2a). Heat hyperalgesia peaked by day 3 following neuritis (Fig. 2b, vehicle group), which was indicated by a reduction in withdrawal latency. Similar to mechanical allodynia, almost complete reversal had occurred by day 7. Treatment with 120 μg/kg ARA290 caused a notable reduction in heat hyperalgesia on day 3 (Fig. 2b). These data clearly demonstrate the positive effects of ARA290 in the treatment of neuritis-induced pain where there is no frank nerve injury.

4 Notes

1. There has been much discussion regarding a suitable negative control for the neuritis. Previous studies have performed the same surgery but replaced the complete Freund's adjuvant with saline. However, it should be noted that surgery alone is sufficient to cause low grade inflammation around the nerve.
2. It is important to monitor anesthesia depth during surgery. For example, the pedal withdrawal reflex should be absent. Visual inspection of chest wall movements will indicate changes in the rate, depth, and pattern of breathing. Respiratory rate should fall during anesthesia.
3. To blunt dissect, insert the scissors (or forceps) closed into the tissue plane that you are trying to separate. Partly open the scissors and slowly remove. Repeat until all tissue has been separated.
4. Make sure you always have a good visual of the surgical site. Never cut through tissue that you cannot see.
5. The surgery is best performed under a dissection microscope with a good light.
6. The complete Freund's adjuvant solution tends to separate. Therefore, it is important to shake the Eppendorf prior to use, between animals.
7. Always feed the Gelfoam under the nerve with both sets of angled forceps. Pulling the Gelfoam with one set of forceps will cause it to break up or stretch/twist the nerve.
8. To reduce the risk of animals chewing stitches, bury the knot below skin. This is best achieved by performing inside-out stitches, whereby the knot will be on the inside of the surgical side.
9. Behavioral testing should always be performed at the same time of each day.

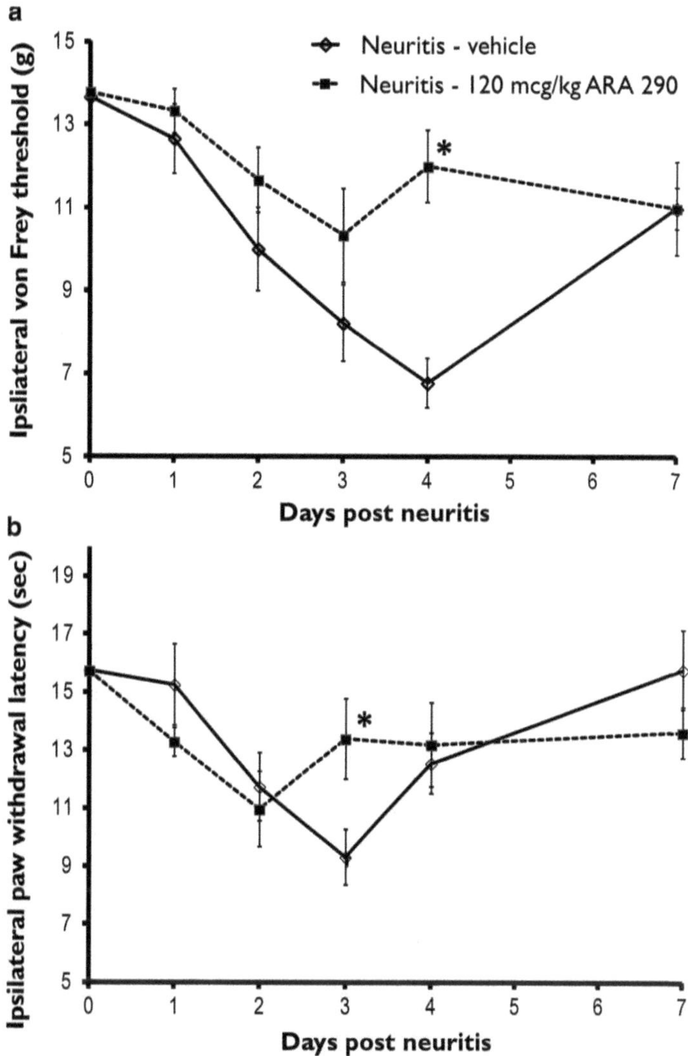

Fig. 2 The effects of ARA290 on the development of (**a**) mechanical allodynia and (**b**) heat hyperalgesia in the neuritis model. Part (**a**) is the mean von Frey thresholds on the ipsilateral side. In the vehicle group ($n=9$), mechanical allodynia peaked on day 4 post neuritis but showed signs of reversal by day 7. Following the i.p. injection of 120 μg/kg ARA290 ($n=9$) there was a clear reduction in the development of mechanical allodynia. Part (**b**) is the mean paw withdrawal latencies in response to the radiant heat source on the ipsilateral side. In the vehicle group ($n=9$), heat hyperalgesia peaked on day 3 post neuritis but showed signs of reversal by day 7. Following injection of 120 μg/kg ARA290 ($n=9$) there was a reduction in the development of heat hyperalgesia on day 3. Error bars = SEM, $^*p<0.05$ compared to the vehicle group at individual time points (unpaired t test) (Data modified from 33)

10. The same experimenter should perform all behavioral testing to avoid variability in technique.
11. The final baseline behavioral test should be performed on the day of surgery.
12. Urination during behavioral testing is rare in well-habituated animals.
13. For both heat hyperalgesia and mechanical allodynia testing, a different part of the rats paw should be tested for each repeat.
14. During the behavioral testing, use a Q-tip to make sure the rat is awake or to induce rat to stand.
15. ARA290 should be made up to a concentration that results in an injection volume of less than 250 μl.
16. Animals should be injected with ARA290 on alternate sides each day.

References

1. Bennett GJ, Xie YK (1988) A peripheral mononeuropathy in rat that produces disorders of pain sensation like those seen in man. Pain 33:87–107
2. Kim SH, Chung JM (1992) An experimental model for peripheral neuropathy produced by segmental spinal nerve ligation in the rat. Pain 50:355–363
3. Decosterd I, Woolf CJ (2000) Spared nerve injury: an animal model of persistent peripheral neuropathic pain. Pain 87:149–158
4. Chen Y, Devor M (1998) Ectopic mechanosensitivity in injured sensory axons arises from the site of spontaneous electrogenesis. Eur J Pain 2:165–178
5. Michaelis M, Blenk KH, Janig W, Vogel C (1995) Development of spontaneous activity and mechanosensitivity in axotomized afferent nerve fibers during the first hours after nerve transection in rats. J Neurophysiol 74:1020–1027
6. Scadding JW (1981) Development of ongoing activity, mechanosensitivity, and adrenaline sensitivity in severed peripheral nerve axons. Exp Neurol 73:345–364
7. Tal M, Eliav E (1996) Abnormal discharge originates at the site of nerve injury in experimental constriction neuropathy (CCI) in the rat. Pain 64:511–518
8. Janig W, Baron R (2003) Complex regional pain syndrome: mystery explained? Lancet Neurol 2:687–697
9. Dilley A (2011) MRI signal hyper-intensity of neural tissues in diffuse chronic pain syndromes—a pilot study. Muscle Nerve 44: 981–984
10. Bove GM, Ransil BJ, Lin HC, Leem JG (2003) Inflammation induces ectopic mechanical sensitivity in axons of nociceptors innervating deep tissues. J Neurophysiol 90:1949–1955
11. Gazda LS, Milligan ED, Hansen MK, Twining CM, Poulos NM, Chacur M, O'Connor KA, Armstrong C, Maier SF, Watkins LR, Myers RR (2001) Sciatic inflammatory neuritis (SIN): behavioral allodynia is paralleled by peri-sciatic proinflammatory cytokine and superoxide production. J Peripher Nerv Syst 6:111–129
12. Dilley A, Lynn B, Pang SJ (2005) Pressure and stretch mechanosensitivity of peripheral nerve fibres following local inflammation of the nerve trunk. Pain 117:462–472
13. Eliav E, Herzberg U, Ruda MA, Bennett GJ (1999) Neuropathic pain from an experimental neuritis of the rat sciatic nerve. Pain 83:169–182
14. Bove GM, Weissner W, Barbe MF (2009) Long lasting recruitment of immune cells and altered epi-perineurial thickness in focal nerve inflammation induced by complete Freund's adjuvant. J Neuroimmunol 213:26–30
15. Meller ST, Gebhart GF (1997) Intraplantar zymosan as a reliable, quantifiable model of thermal and mechanical hyperalgesia in the rat. Eur J Pain 1:43–52
16. Chacur M, Milligan ED, Gazda LS, Armstrong C, Wang HC, Tracey KJ, Maier SF, Watkins LR (2001) A new model of sciatic inflammatory neuritis (SIN): induction of unilateral and bilateral mechanical allodynia following acute unilateral peri-sciatic immune activation in rats. Pain 94:231–244
17. Eliav E, Benoliel R, Tal M (2001) Inflammation with no axonal damage of the rat saphenous nerve trunk induces ectopic discharge and mechanosensitivity in myelinated axons. Neurosci Lett 311:49–52

18. Bove GM, Dilley A (2010) The conundrum of sensitization when recording from nociceptors. J Neurosci Methods 188:213–218
19. Richards N, Batty T, Dilley A (2011) CCL2 has similar excitatory effects to TNF-alpha in a subgroup of inflamed C-fiber axons. J Neurophysiol 106:2838–2848
20. Campbell JN, Meyer RA (2006) Mechanisms of neuropathic pain. Neuron 52:77–92
21. Gracely RH, Lynch SA, Bennett GJ (1992) Painful neuropathy: altered central processing maintained dynamically by peripheral input. Pain 51:175–194
22. LaMotte RH, Shain CN, Simone DA, Tsai EP (1991) Neurogenic hyperalgesia: psychophysical studies of underlying mechanisms. J Neurophysiol 66:190–211
23. Eliav E, Benoliel R, Herzberg U, Kalladka M, Tal M (2009) The role of IL-6 and IL-1beta in painful perineural inflammatory neuritis. Brain Behav Immun 23:474–484
24. Twining CM, Sloane EM, Milligan ED, Chacur M, Martin D, Poole S, Marsh H, Maier SF, Watkins LR (2004) Peri-sciatic proinflammatory cytokines, reactive oxygen species, and complement induce mirror-image neuropathic pain in rats. Pain 110:299–309
25. Leem JG, Bove GM (2002) Mid-axonal tumor necrosis factor-alpha induces ectopic activity in a subset of slowly conducting cutaneous and deep afferent neurons. J Pain 3:45–49
26. Sorkin LS, Xiao WH, Wagner R, Myers RR (1997) Tumour necrosis factor-alpha induces ectopic activity in nociceptive primary afferent fibres. Neuroscience 81:255–262
27. Ozaktay AC, Kallakuri S, Takebayashi T, Cavanaugh JM, Asik I, DeLeo JA, Weinstein JN (2006) Effects of interleukin-1 beta, interleukin-6, and tumor necrosis factor on sensitivity of dorsal root ganglion and peripheral receptive fields in rats. Eur Spine J 15: 1529–1537
28. Dilley A, Bove GM (2008) Resolution of inflammation-induced axonal mechanical sensitivity and conduction slowing in C-fiber nociceptors. J Pain 9:185–192
29. Dilley A, Bove GM (2008) Disruption of axoplasmic transport induces mechanical sensitivity in intact rat C-fibre nociceptor axons. J Physiol 586:593–604
30. Campana WM, Li X, Shubayev VI, Angert M, Cai K, Myers RR (2006) Erythropoietin reduces Schwann cell TNF-alpha, Wallerian degeneration and pain-related behaviors after peripheral nerve injury. Eur J Neurosci 23:617–626
31. Campana WM, Myers RR (2003) Exogenous erythropoietin protects against dorsal root ganglion apoptosis and pain following peripheral nerve injury. Eur J Neurosci 18: 1497–1506
32. Swartjes M, Morariu A, Niesters M, Brines M, Cerami A, Aarts L, Dahan A (2011) ARA290, a peptide derived from the tertiary structure of erythropoietin, produces long-term relief of neuropathic pain: an experimental study in rats and beta-common receptor knockout mice. Anesthesiology 115:1084–1092
33. Pulman KG, Smith M, Mengozzi M, Ghezzi P, Dilley A (2012) The erythropoietin-derived peptide ARA290 reverses mechanical allodynia in the neuritis model. Neuroscience http://dx.doi.org/10.1016/j.neuroscience.2012.12.022 [Epub ahead of print]

Chapter 15

In Vivo Angiogenic Activity of Erythropoietin

Domenico Ribatti

Abstract

The role of erythropoietin (Epo) has been demonstrated in tissues outside the hematopoietic system, including the cardiovascular system, where Epo promotes various effects in endothelial cells. Here, we have demonstrated the angiogenic capacity of recombinant human Epo (rhuEpo) in vivo, by means of the chick embryo chorioallantoic membrane (CAM) assay, a well-established in vivo assay to study angiogenesis and antiangiogenesis.

Key words Angiogenesis, Chorioallantoic membrane, Erythropoietin

1 Introduction

The role of erythropoietin (Epo) has been demonstrated in tissues outside the hematopoietic system, including the cardiovascular system, where Epo promotes various effects in endothelial cells. Epo stimulates both proliferation and migration of human and bovine endothelial cells in vitro and of endothelial cells isolated from rat mesentery (1–4), as well as in the rat aortic ring model (5). Moreover, Epo induces endothelin-1 (ET-1) expression in endothelial cell cultures (5, 6). Recombinant human Epo (rHuEpo) induces an increased cell proliferation, matrix metalloproteinase-2 (MMP-2) expression, and differentiation into vascular tubes of human endothelial cells in vitro (1). Epo receptor (EpoR) mRNA is expressed in human umbilical vein endothelial cells (HUVEC), bovine adrenal capillaries, and rat brain capillaries (2, 3).

Epo and EpoR are expressed in the vasculature during embryogenesis and their deletion in null embryos leads to angiogenic defects, whereas vasculogenesis was not affected, consistent with the differential expression of Epo and EpoR during the early stages of embryonic development (7).

RhuEpo increases also the number of circulating endothelial cells and endothelial precursor cells (EPC) in preclinical models (8)

Pietro Ghezzi and Anthony Cerami (eds.), *Tissue-Protective Cytokines: Methods and Protocols*, Methods in Molecular Biology, vol. 982, DOI 10.1007/978-1-62703-308-4_15, © Springer Science+Business Media, LLC 2013

Table 1
Human tumors which express Epo and/or EpoR

Tumor	Reference
Breast carcinoma	(22)
Lung carcinoma	(12)
Gastric carcinoma	(25)
Hepatocellular carcinoma	(26)
Uterine cervix carcinoma	(27)
Renal carcinoma	(28)
Neuroblastoma	(29)
Pediatric tumors	(30)
Hemangioblastoma	(31)
Lymphoid malignancies	(32)
Glioma	(33)

and human breast cancer and lymphoma models (9). Moreover, Epo stimulates postnatal neovascularization by increasing EPC mobilization from the bone marrow (10).

Several tumor cell lines express Epo and EpoR regardless that they secrete a very small amount of Epo individually and that most of them respond to hypoxic stimuli by enhanced secretion of Epo (11). Kayser and Gabius (12) first suggested that human tumors may express EpoR. In their study 81% of human lung carcinoma tissues possessed Epo-binding site, detected by the use of biotinylated rhEpo. Epo and EpoR mRNA and/or EpoR protein have been detected in several human tumors (Table 1).

Epo favors tumor progression through effects on angiogenesis and it may be considered as an endogenous stimulator of vessel growth during tumor progression through an autocrine/paracrine loop (13). Tumor cell can directly release increasing amounts of vascular endothelial growth factor (VEGF) and placental growth factor (PlGF) in response to Epo (14). In cerebral endothelial cell cultures, rhuEPO enhanced angiogenesis by increasing VEGF levels and this effect was inhibited by anti-Epo antibody and a specific VEGF receptor (VEGFR) antagonist (15). In an experimental model of femoral artery legation using EpoR-null mice, blood flow recovery, activation of VEGF/VEGFR system, and mobilization of EPC were all impaired in EpoR-null mice as compared with wild-type mice (16).

In a Lewis lung carcinoma xenograft model, subcutaneous administration of Epo promoted tumor growth through enhancement of angiogenesis (17). In the dorsal skin-fold window chambers, the co-injection of Epo with labeled rodent mammary

Table 2
Direct and indirect effects of Epo on tumor growth and angiogenesis

Effect	Reference
Direct	
Increased expression of Ki67 proliferation antigen	(18)
Enhanced microvessel density	(18)
Decreased tumor hypoxia	(18)
Increased mobilization of EPC	(16)
Indirect	
Increased release of VEGF and PlGF by tumor cells	(14)

EPC endothelial precursor cells; *VEGF* vascular endothelial growth factor; *PlGF* placental growth factor

carcinoma cells stimulated tumor neovascularization and growth (18). Epo administration to patients with hematological malignancies induced bone marrow angiogenesis and further malignant transformation in plasma cell leukemia and acute monoblastic leukemia, respectively (19–21).

The administration of anti-Epo-antibody, soluble EpoR, or an inhibitor of JAK2 resulted in a delay in tumor growth with 45% reduction in maximal tumor depth in a tumor Z chambers model with rat mammary adenocarcinoma cells (22). On the contrary, the injection of an antibody anti-Epo or the soluble form of EpoR into in malignant tumors of the female reproductive organs reduces capillaries and causes tumor destruction (23). Moreover, blockade of Epo signalling on xenografts of uterine and ovarian cancer leads to the destruction of tumors in nude mice.

The mechanism of tumor growth in the context of Epo is not completely clarified, and it is not still clear whether there is a direct effect of Epo in tumor cells as opposed to exogenous effect on angiogenesis (Table 2). It is also possible that the effect of Epo is multifactorial depending on the type of tumor and level or functionality of EpoR expression in tumor cells as well as other variables such as hypoxic stress, degree of anemia, chemotherapy, radiotherapy, or surgical intervention.

The chick embryo chorioallantoic membrane (CAM) is an extraembryonic membrane which serves as a gas exchange surface and it function is supported by a dense capillary network (24). Because of its extensive vascularization and easy accessibility, the CAM has been broadly used to study the morphofunctional aspects of the angiogenesis process in vivo and to investigate the efficacy and mechanisms of action of pro-angiogenic and anti-angiogenic natural and synthetic molecules.

Here, we have used the CAM assay to demonstrate the angiogenic activity in vivo of rHuEpo.

2 Materials

2.1 Reagents

1. White fertile chicken eggs obtained at day 1–2 postlaying (see Note 1).
2. 70% ethanol in H_2O.
3. Sterile routine tissue culture medium: minimum essential medium (MEM) with nonessential amino acids.
4. Sterilized gelatin sponges (Gelfoam, Upjohn Company, Kalamazoo, MI, USA).
5. Bouin's solution.
6. Toluidine blue 0. Use a 0.5% aqueous solution.

2.2 Equipment

1. Egg incubator (Kemps Koops—Eugene, Oregon, USA).
2. 25–26 gauge hypodermic needles and 1 ml syringes.
3. Pasteur pipets.
4. Curved-tip forceps.
5. Straight-tip forceps.
6. Small dissecting scissors.
7. Stereomicroscope (SXX9, Olympus Italia).
8. Light microscope (BX51, Olympus Italia).

3 Methods

1. Sterilize all instruments in 70% ethanol prior to use.
2. Clean the fertilized white chick eggs with 70% ethanol and incubate at 37°C and 60% humidity in an egg incubator for 48 h (see Note 2).
3. Aspirate 2–3 ml albumen from the egg using 25–26 gauge hypodermic needle and 1 ml syringe at the pin hole of the egg on day 3 of incubation. This procedure creates a false air sac directly over the CAM, allowing its dissociation from the egg shell membrane (see Note 3).
4. After albumen removal, cut a square window into the shell with the aid of scissors. This window can be enlarged to approximately 10 × 10 mm. The underlying embryo and CAM vessels are revealed. Seal the windows with transparent tape or a glass coverslip of the same dimension, and return the egg to the incubator. This step should be done in an enclosed area

such as a laminar flow hood to minimize the risk of infection (see Note 4).

5. On day 8 of incubation, open the window under sterile conditions within a laminar-flow hood and implant a 1-mm^3 sterilized gelatin sponge onto the CAM. For this purpose, the sponge is cut by hand with a blade and gently placed on top of the growing CAM. Use sterile gloves to minimize the risk of infection (see Note 5).
6. Pipette rhEpo (dissolved in 1–3 μl of sterile routine tissue culture medium at doses ranging between 10 and 50 U/implant) onto the sponge. Sponge containing vehicle alone is used as negative control.
7. On incubation day 12, macroscopic observation shows that the gelatin sponge treated with rhuEpo is surrounded by allantoic vessels that develop radially towards the implant in a "spoked-wheel" pattern. Macroscopic evaluation of the vasoproliferative response can be performed as follows: for each egg, the total number of vessels converging towards the graft can be counted under the stereomicroscope at 10× magnification at different time points after implantation from day 8 to day 12. Then, for each experimental group ($n \geq 10$) data are expressed as mean ± SD and kinetics curves can be obtained for proangiogenic stimulus compared to controls (see Note 6).
8. Fix the embryos and their membranes in ovo by pipetting 10 μl of Bouin's fluid solution on the CAM surface and fix for 3 h at room temperature.
9. Cut the sponges, the underlying and the immediately adjacent CAM portions with curved tip-forceps and process the specimen for paraffin embedding.
10. Deydrate tissue samples in ethanol, cleared in toluene and immerse in the embedding paraffin for 2 h, according to standard procedure.
11. Cut eight-micrometer serial sections of paraffin-embedded CAMs in parallel to the surface of the membrane, stain sections with a 0.5% aqueous solution of toluidine blue for 1 min at room temperature and observe under a light microscope.
12. Microscopic evaluation of the angiogenic response. Quantitative evaluation of the angiogenic response, expressed as microvessel density, can be obtained by applying a morphometric method of "point counting" on histological CAM sections. Briefly, two investigators simultaneously identify transversally cut microvessels (diameter ranging from 3 to 10 μm) among the gelatin sponge trabeculae with a double headed photomicroscope at 250× magnification. A square mesh consisting of 12 lines per side, giving 144 intersection points, is inserted in the eyepiece.

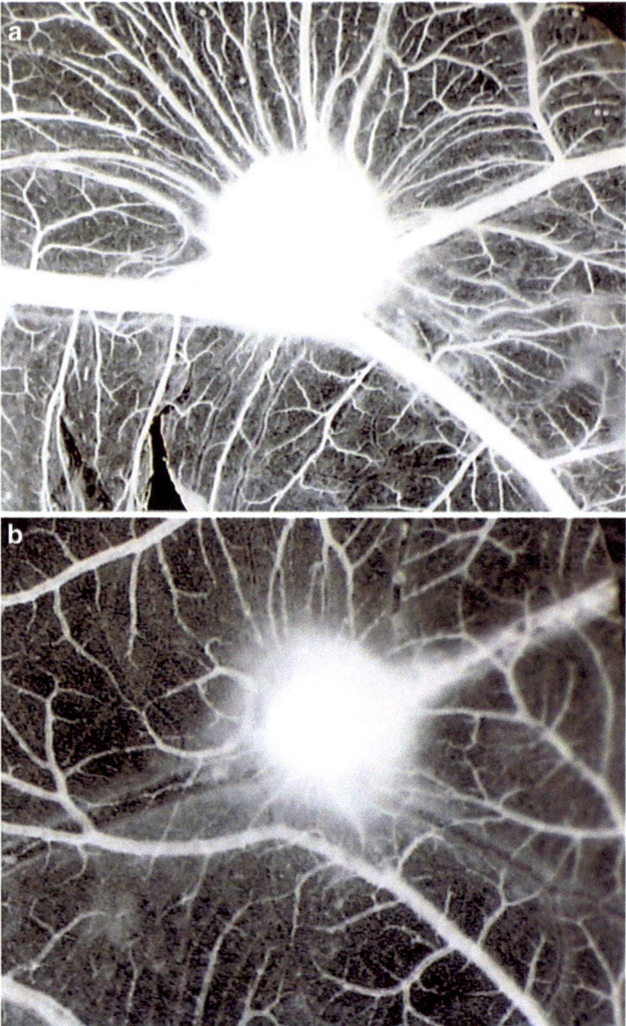

Fig. 1 EPO stimulates angiogenesis in the chick embryo CAM. CAM of 12-day-old chick embryo incubated for 4 days with a gelatin sponge adsorbed with 10 U of rHuEpo (**a**) compared to vehicle (**b**). Note the presence of an increased number of blood vessels with a radially arranged spoked wheel pattern around the implant

Ten sections are analyzed for each CAM specimen by observing every third section within 30 serial slides. For each section, six randomly chosen microscopic fields are evaluated for the number of intersection points occupied by microvessels. Then, mean values ±1 standard deviation are determined for each CAM specimen. The microvessel density is expressed as percentage of intersection points occupied by microvessels. Statistically significant differences between the mean values of the intersection points in the experimental CAMs and control ones are determined by Student's *t*-test for unpaired data.

At microscopic level, a highly vascularized tissue is recognizable among the trabeculae of the sponges treated with rhuEpo. The tissue consists of newly formed blood vessels growing perpendicularly to the plane of the CAM, mainly capillaries within an abundant network of collagen fibers. In contrast, no blood vessels are present among trabeculae of the implants treated with vehicle alone (Fig. 1).

4 Notes

1. Chick embryos from different vendors can vary significantly in their degree of vascularization and developmental status. Thus, consistent use of the same vender can decrease experimental variability.
2. Cleaning the egg shell before incubation will remove any debris associated with the outer surface and decrease the risk of infection. Furthermore, specific incubation conditions, including constant temperature and humidity, are of critical importance for proper vascularization and embryo survival.
3. Make sure the needle opening is pointing away from the embryo during albumen aspiration. In addition, regular changes of needle and syringe limit the carryover of infection from egg to egg.
4. If large pieces of shell fall onto the CAM it may possible to remove these using fine forceps. It is probably best to discard eggs where pieces of shell have fallen onto the CAM and have not been removed very easily, as these eggs may develop false positive response due to inflammation. Reject eggs with an excessively humid CAM; otherwise, sponges may float off during the incubation period.
5. The CAM is an expanding membrane with vessels developing over its entire surface. It is preferable not place the sponges on its outer edges.
6. Two operators, preferably blinded to the sample identity should grade the angiogenic response. Different test substances can produce a range of different types of angiogenic response, such as a mixed response of microvascular growth and large vessel deformation/growth towards the point of application. Samples may also induce local bleeding and the presence and severity of these reactions should be noted, as the response may be secondary to the bleeding or inflammation.

Acknowledgement

Supported by European Seventh Framework Programme (FTP/2007-2013) under grant agreement No 278570 to DR.

References

1. Ribatti D, Presta M, Vacca A, Ria R, Giuliani R, Dell'Era P, Nico B, Roncali L, Dammacco F (1999) Human erythropoietin induces a pro-angiogenic phenotype in cultured endothelial cells and stimulates neovascularization in vivo. Blood 93:2627–2636
2. Anagnostou A, Lee ES, Kessimian N, Levinson R, Steiner M (1990) Erythropoietin has a mitogenic and positive chemotactic effect on endothelial cells. Proc Natl Acad Sci U S A 87: 5982–5987
3. Yamaji R, Okada T, Moriya M, Naito M, Tsuruo T, Miyatake K, Nakano Y (1996) Brain capillary endothelial cells express two forms of erythropoietin receptor mRNA. Eur J Biochem 239:494–500
4. Ashley RA, Dubuque SH, Dvorak B, Woodward SS, Williams SK, Kling PJ (2002) Erythropoietin stimulates vasculogenesis in neonatal rat mesenteric microvascular endothelial cells. Pediatr Res 51:472–478
5. Carlini RG, Reyes AA, Rothstein M (1995) Recombinant human erythropoietin stimulates angiogenesis in vitro. Kidney Int 55:546–553
6. Vogel V, Kramer HJ, Bäcker A, Meyer-Lehnert H, Jelkmann W, Fandrey J (1997) Effects of erythropoietin on endothelin-1 synthesis and cellular calcium messanger system in vascular endothelial cells. Am J Hypertens 10:289–297
7. Kertesz N, Wu J, Chen TH, Sucov HM, Wu H (2004) The role of erythropoietin in regulating angiogenesis. Dev Biol 276:101–110
8. Monestiroli S, Mancuso P, Burlini A, Pruneri G, Dell'Agnola C, Gobbi A, Martinelli G, Bertolini F (2001) Kinetics and viability of circulating endothelial cells as surrogate angiogenesis marker in an animal model of human lymphoma. Cancer Res 61:4341–4344
9. Mancuso P, Burlini A, Pruneri G, Goldhirsch A, Martinelli G, Bertolini F (2001) Resting and activated endothelial cells are increased in the peripheral blood of cancer patients. Blood 97:3658–3661
10. Heeschen C, Aicher A, Lehmann R, Fichtlscherer S, Vasa M, Urbich C, Mildner-Rihm C, Martin H, Zeiher AM, Dimmeler S (2003) Erythropoietin is a potent physiologic stimulus for endothelial progenitor cell mobilization. Blood 102:1340–1346
11. Yasuda Y, Fujita Y, Matsuo T, Koinuma S, Hara S, Tazaki A, Onozaki M, Hashimoto M, Musha T, Ogawa K, Fujita H, Nakamura Y, Shiozaki H, Utsumi H (2003) Erythropoietin regulates tumour growth of human malignancies. Carcinogenesis 24:1021–1029
12. Kayser K, Gabius HJ (1992) Analysis of expression of erythropoietin-binding sites in human lung carcinoma by the biotinylated ligand. Zentralbl Pathol 138:266–270
13. Ribatti D, Vacca A, Roccaro AM, Crivellato E, Presta M (2003) Erythropoietin as an angiogenic factor. Eur J Clin Invest 33:891–896
14. Perelman N, Selvaraj SK, Batra S, Luck LR, Erdreich-Epstein A, Coates TD, Kalra VK, Malik P (2003) Placenta growth factor activates monocytes and correlates with sickle cell disease severity. Blood 102:1506–1514
15. Wang L, Zhang Z, Wang Y, Zhang R, Chopp M (2004) Treatment of stroke with erythropoietin enhances neurogenesis and angiogenesis and improves neurological function in rats. Stroke 35:1732–1737
16. Nakano M, Satoh K, Fukumoto Y, Ito Y, Kagaya Y, Ishii N, Sugamura K, Shimokawa H (2007) Important role of erythropoietin receptor to promote VEGF expression and angiogenesis in peripheral ischemia in mice. Circ Res 100:662–669
17. Okazaki T, Ebihara S, Asada M, Yamanda S, Niu K, Arai H (2008) Erythropoietin promotes the growth of tumors lacking its receptor and decreases survival of tumor-bearing mice by enhancing angiogenesis. Neoplasia 10:932–939
18. Hardee ME, Cao Y, Fu P, Jiang X, Zhao Y, Rabbani ZN, Vujaskovic Z, Dewhirst MW, Arcasoy MO (2007) Erythropoietin blockade inhibits the induction of tumor angiogenesis and progression. PLoS One 2:e549
19. Olujohungbe A, Handa S, Holmes J (1987) Do erythropoietin accelerate malignant transformation in multiple myeloma? Postgrad Med J 73:163–164
20. Bunworasate U, Arnouk H, Minderman H, O'Loughlin KL, Sait SN, Barcos M, Stewart CC, Baer MR (2001) Erythropoietin-dependent transformation of myelodysplastic syndrome to acute monoblastic leukemia. Blood 98:3492–3494
21. Ribatti D (2002) A potential role of erythropoietin in angiogenesis associated with myelodysplastic syndromes. Leukemia 16:1890
22. Arcasoy MO, Amin K, Karayal AF, Chou SC, Raleigh JA, Varia MA, Haroon ZA (2002) Functional significance of erythropoietin receptor expression in breast cancer. Lab Invest 82:911–918
23. Yasuda Y, Fujita Y, Masuda S, Musha T, Ueda K, Tanaka H, Fujita H, Matsuo T, Nagao M, Sasaki R, Nakamura Y (2002) Erythropoietin is involved in growth and angiogenesis in malignant tumours of female reproductive organs. Carcinogenesis 23:1797–1805
24. Ribatti D (2010) The chick embryo chorioallantoic membrane in the study of angiogenesis

and metastasis. Springer Science+Business Media, Dordrecht

25. Ribatti D, Marzullo A, Nico B, Crivellato E, Ria R, Vacca A (2003) Erythropoietin as an angiogenic factor in gastric carcinoma. Histopathology 42:246–250
26. Ribatti D, Marzullo A, Gentile A, Longo V, Nico B, Vacca A, Dammacco F (2007) Erythropoietin/erythropoietin-receptor system is involved in angiogenesis in human hepatocellular carcinoma. Histopathology 50:591–596
27. Acs G, Zhang PJ, McGrath CM, Acs P, McBroom J, Mohyeldin A, Liu S, Lu H, Verma A (2003) Hypoxia-inducible erythropoietin signaling in squamous dysplasia and squamous cell carcinoma of the uterine cervix and its potential role in cervical carcinogenesis and tumor progression. Am J Pathol 162:1789–1806
28. Westenfelder C, Baranowski RL (2000) Erythropoietin stimulates proliferation of human renal carcinoma cells. Kidney Int 58:647–657
29. Ribatti D, Poliani PL, Longo V, Mangieri D, Nico B, Vacca A (2007) Erythropoietin/erythropoietin receptor system is involved in angiogenesis in human neuroblastoma. Histopathology 50:636–641
30. Batra S, Perelman N, Luck LR, Shimada H, Malik P (2003) Pediatric tumor cells express erythropoietin and a functional erythropoietin receptor that promotes angiogenesis and tumor cell survival. Lab Invest 83:1477–1487
31. Krieg M, Marti HH, Plate KH (1998) Coexpression of erythropoietin and vascular endothelial growth factor in nervous system tumors associated with von Hippel-Lindau tumor suppressor gene loss of function. Blood 92:3388–3393
32. Kokhaei P, Abdalla AO, Hansson L, Mikaelsson E, Kubbies M, Haselbeck A, Jernberg-Wiklund H, Mellstedt H, Osterborg A (2007) Expression of erythropoietin receptor and in vitro functional effects of epoetins in B-cell malignancies. Clin Cancer Res 13:3536–3544
33. Nico B, Annese T, Guidolin D, Finato N, Crivellato E, Ribatti D (2011) Epo is involved in angiogenesis in human glioma. J Neuroncol 102:51–58

Chapter 16

Photoreceptor Degeneration in Mice: Adeno-Associated Viral Vector-Mediated Delivery of Erythropoietin

Pasqualina Colella and Alberto Auricchio

Abstract

The exogenous delivery of erythropoietin (EPO) and EPO derivatives (EPO-Ds) represents a valuable strategy to protect the retina from degeneration. In this chapter we describe a method to deliver EPO and the EPO derivative S100E in the light-damage model of induced retinal degeneration using adeno-associated viral (AAV) vectors and to evaluate the functional and morphological protection of the retina from light damage.

Key words Retina, Photoreceptor degeneration, Adeno-associated viral vectors, Erythropoietin derivatives, Light damage

1 Introduction

Inherited retinal degenerations (IRDs) are common untreatable conditions leading to blindness that include retinitis pigmentosa (RP) (1), Leber congenital amaurosis (LCA) (2), and cone-rod dystrophies (3). IRDs are characterized by high genetic heterogeneity. Nonetheless, they show a common phenotype consisting in photoreceptor (PR) dysfunction and progressive PR cell loss (4). Based on this, the delivery of neurotrophic/antiapoptotic genes, proteins, or compounds may represent a valuable strategy to either inhibit/slow down PR cell death or to sustain PR function and/or survival independently of the mutated gene (5–8). Notably, many studies have reported that EPO exerts a protective function on retinal photoreceptors (PRs) (9–12), ganglion cells (GCs) (13–17), and on the retinal pigment epithelium (RPE) (18–20) (Fig. 1). The delivery of EPO through viral vectors allows sustained and/or regulatable expression (21, 22) and would be advantageous to avoid repeated systemic or intraocular administrations that would be required to treat chronic and progressive diseases like IRDs. Vectors derived from the adeno-associated virus (AAV) are the most efficient

Pietro Ghezzi and Anthony Cerami (eds.), *Tissue-Protective Cytokines: Methods and Protocols*, Methods in Molecular Biology, vol. 982, DOI 10.1007/978-1-62703-308-4_16, © Springer Science+Business Media, LLC 2013

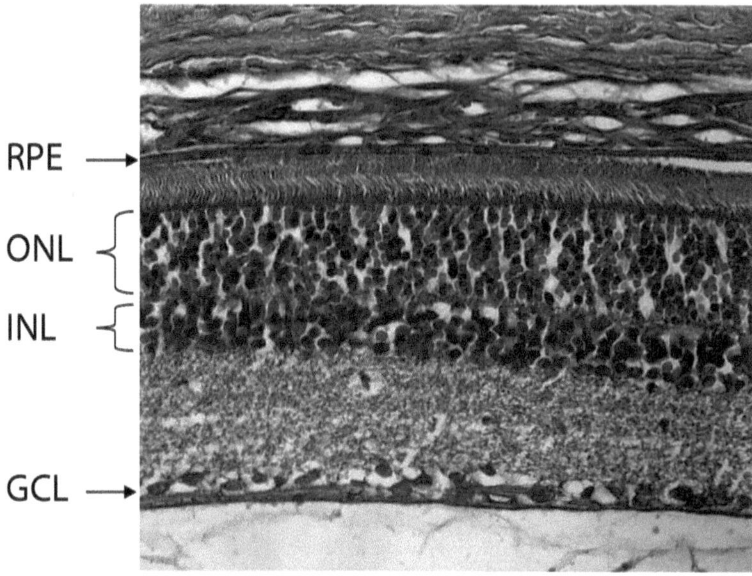

Fig. 1 Representative picture of the rat retina. The retina is organized into three layers of cells: (1) the outer nuclear layer (ONL) composed by the photoreceptor nuclei; (2) the inner nuclear layer (INL) composed by nuclei of amacrine, Muller, bipolar, and horizontal cells, and (3) the ganglion cell layer (GCL) containing ganglion and displaced amacrine cells. The retinal pigment epithelium (RPE) lies above the retina. The depicted retinal section is taken from a Lewis rat at postnatal day 65 and is stained with hematoxylin and eosin. The picture magnification is 20×

tools for retinal gene transfer given their capability to selectively tranduce different retinal cell types and to ensure long-term gene expression after a single vector administration (8, 23). In addition, the safety and efficacy of AAV-mediated gene delivery has been proven in preclinical models (8, 24) and clinical trials (25–33) aimed at treating IRDs. Notably, the systemic AAV-mediated delivery of EPO has resulted to be effective at protecting PRs in animal models of induced and inherited retinal degeneration (34). However, the systemic delivery of EPO (either as protein or gene) is associated with undesirable side effects such as significant hematocrit increase (34, 35), increased thrombotic risk (36), and platelet hyperreactivity (37, 38) that could lead to serious life-threatening consequences (39, 40). EPO side effects may be particularly relevant considering that the EPO doses required to achieve tissue-protection are higher than those required for erythropoiesis (i.e., for the treatment of anemia) (41). Therefore, the treatment of chronic and/or progressive conditions such as IRDs would require continuous EPO delivery resulting in deleterious effects. To overcome EPO side effects due to its function in the hematopoietic system several EPO derivatives (EPO-Ds) have been generated that should retain tissue protective functions while avoiding the EPO-mediated erythropoiesis (42–45) as well as other hormone side effects (38). Notably, we and others have showed that AAV-mediated delivery of non-erythropoietic EPO-Ds protects PRs from induced (46) and inherited (46, 47) degeneration.

Based on our study (46) we describe a method to deliver EPO and the EPO derivative S100E (S100E) in the rat model of light damage by AAV vectors and to evaluate the protective effects of EPO and S100E upon PR function and survival.

In addition, this method may be useful to study the mechanisms responsible for the protective effects induced by EPO and EPO-Ds in the retina.

2 Materials

2.1 Generation of Plasmids for AAV Vector Production Expressing EPO and S100E

1. A plasmid for AAV vector production containing the AAV2 inverted terminal repeats (ITRs) (e.g., the pAAV2.1-CMV-EGFP (48) that can be supplied by the author) (see Note 1).
2. Murine cDNA (e.g., from kidney, retina, or other tissues expressing EPO).
3. Sequence-specific forward and reverse oligonucleotides for the amplification of the EPO coding sequence (CDS). The oligonucleotides must contain the restriction sites for the directional cloning of the EPO CDS into the pAAV2.1-CMV-EGFP plasmid (e.g., NotI and BamHI).
4. Standard equipment and reagents for PCR amplification.
5. Standard equipment for DNA electrophoresis.
6. Standard equipment and reagents for DNA gel extraction and PCR purification.
7. Standard equipment and reagents for: molecular cloning (e.g., restriction enzymes and DNA ligase); bacterial transformation (e.g., *Escherichia coli* DH5α cells), growth (e.g., LB medium), and selection (e.g., Ampicillin); low-scale plasmid purification reagents.
8. Quick Change II XL Site-Directed Mutagenesis kit (Agilent Technologies, USA).
9. Sequence-specific forward and reverse oligonucleotides to generate the S100E mutant by EPO mutagenesis (see Note 2).
10. SmaI and SnabI restriction enzymes.
11. Endotoxin-free plasmid purification materials (e.g., Endofree Plasmid Giga kit, Qiagen, USA).

2.2 EPO and S100E Protein Expression In Vitro

2.2.1 Cell Culture and Transfection

1. COS7 cells (see Note 3).
2. Standard equipment and reagents for cell cultures.
3. Culture medium: Dulbecco's Modified Eagle's Medium (DMEM, high glucose) containing 10% of Fetal Bovine Serum (FBS), 2 mM l-Glutamine and 1× antibiotic-antimycotic solution.
4. PolyFect Transfection Reagent (Qiagen).

2.2.2 Protein Extraction, Quantification, and Western Blot Analysis

1. RIPA buffer: 50 mM Tris–HCl pH 8.0, 150 mM NaCl, 1% NP40, 0.5% Na-Deoxycholate, 1 mM EDTA pH 8.0, 0.1% SDS dissolved in H_2O.
2. Protease inhibitors (e. g. Complete Protease inhibitor cocktail tablets, Roche; dissolve 1 tablet in 500 μL of H_2O to achieve a 100× stock solution, store at −20°C).
3. 200 mM (200×) Phenylmethanesulfonylfluoride (PMSF) dissolved in isopropyl alcohol (store the solution at −20°C).
4. Standard equipment and reagents for protein quantification: a colorimetric assay (e.g., BCA protein assay reagent, Thermo fisher scientific, USA or BIO-RAD protein assay, Bio-Rad, USA); a spectrophotometer or a microplate reader (supplied with 560 and 595 nm filters).
5. Standard equipment and reagents for Western Blot analysis (SDS-PAGE and immunoblot).
6. 1× Tris–HCl Buffered Saline-Tween (1× TBS-T): 50 mM Tris–HCl, 150 mM NaCl, 0.05% Tween 20.
7. 5% nonfat dry milk dissolved in 1× TBS-T.
8. Anti EPO primary antibody (EPO N-19, sc-1310 Santa Cruz Biotechnology Inc., USA).
9. Secondary antibody (e.g., Anti-Goat IgG HRP Affinity Purified, HAF109, R&D Systems, USA).
10. Enhanced chemiluminescence (ECL) substrate.

2.3 Subretinal and Intramuscular AAV Injections

2.3.1 Intramuscular and Subretinal Vector Administration

1. Purified AAV vectors encoding EPO, S100E, or EGFP as negative control (see Notes 4–6).
2. Sterile 1× phosphate buffered saline pH 7.4 (1× PBS): 137 mM NaCl, 2.7 mM KCl, 4.3 mM Na_2HPO_4, 1.47 mM KH_2PO_4 dissolved in H_2O.
3. H_2O.
4. 70% Ethanol.
5. Avertin (Anesthetic): 1.25% w/v of 2,2,2-tribromoethanol and 2.5% v/v of 2-methyl-2-butanol. To prepare a working solution mix 2.5 g of 2,2,2-tribromoethanol in 5 mL of 2-methyl-2-butanol in the dark (i.e., use a bottle wrapped in foil), let the powder dissolve heating it at 50°C on a magnetic stirrer. Once the powder is completely dissolved add 200 mL of an isotonic saline solution (0.9% NaCl in H_2O) and let the solution mix overnight (about 12 h) at room temperature using a magnetic stirrer. Filter the solution through a 0.22 μm filter. Store at 4°C wrapped in foil.

2.3.2 Subretinal Injections

1. Hamilton syringe 95RN 5 μL (Hamilton Company, USA).
2. Dumont tweezers #5, 11 cm long, 0.05 × 0.01 mm bent (45°) tips.
3. Vannas scissors, 82 mm long, 9.5 mm sharp straight tips.
4. 30 gauge needles (to be mounted on 1 mL syringes).
5. Removable 33 gauge needle to be mounted on the Hamilton syringe (Hamilton Company, USA).
6. Dissecting microscope.
7. Fiber optic light sources.
8. Ophthalmic antibiotic ointment.

2.3.3 Intramuscular Injections

1. Hamilton syringe 710RN 100 μL (Hamilton Company, USA).
2. Sterile stiff support.
3. Removable 30 gauge needles to be mounted on the 710RN Hamilton syringe (Hamilton Company, USA).
4. Dumont tweezers #5, 11.5 cm long, 0.17 × 0.10 mm serrated tips.
5. Mini dissecting scissors 9.5 cm long, with sharp fine tips.
6. Sterile razor blades.
7. A topical antibiotic ointment.
8. Catgut (monofilament polyglyconate synthetic absorbable suture).
9. Adhesive gauze.

2.4 Light Damage

1. Lewis rats (see Note 7).
2. A custom-made apparatus for light damage composed of: eight 36-W white bulbs, two transparent Plexiglas sheets, four transparent Plexiglas cages, a wired support to lodge the cages (see Note 8).
3. Luxometer (see Notes 9 and 10).

2.5 Electroretinogram Recordings

1. A room to work under red dim light.
2. A completely obscured and highly ventilated box to adapt the rats to the dark.
3. A stereotaxic support.
4. Avertin (see Subheading 2.3.1).
5. A heating pad.
6. Two reference needle electrodes.
7. One ground needle electrode.
8. Two gold-plated electrodes.

9. A differential amplifier band pass-filtered of 0.1–1,000 Hz (CSO, Italy).
10. A rack-mounted terminal block (BNC 2090, National Instruments, USA).
11. A Ganzfeld globe (CSO, Italy).
12. Software to control the light stimuli (e.g., Ganzfeld stimulator, custom), to register and analyze the Electroretinogram (ERG) responses (e.g., ERG2Ch.vi, custom).

2.6 Sample Collection and Hematocrit Measurement

2.6.1 Serum and Blood Collection and Hematocrit Measurement

1. Avertin (see Subheading 2.3.1).
2. Sterile stiff support.
3. Sterile razor blade.
4. Heparinized capillaries.
5. 0.5 M EDTA pH 8.0.
6. Microfuge (for 1.5/2 mL test tubes).
7. Haematocrit 210 microfuge (24 hematocrit capillaries).
8. Clay sealant and specimen holding tray (24 hematocrit capillaries).
9. Ruler.

2.6.2 Anterior Chamber Fluid Collection

1. Hamilton syringe 1702RN 25 μL (Hamilton Company, USA).
2. Removable 33 gauge needles: RN NDL 6/PK (33/*/*) (to be mounted on the Hamilton syringe 1702RN) (Hamilton Company, USA).

2.6.3 Eye Harvesting and Embedding

1. Cauterizer: low-temperature cautery (World Precision Instruments, Germany).
2. Mini dissecting scissors, 9.5 cm long, sharp fine tips.
3. Dumont tweezers #5, 11.5 cm long, 0.17 × 0.10 mm serrated tips.
4. 1× PBS pH 7.4.
5. 4% Paraformaldehyde in 1× PBS.
6. 30% Sucrose in 1× PBS.
7. Tissue freezing medium: Optimum cutting temperature (O.C.T.) formulation of water-soluble glycols and resins.
8. Disposable embedding molds peel-a-way truncated-T12 (Polysciences Europe GmbH, Germany).
9. Dry-ice denatured alcohol bath.

2.7 Quantification of EPO and S100E Proteins by ELISA Assay

1. Microplate ELISA reader equipped with the 450 and 540 (or 570) nm filters.
2. Mouse/rat erythropoietin ELISA (e.g., Quantikine, R&D Systems, USA).
3. Horizontal orbital microplate shaker.

2.8 Histological Analysis

1. Cryostat.
2. Round cryostat chuck 1–1/2″ (40 mm).
3. Sectioning Blades.
4. Trimming Blades.
5. Glass slides.
6. Dissecting microscope.
7. 4′,6-Diamidino-2-phenylindole (DAPI).
8. Rectangular coverslips (24 × 60).
9. Transparent nail-polish.
10. Fluorescence microscope.

3 Methods

The methods described in this chapter include: Subheading 3.1 the generation of plasmids for AAV vector production including expression cassettes encoding for EPO and the EPO derivative S100E. Subheading 3.2 the expression analysis of the EPO and S100E proteins in vitro; Subheading 3.3 the subretinal and intramuscular injections of AAV vectors encoding EPO and S100E in rats; Subheading 3.4 the setup of the light-damage model of induced retinal degeneration in rats; Subheading 3.5 the evaluation of retinal function by full-field electroretinogram (ERG) recordings; Subheading 3.6 biological sample (blood, serum, anterior chamber fluid (ACF), and eye) collection and hematocrit measurement; Subheading 3.7 the quantification of EPO and S100E protein levels in biological fluids of rats; Subheading 3.8 the evaluation of rat photoreceptor survival by retinal histology.

3.1 The Generation of Plasmids for AAV Vector Production Including Expression Cassettes Encoding for EPO and the EPO Derivative S100E

3.1.1 Basics of AAV Vector Production

The detailed methods and protocols for AAV vector production are described in "*Viral Vectors for Gene Therapy. Methods and Protocols*" (49). Briefly, AAV vectors can be produced at high yields by transient triple transfection of mammalian cells or infection of packaging eukaryotic and insect cells (49). Independently of the specific method used for AAV vector production, it is necessary to generate a *cis* plasmid that contains the expression cassette with the gene of interest flanked by the viral inverted terminal repeats (ITRs, 145 bp). The ITRs are the only viral sequences required in *cis* to allow the encapsidation of the AAV vector genomes for the production of viable vector particles during the AAV vector production process (Fig. 2).

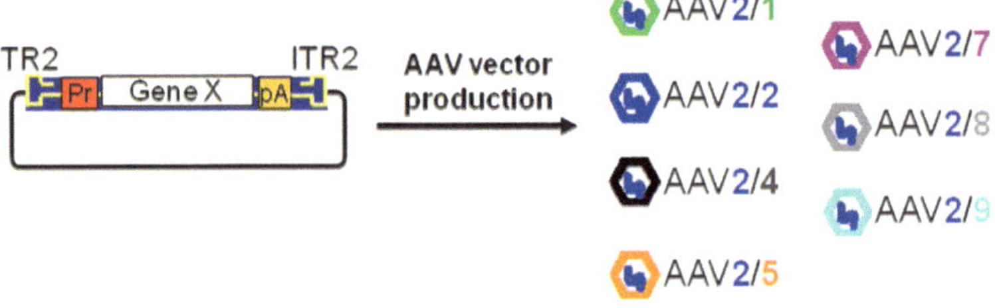

Fig. 2 Versatility of AAV vector production. The picture shows a schematic representation of a *cis* plasmid required for AAV vector production. The *cis* plasmid contains the expression cassette encoding for the gene of interest flanked by the ITRs from AAV2. The availability of a different capsid from each serotype and the versatility of the AAV production system allow to easily exchange capsids among various AAV serotypes thus creating hybrid vectors containing a genome with the same ITRs (e.g., from AAV2) and the capsid from a different serotype (54). Vectors obtained through this trans-capsidation system are named AAV2/*n*, where the first number refers to the ITRs and the second to the capsid. Each AAV serotype has its own transduction characteristics (i.e., target cells, kinetics of transgene expression) and the user can choose the most appropriate for the cell/organ of interest (55). *Pr* promoter, *pA* polyadenylation signal sequence, *gene X* gene of interest

3.1.2 Generation of the pAAV2.1-CMV-EPO Plasmid

1. PCR-amplify the full-length coding sequence (CDS) of the murine EPO gene using as template murine cDNA, a forward oligonucleotide containing the NotI restriction site and a reverse oligonucleotide containing the BamHI restriction site.
2. Recover the amplicon of the expected size by gel extraction.
3. Sequence the amplicon to be sure that it corresponds to the EPO CDS and it does not contain mutations.
4. Digest the pAAV2.1-CMV-EGFP3 plasmid (48) using NotI and BamHI restriction enzymes to remove the EGFP CDS and purify the fragment of interest by gel extraction.
5. Digest the amplified EPO CDS using NotI and BamHI restriction enzymes and purify the fragment.
6. Ligate the EPO CDS and the digested pAAV2.1-CMV-EGFP3 backbone fragments following standard procedures.
7. Select the clone of interest using standard bacterial transformation and selection methods.
8. Once the clone pAAV2.1-CMV-EPO has been obtained, make a large plasmid preparation using endotoxin-free solutions and materials (e.g., Endofree Plasmid Giga kit, Qiagen, S.P.A., Italy).

3.1.3 Generation of the pAAV2.1-CMV-S100E Plasmid

1. Mutagenize the EPO CDS following the protocol contained in the Quick Change II XL Site-Directed Mutagenesis (Agilent Technologies, USA) datasheet. Use the pAAV2.1-CMV-EPO

plasmid as template and the oligonucleotides designed to generate the S100E substitution (see Note 2).

2. Confirm the mutagenesis by sequencing the *pAAV2.1-CMV-S100E* plasmid.
3. Make a large plasmid preparation of the *pAAV2.1-CMV-S100E* using endotoxin-free solutions and materials.

3.2 Expression Analysis of the EPO and the S100E Proteins In Vitro

The proper expression of the EPO and the S100E proteins from the pAAV2.1 plasmids can be assessed in vitro prior to the generation of AAV vectors.

3.2.1 COS7 Cells Transfection

1. Maintain COS7 cells in DMEM containing 10% FBS and 2 mM l-Glutamine and 1× antibiotic-antimycotic solution.
2. Plate the cells in 6-well-plates at a density of 1×10^5 cell/well.
3. Twenty-four hours later, transfect the cells with 2 μg of either pAAV2.1-CMV-EPO, pAAV2.1-CMV-S100E or pAAV2.1-CMV-EGFP as negative control, using the PolyFect Transfection Reagent (follow the protocol supplied by the provider).
4. Thirty-six hours later incubate the cells in serum-free DMEM.
5. Harvest the media and the cells 12 h later.
6. Collect the media and spin at 4000 × *g* for 10 min at 4°C in a microfuge to remove any cellular debris.
7. Store the media at 4°C (or −80°C for long storage).
8. Scrape the adherent cells in 1 mL of 1× PBS, spin the cells at 4000 × *g* for 10 min at 4°C in a microfuge.
9. Lyse the cells in RIPA buffer (50 μL) in the presence of 1× protease inhibitors and 1 mM PMSF for 30 min at 4°C.
10. Spin the cell lysates at 16,000 × *g* for 15 min at 4°C in a microfuge to remove the cellular debris and membranes.
11. Quantify both the cell lysates and the media using standard procedures.
12. Separate the medium and lysate samples by 12% SDS-PAGE and blot on a membrane following standard procedures (see Note 11).
13. Block the membrane in 5% milk/TBS-T for 1 h on a shaker.
14. Incubate the membrane with the anti-EPO antibody diluted 1:400 in 5% milk/TBS-T for 2 h at room temperature on a shaker.
15. Wash the membrane three times in TBS-T for 5 min each time at room temperature on a shaker.

Incubate the membrane in the secondary antibody diluted 1:3,000 in 5% milk/TBS-T at room temperature for 45 min on a shaker.

17. Wash the membrane three times in TBS-T (10 min each time) at room temperature on a shaker.
18. Incubate the membrane in ECL for 5 min at room temperature.
19. Visualize the protein bands by juxtaposing the filter to X-ray films and by developing X-ray films under red dim light following standard procedures (see Note 12).
20. Once the expression of the EPO and S100E proteins has been confirmed generate AAV vectors using either the *pAAV2.1-CMV-EPO* or the *pAAV2.1-CMV-S100E* (see Notes 13–15).

3.3 Subretinal and Intramuscular Injection of AAV Vectors in Albino Rats

3.3.1 AAV Vector Preparation

1. Thaw the required number of AAV vector aliquots; the AAV vectors required are the AAV-EPO, the AAV-S100E and the AAV-EGFP as negative control (see Notes 4–6).
2. If necessary, dilute AAV vectors using 1× PBS as follows (steps 3 and 4).
3. For the subretinal injections dilute each AAV vector to 1×10^9 GC (genome copies)/μL considering that 2 μL will be injected in each eye (thus, the dose will be 2×10^9 GC/eye).
4. For the intramuscular injections dilute:
   (a) AAV-EPO to 1×10^8 GC/μL, considering that 100 μL per rat are required (thus, the dose will be 1×10^{10} GC/rat) (see Note 16).
   (b) AAV-S100E to 1×10^9 GC/μL considering that 100 μL per rat are required (thus, the dose will be 1×10^{11} GC/rat).
   (c) The control vector AAV-EGFP to 1×10^9 GC/μL considering that 100 μL per rat are required (thus, the dose will be 1×10^{11} GC/rat).
5. Put the working dilution on ice, store the leftover of the AAV vector stocks at −80°C.

3.3.2 Subretinal Injections (See Notes 17–20)

1. Anesthetize the 4 week-old rat with an intraperitoneal injection of 2 mL/g of body weight of Avertin (prewarmed at 37°C). Wait until the rat is properly anesthetized.
2. Place the anesthetized rat under a dissecting microscope and position the light fibers as preferred. Make sure that the eye to be injected is clearly observable.
3. Delicately grab the temporal conjunctiva using tweezers.
4. Using scissors carefully perform a single cut of the conjunctiva at the level of the pars plana (Fig. 3).

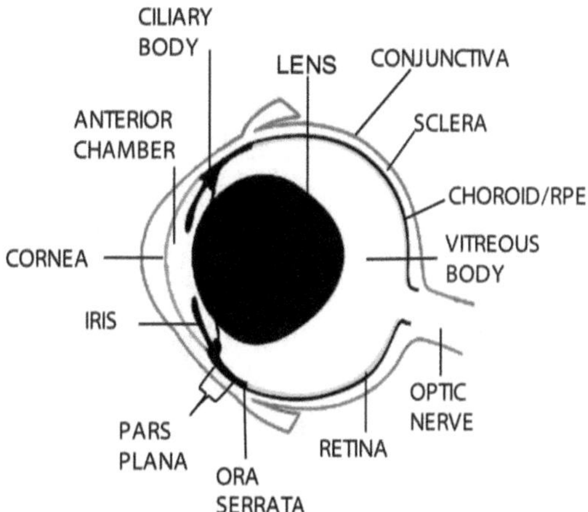

Fig. 3 Schematic representation of eye. Transverse section of the murine eye

5. Introduce the lower blade of the scissors into the cut and perform a 180° conjunctival peritomy extending superiorly and inferiorly. You have thus created a conjunctival flap to which you can grab to by using the tweezers in the following steps of the procedure.
6. Using the tweezers grab onto the conjunctival flap and pull it towards the nasal side of the eye to expose its temporal posterior part.
7. Punch through the sclera and choroid at the level of the pars plana posteriorly to the sclera-corneal margin using the tip of a sterile 30 gauge needle. Usually a drop of vitreous fluid leaks from the punctured site.
8. Using the tweezers grab onto the conjunctival flap and pull it towards the nasal side of the eye to expose its temporal posterior part and the puncture.
9. An assistant has previously loaded the Hamilton syringe with 2 μL (see Subheading 3.3.1, step 3) of the AAV solution. Avoid making air bubbles when aspirating the vector solution.
10. Insert the tip of the 33 gauge needle (mounted on the Hamilton syringe) through the puncture into the subretinal space (the area between the RPE/choroid and the retina). To maximize the chances of applying the vector-containing solution to the subretinal space the needle needs to be oriented tangentially to the globe and with the hub turned towards the operator.
11. Keep the hand holding the syringe steady while the assistant presses the piston to deliver the injection solution. Notably, in albino animals the operator should be able to clearly see the needle through the transparent sclera and to monitor the liquid delivery. A proper subretinal injection would result in a local

retinal detachment that can be visualized by the operator as a bleb.

12. Keep maintaining the conjunctival flap using tweezers while slowly and carefully removing the needle.
13. Let the conjunctival flap go.
14. Gently apply a thin layer of ointment containing antibiotics and steroids to the operated area.
15. Starting from step 1, repeat the procedure to inject the contralateral eye of the same rat.

    Put the cage containing the anesthetized animal on a hot plate set at 37°C to limit the anesthesia-induced hypothermia and to speed up the recovery from anesthesia (the complete recovery usually takes 2–4 h) (see Notes 21–24).

3.3.3 Intramuscular Injections

1. Anesthetize the 4 week-old rat (see Subheading 3.4, step 1).
2. Gently lay the rat flat on its stomach.
3. Gently secure the rat's legs to the support using an adhesive gauze and sterilize the area of the leg to be operated using a disinfectant.
4. Carefully shave the area to be injected.
5. Using tweezers grip the skin and carefully make a cut to expose the underlying gastrocnemius muscle (see Note 25).
6. Keep the skin layer steady with the tweezers.
7. An assistant has previously loaded the Hamilton syringe (armed with a 30 Gauge needle) with 100 μL (see Subheading 3.3.1, step 4) of the AAV vector solution. Avoid making air bubbles when aspirating the vector solution.
8. Puncture the gastrocnemius allowing only the tip of the needle to enter the muscle.
9. Hold the syringe steadily while the assistant presses the piston to deliver the injection volume (up to 40 μL). A proper intramuscular injection results in a local muscular lump.
10. Inject 30 μL in two additional locations of the gastrocnemius to evenly distribute the dose of vector in the muscle.
11. Slowly and carefully remove the needle.
12. Suture the skin using the catgut.
13. Gently apply a thin layer of antibiotic ointment to the operated area.
14. Relocate the rat into its cage.
15. Put the cage containing the anesthetized animal on a hot plate set at 37°C to limit the anesthesia-induced hypothermia and to speed up the recovery from anesthesia (see Notes 21–24).

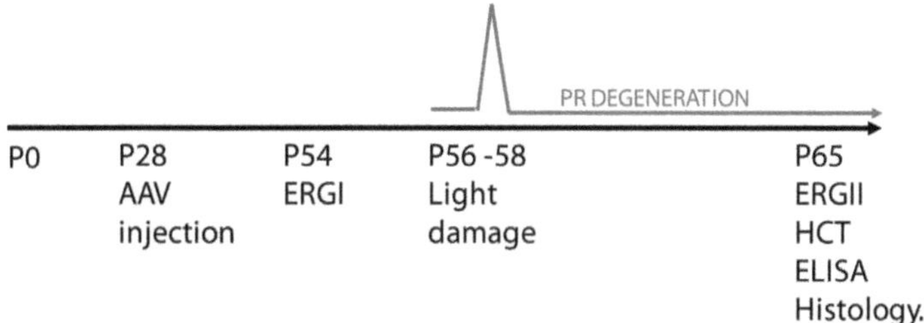

Fig. 4 Schematic representation of the light damage protocol. Setup of the light damage experiment: (1) Lewis rats are injected systemically or subretinally with AAV vectors at postnatal day (P) 28; (2) the rats are reared under physiological 12 h light/dark cycle for 4 weeks; (3) ERGs are recorded before light damage to evaluate retinal function in treated and untreated eyes; (4) the rats are submitted to light damage for 48 h; (5) the rats are reared for 7 days under physiological 12 h light/dark cycle for photoreceptor degeneration to occur; (6) ERGs are recorded at P65 to evaluate the protection from light damage; (7) the rats are sacrificed and hematocrit variations, EPO and S100E expression levels, and PR survival are assessed. The *gray line* represents the acute damage induced by light that causes photoreceptor (PR) degeneration

3.4 Setup of the Light-Damage Model of Induced Retinal Degeneration in Albino Rats

1. Following AAV injection (see Subheading 3.3) rear the rats in a physiological 12 h light/dark cycle for 4 weeks (Fig. 4).
2. At around postnatal day (P) 56 put the rats separately (1 rat/cage, 4 total rats-cages/light damage cycle) in clear Plexiglas cages and insert the cages in the light damage apparatus (see Notes 8–10, 26).
3. House the rats under continuous light exposure for 48 h (see Notes 27 and 28) (Fig. 4).
4. After light damage put the rats back under a physiological 12 h light–dark cycle for 1 week.
5. Assess retinal function by electrophysiological analyses; then, sacrifice the rats to collect biological samples for further analysis including assessing AAV-mediated gene expression and performing histological analyses (see Subheadings 3.5–3.8) (Fig. 4).

3.5 Evaluation of Retinal Function by Full-Field Electroretinogram Recordings See Note 29

1. Adapt the rat to the dark by placing it in a darkened ventilated box for at least 3 h.
2. Perform all the following steps under dim red light.
3. Anesthetize the rat (see Subheading 3.3.2, step 1).
4. Lay the rat on its stomach on a sterotaxic support with a heating pad (37.5°C) to maintain it warm during the recordings.
5. Secure the rat using an adhesive gauze.
6. Insert two inactive reference needle electrodes subcutaneously at the center of the scalp.
7. Insert the ground needle electrode subcutaneously in the tail.

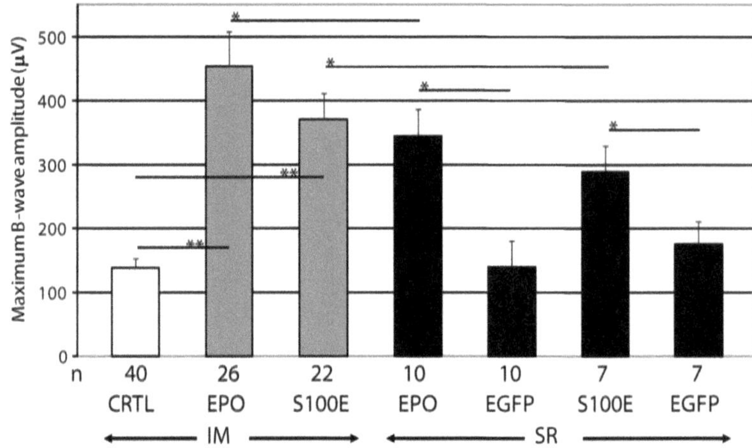

Fig. 5 AAV-mediated delivery of EPO and S100E protects PR function in light-damaged rats. The chart shows the scotopic b-wave amplitudes recorded by ERGs at the highest luminance (20 cd s/m^2). The scotopic b-wave amplitude from age-matched, non-damaged Lewis rats at 20 cd s/m^2 is 1,291.3 ± 92 μV (n = 8). The values are means ± SE. The number (n) of eyes in each group is depicted under the corresponding bar. *CTRL* uninjected rats, *IM* intramuscular AAV injection, *SR* subretinal AAV injection. The statistical significance is calculated by ANOVA. *p value ≤ 0.05. **p value ≤ 0.001. In addition to those depicted in the chart significant differences (p value < 0.05) were found between the following groups of rats: CTRL vs. SR EPO; CTRL vs. SR S100E; IM EPO vs. SR EGFP; IM S100E vs. SR EGFP; IM EPO vs. SR S100E (46). Reproduced from Colella P. 2011 (46) by permission of Oxford University Press

8. Place the active gold-plated electrode on the cornea close to the lower eyelids (one electrode per eye).
9. Place the rat head in the Ganzfeld globe.
10. Set the parameters for stimulation and recording using the appropriate software (see Notes 30–34).
11. Start the stimulation, which, from now on, is completely automated and managed by the ERG software.
12. Once the recordings are completed put the rat back in its cage.
13. Put the cage containing the anesthetized animal on a hot plate set at 37°C to limit the anesthesia-induced hypothermia and to speed up the recovery from anesthesia (the complete recovery usually takes 2–4 h).
14. Analyze and compare the ERGs between treated and untreated eyes (see Note 35).

Figure 5 shows the ERG data obtained in rats treated with AAV and light damaged or not (46).

3.6 Sample Collection and Hematocrit Measurement

3.6.1 Serum Collection

1. Anesthetize the rat (see Subheading 3.3.2, step 1).
2. Gently lay the rat flat on its stomach upon a sterile support.
3. Using a sterile razor blade make a short transversal cut upon the caudal vein (see Note 36).
4. Collect the leaking blood using heparinized capillary tubes.
5. Transfer the blood from two capillaries to a test tube containing 15 μL of 0.5 M EDTA pH 8.0.
6. Gently flip the test tube containing EDTA and the collected blood once to avoid clotting.
7. Centrifuge the blood sample at 2,000 × *g* for 5 min in a microfuge at 4°C.
8. Transfer the serum (the upper phase) to a new test tube (short-term storage at 4°C; long-term storage at −80°C).

3.6.2 Hematocrit Measurement

1. Collect one capillary of blood from the rat (see Subheading 3.6.1, steps 1–4).
2. Seal one end of the capillary with the sealant and place it on the holding tray.
3. Centrifuge the capillary for 3 min at 16,000 × *g* at room temperature using a microfuge equipped with a capillary rotor. Following centrifugation the blood is separated into two distinct volumes: the packed cell volume (PCV) and the serum volume.
4. Measure the length of the PCV fraction and the length of the total blood volume contained in the capillary using a ruler.
5. Calculate the hematocrit dividing the length of the PCV by the total length of the blood sample and express the measurement as percentage.

Table 1 shows the hematocrit values measured in rats treated with AAV (46).

3.6.3 Anterior Chamber Fluid Collection

1. Anesthetize the rat (see Subheading 3.3.2, step 1).
2. Lay the rat on his stomach and place it with the eyes under an adequate light source.
3. Keep the rat head steady with one hand.
4. Insert a 33 gauge needle (mounted on the Hamilton syringe) into the anterior chamber of the eye. The anterior chamber is visible like a clear transparent ring (see Note 37).
5. Aspirate the ACF (about 10 μL) with the syringe.
6. Remove the needle slowly and carefully and transfer the ACF into a test tube. Store the ACF at 4°C or −80°C for short- and long-term storage, respectively.

Table 1
Hematocrit levels in light-damaged rats

	HCT (%)	
LD Ctrl	47.2±1.3 (24)	
	EPO	S100E
LD IM	71.5±5.7 (10)*	52.3±5.2 (10)*
LD SR	57.8±7.9 (10)*	49.8±3.7 (10)

Ctrl uninjected rats, *HCT* hematocrit, *LD* light-damaged rats, *IM* intramuscular injection, *SR* subretinal injection. The measurements were performed at P65. The values are means ± SD, the number of eyes analyzed is shown in brackects (46). The asterisk indicates hematocrits significantly increased (t-test $p<0.05$) compared with controls. Reproduced from Colella P. 2011 (46) by permission of Oxford University Press

3.6.4 Eye Harvesting and Embedding

1. Following blood and ACF collection (see Subheadings 3.6.2 and 3.6.3), sacrifice the rat using a lethal dose of Avertin or by cervical dislocation.
2. Mark the temporal side of the eye searing the sclera with a cauterizer (see Note 38).
3. Enucleate the eye making several incisions around the globe with scissors. Be gentle to avoid damaging the eye.
4. Put the eye in a test 2 mL tube containing 4% PFA for 12 h at 4°C.
5. Wash the eye for 5 min twice in 1× PBS.
6. Dehydrate the eye in 30% sucrose until the eye drops to the bottom of the 2 mL tube (usually about 4–8 h for a rat eye).
7. Take the eye with tweezers and lean it on a paper to remove dripping sucrose.
8. Fill in a plastic mold with tissue freezing medium (O.C.T.).
9. Lay the eye on one side at the bottom of the mold. Lay the eye so that neither the anterior nor posterior part of the eye faces the bottom of the mold (see Note 39).
10. Mark the side of the mold (left or right) that corresponds to the temporal side of the eye (see Note 40).
11. Allow the O.C.T. to solidify putting the mold in a bath with dry ice and denatured ethanol.
12. Store the mold at –80°C.

3.7 Quantification of EPO and S100E Protein Levels in the Serum and ACF of Rats

1. Thaw the serum and ACF samples (if stored at -80°C).
2. Use the Quantikine Mouse/Rat Erythropoietin ELISA.
3. Prepare all reagents, standard dilutions, and positive control as described in the ELISA datasheet.
4. Measure two different dilutions of each sample.
5. Use 10–50 μL of serum and 0.1–10 μL of ACF (see Notes 41–44).

Table 2 shows the EPO and the S100E protein levels measured in rats injected with AAV (46).

3.8 Evaluation of Photoreceptor Survival by Retinal Histology

3.8.1 Eye Cryosectioning

1. Set the temperature of the main cryostat chamber at -26°C and that of the specimen at -23°C.
2. Keep the mold containing the O.C.T.-embedded eye in the cryostat chamber for 15 min to let the mold reach -26°C.
3. Remove the O.C.T. block from the mold, put a thin layer of fresh O.C.T. on the specimen chuck, put the block on the chuck so that the bottom of the block (where the eye is located) is in front of you and is parallel to the chuck base (see Note 45).
4. Let the thin layer of O.C.T. solidify so that the block is firmly fixed on the chuck.
5. Insert the chuck in the clamp and tighten the wing nuts to hold it firmly.
6. Insert the blade into the blade holder.
7. Move the cryostat arm backwards with the wheel so that the block is located behind the blade.
8. Try to align the flat bottom side of the block and the blade edge so that they are parallel.
9. Remove the excess of O.C.T. surrounding the embedded eye by hand using a trimming blade.
10. Set the section thickness at 20–30 μm and start trimming.
11. Pick up each section by leaning the slide towards the polarized side of the section. The section will immediately stick to the slide.
12. Observe the section under a dissecting microscope (see Note 46).
13. Once you achieve an eye section in which it is possible to distinguish the anterior (e. g. cornea and ciliary epithelium) from the posterior (retina and RPE) part of the eye, you can stop trimming and set the section thickness to 10 μm.
14. Number 30 slides (from 1 to 30) and mark the side of the slide corresponding to the temporal side of the section with a pencil.
15. Place the first series (1–15) on the cryostat in front of you.

Table 2
Levels of EPO and S100E proteins in the serum and ACF of light-damaged rats

Serum levels (pg/mL)			ACF levels (pg/mL)	
	EPO	**S100E**	**EPO**	**S100E**
LD Ctrl	U (20)		U (10)	
LD IM	90±16 (10)	261±24 (10)	12±6 (10)	17±2 (10)
LD SR	4±2 (4/10)	6±2 (3/10)	2400±750 (9)	4550±1370 (8)

ACF anterior chamber fluid, *Ctrl* uninjected rats, *LD* light-damaged rats, *IM* intramuscular injection, *SR* subretinal injection, *U* levels below the ELISA sensitivity which is 0.65 pg/mL for murine EPO and 0.33 pg/mL for rat EPO. The measurements were performed at P65. The values are means ± SE, the number of eyes analyzed is shown in brackects (46). Reproduced from Colella P. 2011 (46) by permission of Oxford University Press

Move the cryostat arm backwards with the wheel so that the following section results 10 μm thick.

17. Continue sectioning and picking up each section in a serial way (put the first section on slide 1, the second on slide 2, the 16th section again on slide 1 and so on) (see Note 47).
18. When the first series of slides is completed, start filling in the second series of slides (slides 15–30).
19. Check the quality of the sections under a dissecting microscope every once in a while.
20. Air-dry the sections 1–3 h at room temperature.
21. Store the slides either at −20°C or at −80°C.

3.8.2 DAPI Staining of the Sections

1. Thaw two slides per eye and air dry them for 15 min at room temperature.
2. Rehydrate the sections rinsing the slides twice in 1× PBS.
3. Put a few drops of 4′,6-diamidino-2-phenylindole (DAPI) per slide.
4. Cover the slides with cover-slips.
5. Fix the cover-slip passing a thin layer of transparent nail polish on the coverslip borders.
6. Store the stained sections either at +4°C (short-term) or at −20°C (long-term) in the dark.

3.8.3 Assessment of Photoreceptor Nuclei Rows

1. Observe the eye sections stained with DAPI under a fluorescence microscope.
2. Observe all the sections from each slide using a 5–10× magnification.
3. Select three sections from the central part of the eye, which have the largest globe diameter and contain the optic nerve head (ONH).

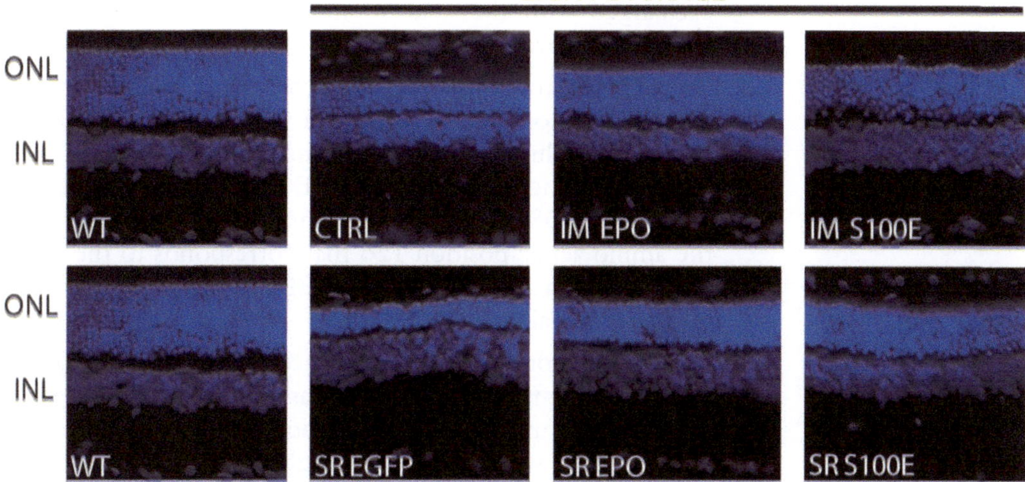

Fig. 6 AAV-mediated delivery of EPO and S100E sustains PR survival in light-damaged rats. The figure shows the representative retinal histology of Lewis rats whether light-damaged or not. *CTRL* uninjected animals, *IM* intramuscular AAV injection, *INL* inner nuclear layer, *ONL* outer nuclear layer, *SR* subretinal AAV injection, *WT* wild type, age-matched non-damaged Lewis rats. The picture magnification is 40×. Retinal sections were stained with DAPI to count the rows of PR nuclei in the ONL (46). Reproduced from Colella P. 2011 (46) by permission of Oxford University Press

4. Identify the outer nuclear layer (ONL) that contains the photoreceptor nuclei.
5. Count the number of rows of photoreceptor nuclei that are present in the ONL in each selected section at 40× magnification. Count the rows in two different locations on each side of the ONH: i.e., the nasal and central on one side, the temporal and central on the other side (see Notes 48 and 49).
6. Average the number of rows of PR nuclei counted in each eye and compare PR survival between treated and untreated rats.

Figure 6 shows representative retinal sections taken from treated and untreated rats (46).

4 Notes

1. The pAAV2.1-CMV-EGFP3 plasmid contains the 5′ ITR from the AAV serotype 2, the Cytomegalovirus (CMV) promoter, the SV40 intron, the CDS of the enhanced green fluorescent protein (EGFP), the bovine growth hormone polyadenylation signal (bGH polyA) and the 3′ ITR from the AAV serotype 2.

The EGFP CDS and the regulatory elements can be easily exchanged with the gene of interest and other regulatory elements, respectively.

2. To perform the S100E site-directed mutagenesis we used primers that contained a nucleotide triplet (e.g., GAA) corresponding to glutamate, thus resulting in the substitution of the triplet encoding serine in the S100E PCR product. In the wild type EPO CDS the AGT triplet encoding serine is located at the amino acidic position 126 that corresponds to the amino acidic position 100 in the mature EPO protein (devoid of the hormone signal peptide).
3. The transfection of the plasmids encoding EPO and S100E in cell lines other than COS7 is also feasible, however it is possible that the EPO and S100E glycosylation profile (and their apparent molecular weight) may vary slightly between different cell lines.
4. AAV vectors are handled as biosafety level 2 genetically-modified micro-organisms in several countries. Please observe the legislation regarding biosafety levels operating procedures of the country where you plan to perform the experiments using AAV vectors.
5. For the optimal storage of AAV vectors store them in aliquots at −80°C in 5% sterile glycerol.
6. No reduction of viral infectivity is observed up to eight sequential freeze and thaw events (Doria M. and Auricchio A, unpublished data). However we suggest using a freshly thawed aliquot each time.
7. It is possible to use albino rat strains other than Lewis (e.g., Sprague Dawley or Wistar) but it is important to consider that the genetics as well as the age and the diet can affect the degree of damage induced by light.
8. During light damage, animals need to have free access to food and water.
9. To compare the protective effect of a specific treatment from light damage it is important to use an adequate number of sham-treated controls. In particular, it is important to have a sham-treated control per light damage cycle to evaluate the variability of the damage that mainly depends on the luminance variability and the rat behavior (e.g., sleeping time and eyes position with respect to the light sources).
10. The retinal light damage varies also with light history, circadian rhythm, and the prior extent of dark-adaptation (50). Therefore, to reduce the light-damage variability between groups of rats undergoing distinct light-damage cycles it is important to start each light damage cycle at a fixed time of day.

11. The amount of cellular lysate to be analyzed by SDS-PAGE may vary from 10 to 50 μg per sample depending on the transfection efficiency. We suggest loading at least two different dilutions of each sample to ensure the clear detection of the EPO and S100E proteins.
12. The murine EPO protein has a molecular mass of 30.4 kDa, however, due to its secondary modifications (glycosylation), it migrates with a higher apparent size of 34–38 kDa on SDS-PAGE.
13. Several academic and nonacademic providers sell AAV vector preparations. The only material that the applicant has to provide is a certain amount of the *cis* AAV plasmid containing the gene of interest (Fig. 2). The AAV vector core facility operating at the Telethon Institute of Genetics and Medicine (TIGEM) requires 1.3 mg of endotoxin-free pAAV2.1 plasmid for the production of an AAV preparation (4 mL of vector preparation containing ≥10e12 AAV genome copies).
14. Before providing the plasmid containing the transgene of interest to the facility it is important to check the integrity of the AAV ITRs that flank the transgene expression cassette. The AAV ITRs are palindromes difficult to sequence unambiguously. Thus, the assessment of ITR integrity is usually based on the analysis of their restriction pattern. In the pAAV2.1 plasmid a fragment of 125 bp containing the AAV2 ITRs can be excised by digestion of pAAV2.1 with the SmaI and SnabI restriction enzymes.
15. In our experiments in the light damage rat model of photoreceptor degeneration we used AAV serotype 1 (AAV2/1) because it is very efficient at transducing both muscle and the retinal pigment epithelium (RPE), which we used for systemic and intraocular expression of EPO and its derivatives, respectively (48, 51).

    The intramuscular injection of AAV-EPO in rats at the dose of 1×10^{11} GC determines significant mortality at about 3 weeks following the injections. This is probably due to the extremely high increase in hematocrits that is caused by the systemic overexpression of EPO. The mortality rate of rats injected intramuscularly with AAV-EPO can be significantly reduced using a tenfold lower AAV-EPO dose (1×10^{10} GC/per rat).
17. The surgical procedures have to be performed in accordance with the Association for Research in Vision and Ophthalmology Statement for the Use of Animals in Ophthalmic and Vision Research and with the specific country regulations.
18. For unambiguous identification number the rats before the injections using ear punches or toe clipping.

19. To properly perform subretinal injection, it is necessary that the novice observes an experienced operator and that he/she performs the injection on his/her own several times to get acquainted before attempting the injections in actual experiments. A good practice is important to efficiently transduce the retina and/or the RPE and to avoid damages to the retina and the lens.
20. The efficiency of retinal and RPE transduction following subretinal injection may vary from one injection to another due to the intrinsic variability of the procedure.
21. The animals that are treated with AAV vectors should be housed in a separate room of the animal facility.
22. The surgical instruments must be cleaned with disinfectant and sterilized under UV light before and after their use.
23. To clean the Hamilton syringes and needles after the injections wash them five times with 70% Ethanol and five times with H_2O, then put the syringes under UV light overnight.
24. When injecting different AAV vectors during the same surgical session (e.g., an AAV vector encoding for EPO, another one encoding for S100E or EGFP) clean the Hamilton syringes as described in Note 11 when switching from one vector to another.
25. If experienced, operators can perform intramuscular injections directly into the gastrocnemius of 4-week old rats without making a skin incision to expose the muscle.
26. The custom device for light damage in rats is made of a wired support for the cages (that allows the cage insertion and removal as a drawer). The wired support we use can accommodate a total of four cages (therefore a total of four rats per light damage cycle, as each rat has to be housed in a single cage to avoid that more animals in the same cage shield each other from the light). A white Plexiglas sheet is located above and below the cages (at a distance of 30 cm). White light bulbs are placed above and below (four bulbs per side) the superior and inferior Plexiglas sheets, respectively. The lights can be turned on and off at will. The bottom of each cage is grated to allow the excrements to pass through the cage in order to be removed by the animal house assistants.
27. The average luminance generated by the lights should be around 2,000 lx as measured using a luxometer before the rats are submitted to light damage. After many hours of use the light bulbs may wear out, therefore if the luminance measured is significantly lower than 2,000 lx the light bulbs need to be replaced.
28. The retinal damage induced by light can be modulated by varying the light intensity as well as the onset and duration of

light exposure (5, 52). In nocturnal animals such as rats the intensity of light required to induce retinal damage needs to be only two to three times above the normal room lighting (200–400 lx) (50).

29. The full-field electroretinogram (ERG) is used to evaluate the electrical response of the retina to light stimulation. The ERG is a test used worldwide to assess the function of the retina in humans and animal models. The method is based on the stimulation of the retina with bright light such as a flash produced by light emitting diodes or lamps. The flash of light evokes a recordable bi-phasic wave composed by a negative a-wave and a positive b-wave. The amplitude of the a- and b-waves is used to evaluate retinal function (53).

    The ERG recording and analysis require practice. In our opinion it is necessary for a novice to follow an experienced operator to set up the whole ERG apparatus, to perform the recordings and to analyze the data. Extensive practice is also important to achieve reproducible results.

30. Use the Ganzfeld software to set: (1) the pattern of light stimuli (that are generated by the Ganzfeld globe); (2) the intensity of the background light. Use the ERG2Ch.vi software to: (1) register and analyze the evoked ERG traces; (2) vary the ERG amplification; (3) control the background noise. The ERG2Ch.vi software computerizes the ERG traces and allows their visualization on the oscilloscope in real time.

31. Amplify the ERG traces 10,000 folds (1K gain), set the band pass-filter at 0.3–1,000 Hz and the notch filter at 50 Hz.

32. Record the scotopic ERG using 12 flashes of increasing intensity starting from 0.0001 to 20.0 cd s/m^2 (candela per second per square meter). The luminance varies from –4 to +1.3 log cd s/m^2 and the gap between flashes of different intensities is 0.6 log. To minimize the noise, the operator averages three ERG traces per stimulus over the indicated luminance range using a temporal frequency of 0.7 Hz.

33. Record photopic ERGs using a constant background illumination (50 cd/m^2) and ten flashes having the same intensity (20 cd s/m^2). To minimize the noise, the operator averages the resulting ten ERG traces at the indicated intensity (20 cd s/m^2).

34. The scotopic and photopic ERGs can be recorded one after the other.

35. The amplitudes of the a- and b-waves generated from a single eye are plotted as a function of the increasing flash intensity.

36. In case a little amount of blood is recovered from the caudal vein it is also possible to collect it from the jugular vein. Briefly, anesthetize the rat and lay it flat on its back, shave the fur

between the sternum and the shoulders and insert a 23 gauge needle (mounted on a 1 mL syringe) in the jugular vein, which should be clearly visible under the neck skin. If you prick the vein correctly, the blood will flow into the hub of the needle. Keep the syringe firm and draw the blood. Once you have obtained the desired amount of blood, slowly pull back the needle and apply a slight amount of pressure to the puncture site. Usually 100–200 μL of blood is sufficient to measure by ELISA the concentration of EPO or S100E in the serum of the AAV injected rats.

37. The ACF can be also collected punching the anterior chamber with a 30 gauge needle mounted on a 1 mL syringe. After punching a hole in the anterior chamber the ACF flows out from the puncture site and can be collected using a 20 μL pipette. However, using this method it is possible that some tear fluid is collected together with the ACF.
38. The eye can be also oriented using a catgut suture. Follow step 1 described in Subheading 3.6.4. Then, pass the catgut superficially through the temporal conjunctiva. Fix the catgut making a double knot and finally cut the catgut in excess using scissors. The suture can be performed under a dissecting microscope, if required.
39. During eye cryosectioning (see Subheading 3.8.1) the lens may tend to fold on the retina wasting the section. To avoid this, it is suggested to embed the optic cup rather than the whole eye by removing the cornea and the lens. To do this, dissect the eye under a dissecting microscope. Briefly, grasp the cornea using tweezers, cut away the cornea using Vannas scissors and finally remove the lens using tweezers.
40. In most cases the cauterized part of the sclera is not visible to the naked eye. Therefore, to embed the eye in the oriented way it is necessary: (1) to observe it under a dissecting microscope; (2) to position it in the desired orientation using tweezers; and (3) to transfer it into the mold preserving the desired orientation.
41. The intramuscular injection of AAV vectors in rats is quite a reproducible method therefore the levels of EPO and S100E in the serum of injected rats should not vary widely. The protein (EPO and S100E) concentrations obtained following AAV delivery should be included in the EPO standard curve range if assaying 10–50 μL of serum by ELISA.
42. Unlike intramuscular injections, the ACF levels of EPO and S100E proteins are highly variable due to the inherent variability of the subretinal injection as well as of the ACF collection procedure. However, 0.1–5 μL of ACF by ELISA should allow measuring EPO and S100E concentrations included in

the EPO standard curve range. Instead, to measure the ACF levels of EPO and S100E of rats injected with AAV systemically it is recommended to assay 2–10 μL of ACF.

43. It is important to underline that in control rats the levels of endogenous EPO (in both serum and ACF) are below the ELISA sensitivity that corresponds to 0.33 pg/mL.
44. The rat and murine EPO proteins are similarly recognized by the Quantikine Mouse/Rat Erythropoietin ELISA as reported by the producer.
45. Before putting the O.C.T. block on the chuck it is important to mark the block side that corresponds to the temporal part of the eye. To do this, make a transverse incision in the block using the trimming blade.
46. Make sure that the eyes are cut along the horizontal meridian.
47. The sections are progressively distributed on slides so that each slide contains representative sections of either the first (slides 1–15) or the second half (sections 16–30) of the eye.
48. In rats the light damage is typically confined to rods and the cell loss is mainly restricted to the outer nuclear layer (ONL) that only contains the PR nuclei. Therefore, the measurement of the ONL thickness or the count of rows of PR nuclei represent appropriate methods to quantify the damaging effect of light. Another method that can be used to estimate PR cell loss after light damage is based on the quantification of rod-specific products (e. g. proteins involved in rod phototransduction).
49. It is important to consider that in rodents retinal light damage results in a graded response with areas containing little or no damage located close to more severely affected regions. Within the superior hemisphere, light damage is particularly severe in a region that is between 1 and 2 mm away from the optic nerve head (50).

References

1. Hartong DT, Berson EL, Dryja TP (2006) Retinitis pigmentosa. Lancet 368:1795–1809
2. den Hollander AI, Roepman R, Koenekoop RK, Cremers FP (2008) Leber congenital amaurosis: genes, proteins and disease mechanisms. Prog Retin Eye Res 27:391–419
3. Hamel CP (2007) Cone rod dystrophies. Orphanet J Rare Dis 2:7
4. Wright AF, Chakarova CF, Abd El-Aziz MM, Bhattacharya SS (2010) Photoreceptor degeneration: genetic and mechanistic dissection of a complex trait. Nat Rev Genet 11:273–284
5. LaVail MM, Yasumura D, Matthes MT et al (1998) Protection of mouse photoreceptors by survival factors in retinal degenerations. Invest Ophthalmol Vis Sci 39:592–602
6. Wenzel A, Grimm C, Samardzija M, Reme CE (2005) Molecular mechanisms of light-induced photoreceptor apoptosis and neuroprotection for retinal degeneration. Prog Retin Eye Res 24:275–306
7. Buch PK, MacLaren RE, Ali RR (2007) Neuroprotective gene therapy for the treatment of inherited retinal degeneration. Curr Gene Ther 7:434–445
8. Colella P, Cotugno G, Auricchio A (2009) Ocular gene therapy: current progress and future prospects. Trends Mol Med 15:23–31

9. Grimm C, Wenzel A, Groszer M et al (2002) HIF-1-induced erythropoietin in the hypoxic retina protects against light-induced retinal degeneration. Nat Med 8:718–724
10. Junk AK, Mammis A, Savitz SI et al (2002) Erythropoietin administration protects retinal neurons from acute ischemia-reperfusion injury. Proc Natl Acad Sci U S A 99:10659–10664
11. Grimm C, Wenzel A, Stanescu D et al (2004) Constitutive overexpression of human erythropoietin protects the mouse retina against induced but not inherited retinal degeneration. J Neurosci 24:5651–5658
12. Rex TS, Wong Y, Kodali K, Merry S (2009) Neuroprotection of photoreceptors by direct delivery of erythropoietin to the retina of the retinal degeneration slow mouse. Exp Eye Res 89:735–740
13. Yamasaki M, Mishima HK, Yamashita H et al (2005) Neuroprotective effects of erythropoietin on glutamate and nitric oxide toxicity in primary cultured retinal ganglion cells. Brain Res 1050:15–26
14. Tsai JC, Wu L, Worgul B et al (2005) Intravitreal administration of erythropoietin and preservation of retinal ganglion cells in an experimental rat model of glaucoma. Curr Eye Res 30:1025–1031
15. Zhong L, Bradley J, Schubert W et al (2007) Erythropoietin promotes survival of retinal ganglion cells in DBA/2J glaucoma mice. Invest Ophthalmol Vis Sci 48:1212–1218
Weishaupt JH, Rohde G, Polking E et al (2004) Effect of erythropoietin axotomy-induced apoptosis in rat retinal ganglion cells. Invest Ophthalmol Vis Sci 45:1514–1522
17. King CE, Rodger J, Bartlett C et al (2007) Erythropoietin is both neuroprotective and neuroregenerative following optic nerve transection. Exp Neurol 205:48–55
18. Wang ZY, Shen LJ, Tu L et al (2009) Erythropoietin protects retinal pigment epithelial cells from oxidative damage. Free Radic Biol Med 46:1032–1041
19. Chung H, Lee H, Lamoke F et al (2009) Neuroprotective role of erythropoietin by antiapoptosis in the retina. J Neurosci Res 87:2365–2374
20. Zhang J, Wu Y, Jin Y et al (2008) Intravitreal injection of erythropoietin protects both retinal vascular and neuronal cells in early diabetes. Invest Ophthalmol Vis Sci 49:732–742
21. Auricchio A, Rivera VM, Clackson T et al (2002) Pharmacological regulation of protein expression from adeno-associated viral vectors in the eye. Mol Ther 6:238–242
22. Lebherz C, Auricchio A, Maguire AM et al (2005) Long-term inducible gene expression in the eye via adeno-associated virus gene transfer in nonhuman primates. Hum Gene Ther 16:178–186
23. Surace EM, Auricchio A (2008) Versatility of AAV vectors for retinal gene transfer. Vision Res 48:353–359
24. Allocca M, Tessitore A, Cotugno G, Auricchio A (2006) AAV-mediated gene transfer for retinal diseases. Expert Opin Biol Ther 6:1279–1294
25. Simonelli F, Ziviello C, Testa F et al (2007) Clinical and molecular genetics of Leber's congenital amaurosis: a multicenter study of Italian patients. Invest Ophthalmol Vis Sci 48:4284–4290
26. Bainbridge JW, Smith AJ, Barker SS et al (2008) Effect of gene therapy on visual function in Leber's congenital amaurosis. N Engl J Med 358:2231–2239
27. Maguire AM, Simonelli F, Pierce EA et al (2008) Safety and efficacy of gene transfer for Leber's congenital amaurosis. N Engl J Med 358:2240–2248
28. Cideciyan AV, Aleman TS, Boye SL et al (2008) Human gene therapy for RPE65 isomerase deficiency activates the retinoid cycle of vision but with slow rod kinetics. Proc Natl Acad Sci U S A 105:15112–15117
29. Hauswirth WW, Aleman TS, Kaushal S et al (2008) Treatment of Leber congenital amaurosis due to RPE65 mutations by ocular subretinal injection of adeno-associated virus gene vector: short-term results of a phase I trial. Hum Gene Ther 19:979–990
30. Cideciyan AV, Hauswirth WW, Aleman TS et al (2009) Vision 1 year after gene therapy for Leber's congenital amaurosis. N Engl J Med 361:725–727
31. Cideciyan AV, Hauswirth WW, Aleman TS, Kaushal S et al (2009) Human RPE65 gene therapy for Leber congenital amaurosis: persistence of early visual improvements and safety at 1 year. Hum Gene Ther 20:999–1004
32. Simonelli F, Maguire AM, Testa F et al (2009) Gene therapy for Leber's congenital amaurosis is safe and effective through 1.5 years after vector administration. Mol Ther 18(3):643–650
33. Stein L, Roy K, Lei L, Kaushal S (2011) Clinical gene therapy for the treatment of RPE65-associated Leber congenital amaurosis. Expert Opin Biol Ther 11:429–439
34. Rex TS, Allocca M, Domenici L et al (2004) Systemic but not intraocular Epo gene transfer protects the retina from light- and genetic-induced degeneration. Mol Ther 10: 855–861
35. Brines M, Cerami A (2005) Emerging biological roles for erythropoietin in the nervous system. Nat Rev Neurosci 6:484–494
36. Rosenzweig MQ, Bender CM, Lucke JP et al (2004) The decision to prematurely terminate

a trial of R-HuEPO due to thrombotic events. J Pain Symptom Manage 27:185–190
37. Stohlawetz PJ, Dzirlo L, Hergovich N et al (2000) Effects of erythropoietin on platelet reactivity and thrombopoiesis in humans. Blood 95:2983–2989
38. Kirkeby A, Torup L, Bochsen L et al (2008) High-dose erythropoietin alters platelet reactivity and bleeding time in rodents in contrast to the neuroprotective variant carbamyl-erythropoietin (CEPO). Thromb Haemost 99:720–728
39. Belonje AM, Voors AA, van Gilst WH, van Veldhuisen DJ (2007) Erythropoietin in chronic heart failure. Congest Heart Fail 13:289–292
40. Ehrenreich H, Weissenborn K, Prange H et al (2009) Recombinant human erythropoietin in the treatment of acute ischemic stroke. Stroke 40:e647–e656
41. Brines M, Cerami A (2008) Erythropoietin-mediated tissue protection: reducing collateral damage from the primary injury response. J Intern Med 264:405–432
42. Campana WM, Misasi R, O'Brien JS (1998) Identification of a neurotrophic sequence in erythropoietin. Int J Mol Med 1:235–241
43. Leist M, Ghezzi P, Grasso G et al (2004) Derivatives of erythropoietin that are tissue protective but not erythropoietic. Science 305:239–242
44. Brines M, Patel NS, Villa P et al (2008) Nonerythropoietic, tissue-protective peptides derived from the tertiary structure of erythropoietin. Proc Natl Acad Sci U S A 105:10925–10930
45. Brines M, Cerami A, Coleman T (2009) Tissue protective peptides and uses thereof. US Patent Application 2009/0221482, Warren Pharmaceuticals, Inc.
46. Colella P, Iodice C, Di Vicino U et al (2011) Non-erythropoietic erythropoietin derivatives protect from light-induced and genetic photoreceptor degeneration. Human Mol Genet 20(11):2251–2262
47. Sullivan T, Kodali K, Rex TS (2011) Systemic gene delivery protects the photoreceptors in the retinal degeneration slow mouse. Neurochem Res 36:613–618
48. Auricchio A, Kobinger G, Anand V et al (2001) Exchange of surface proteins impacts on viral vector cellular specificity and transduction characteristics: the retina as a model. Hum Mol Genet 10:3075–3081
49. Merten OW, Al-Rubeai M (2011) Viral vectors for gene therapy. Methods and protocols, vol 737, 1st edn. Humana, New York
50. Organisciak DT, Vaughan DK (2010) Retinal light damage: mechanisms and protection. Prog Retin Eye Res 29:113–134
51. Xiao W, Chirmule N, Berta SC et al (1999) Gene therapy vectors based on adeno-associated virus type 1. J Virol 73:3994–4003
52. LaVail MM, Unoki K, Yasumura D et al (1992) Multiple growth factors, cytokines, and neurotrophins rescue photoreceptors from the damaging effects of constant light. Proc Natl Acad Sci U S A 89:11249–11253
53. Weymouth AE, Vingrys AJ (2008) Rodent electroretinography: methods for extraction and interpretation of rod and cone responses. Prog Retin Eye Res 27:1–44
54. Auricchio A (2003) Pseudotyped AAV vectors for constitutive and regulated gene expression in the eye. Vision Res 43: 913–918
55. Vandenberghe LH, Wilson JM, Gao G (2009) Tailoring the AAV vector capsid for gene therapy. Gene Ther 16:311–319

Chapter 17

Myocardial Infarction: Cardioprotection by Erythropoietin

Mark I. Talan and Roberto Latini

Abstract

Extensive research during the last decade demonstrated that a single systemic administration of erythropoietin (EPO) lead to significant attenuation of myocardial infarction (MI) induced in animals, mostly small rodents, either by a myocardial ischemia followed by reperfusion or by a permanent ligation of a coronary artery. Both methods are critically reviewed with the aim of helping the reader in appreciating key issues in the translation of experimental results to the clinic. Results of several clinical trials in patients with acute MI completed to date failed to demonstrate beneficial effects of EPO, and thus put into question the validity of results obtained in animal models. Comprehensive review of design and results of animal experiments and clinical trials presented here allowed authors to postulate that therapeutic window for EPO during developing MI is very narrow and was possibly missed in negative clinical trials. This point was illustrated by the negative outcome of experiment in the rat model of MI in which timing of EPO administration was similar to that in clinical trials. The design of future clinical trials should allow for a narrow therapeutic window of EPO. Given current standards for onset-to-door and door-to-balloon time the optimal time for EPO administration should be just prior to PCI.

Key words Heart, Cardiac, Ischemia, Ischemia-reperfusion, Cardiomyocytes, Clinical trials

1 Introduction

Chronic Heart Failure (CHF)—a debilitating condition that reached epidemic proportion in the Western society and responsible for probably the largest portion of health care budget predominantly affecting the aging population. The major cause for severity of developing CHF is the extent of myocardial damage during an ischemic episode leading to a myocardial infarction (MI). Thus, one strategic approach to reduce the severity of CHF is to limit the extent of myocardial damage during an ischemic event. In spite of a great progress achieved during the last 20 years to treat an acute MI, mainly using timely revascularization of myocardium, the need for additional therapies to limit myocardial damage remains on the forefront of cardiovascular research.

The first experimental study demonstrating that exogenous recombinant human erythropoietin rhEPO protects myocardium

Pietro Ghezzi and Anthony Cerami (eds.), *Tissue-Protective Cytokines: Methods and Protocols*, Methods in Molecular Biology, vol. 982, DOI 10.1007/978-1-62703-308-4_17,

from ischemic damage had been published in 2003. Since then, numerous successful studies in different experimental models had been reported and summarized in several review papers (1–6). Now, almost a decade and several phase I and II clinical trials later, we still failed to translate these highly promising experimental findings into clinical advances. In this review we reassess the wells of experimental data and outcomes of clinical trials and attempt to analyze the reason of a chasm between them. Should we write off the decade of intensive experimental work as yet another example of biological differences between species or is there another, currently overlooked, reason for the negative outcomes of clinical trials? To this effect we systematically review all available in vivo experimental studies, conducted in two major experimental models, MI induced by a permanent ligation of a coronary artery or by an ischemia–reperfusion method. We put an emphasis of this review on the experimental details such as effective doses and therapeutic window. Then, in the light of these experimental details, we reassess the design and clinical outcomes of reported clinical trials.

1.1 Review of published studies

1.1.1 Permanent occlusion of a coronary artery model

Animal experiments in the model of permanent occlusion of a coronary artery are summarized in the Table 1. The first experiment had been published in 2003 (7). Rats were subjected to a surgical occlusion of the left descending coronary artery and were treated either with intraperitoneal injection of 5,000 IU/kg of erythropoietin (type is not reported) or with saline. After 60 min of occlusion rats were sacrificed and apoptosis was assessed in the area at risk via TUNEL staining. The number of TUNEL positive nuclei in the AAR were dramatically reduced ($p < 0.0001$) in EPO-treated rats.

Two more studies were published at the same time. In one (8) rats were treated with a single systemic (intraperitoneal) injection of Epoetin-alfa (3,000 IU/kg) immediately following a permanent occlusion of an anterior descending coronary artery. Apoptosis in the area at risk (peri-infarct area) was assessed in the subgroup of rats 24 h after coronary occlusion and was found to be reduced by 50%. The rest of the animals were followed up with serial echocardiography during 8 weeks. Experiment was concluded by a histological evaluation of the heart. At the end of experiment left ventricular (LV) volumes were significantly smaller and EF was significantly higher in the EPO treated group comparing with control, untreated animals. Moreover, histologically measured MI size was fourfold smaller in the EPO-treated animals. Interestingly, hematocrit level measured during the first 3 weeks after surgery was not elevated in the EPO-treated group. Results of this first study had been consistently reproduced by the same group later on in 2005, 2006, and 2011 in different experiments in which a single injection of 3,000 IU/kg of EPO immediately after coronary occlusion served to elicit a standard effect to compare with other types of interventions (9–12).

Table 1
Animal experiments (permanent occlusion)

First author and references	Date	Species	EPO, type and doses (IU/kg)	Time of EPO: Relative to occlusion	Time of EPO: Relative to reperfusion	End-points: End-points	End-points: Outcomes
Tramontano (7)	2003	Rat	EPO, 3,000 single dose	0 min	NA	At 60 min: apoptosis in the AAR	↓ $p<0.001$
Moon (8)	2003	Rat	Epoetin alfa, 3,000 single dose	0 min	NA	At 24 h: apoptosis in the AAR 8 weeks: MI size LVEDV LVESV LVEF	↓ 50% $p<0.001$ ↓ 80% $p<0.001$ ↓ $p<0.002$ ↓ $p<0.005$ ↑ $p<0.001$
Parsa (13)	2003	Rabbits	Epoetin alfa, 5,000 single dose	0 min	NA	At 6 h: apoptosis At 4 days: AAR MI size	↓ $p<0.001$ ns ↓ $p<0.1$
Van der Meer (14)	2005	Rat	Darbepoetin alfa, 8,000 single dose or repeated	(1) 0 min or (2) 0 min + every 3 weeks or (3) +3 weeks + every 3 weeks	NA	At 9 weeks: MI size Capillary density	↓ (1) $p<0.01$ ↓ (2) $p<0.01$ ns (3) ↑ (1, 2, 3) $p<0.01$

(continued)

Table 1 (continued)

First author and references	Date	Species	EPO, type and doses (IU/kg)	Time of EPO		End-points	
				Relative to occlusion	Relative to reperfusion	End-points	Outcomes
Moon (9)	2005	Rat	Epoetin alfa 3,000 or 1,000 or 500 or 150 or 50, single dose	0 min	NA	At 24 h: apoptosis	0 min groups: ↓ 3,000 $p<0.05$ ↓ 150 $p<0.05$ 50 ns
			3,000 or 150, single dose	+4 h or +8 h or + 12 h or + 24 h		At 4 weeks: MI size	0 min groups: ↓ 3,000–150 $p<0.05$ 50 ns Delayed: 3,000: ↓ +4,+8,+12 h 150: ↓ +4 h
Hale (15)	2005	Rat	EPO 5,000 repeated doses	−24 h, 0 min, +24 h and 4 more times daily	NA	At 6 weeks: MI size LV volume LVEF	 ns ns ns
Moon (10)	2006	Rat	Epoetin alfa 3,000 single dose and repeated doses	0 min or 0 min,+24 and 6 more times daily	NA	At 4 weeks: MI size LV volume LVEF	 Single ↓ $p<0.05$ Repeated ↓ ns ↓ $p<0.05$ ↑ $p<0.05$
Moon (11)	2006	Rat	Epoetin alfa 3,000 single dose	0 min	NA	At 4 weeks: MI size LVEDV LVESV LVEF	 ↓ $p<0.05$ ↓ $p<0.05$ ↓ $p<0.05$ ↑ $p<0.05$

Hirata (16)	2006	Dog	EPO 1,000 single dose	0 min (0 min group) or +6 h (6 h group) or +1 week (1 week group)	NA	At 6 h: MI size At 1 week: CD34+ cells At 4 weeks: MI size LVEF Capillary density	↓ $p<0.05$ ↑ $p<0.05$ ↓ in 0 min group ns in 6 h group and 1 week group ↑ in 0 min and 6 h ns in 1 week ↑ in 0 min and 6 h ns in 1 week
Chua (17)	2011	Mini-pigs	EPO 7,500 repeated doses	+30 min and +24 h	NA	At 14 days: Apoptosis Fibrosis Capillary density MI size LVEF	 ↓ $p<0.001$ ↓ $p<0.01$ ↓ $p<0.05$ ↓ $p<0.05$ ↓ $p<0.05$
Ahmet (12)	2011	Rat	Epoetin alfa 3,000 single dose	0 min	NA	At 24 h: Apoptosis Inflammation MI size At 8weeks: MI size LVEDV LVESV LVEF	 ↓ (80%) $p<0.05$ ↓ (30%) $p<0.05$ ↓ (50%) $p<0.05$ ↓ $p<0.05$ ↓ $p<0.05$ ↓ $p<0.05$ ↑ $p<0.05$

In another study employing a permanent ligation model in rabbits (13) a single dose of 5,000 IU/kg of epoetin-alfa was injected immediately after ligation (the route of injection was not reported) and animals were sacrificed 4 days after surgery. AAR was not different between groups. While the number of TUNEL-positive cells in the AAR at 6 h after ligation was 50% less in the EPO-treated animals, the MI size at day 4 had only a tendency to be smaller ($p < 0.1$).

In 2005, in experiments on rats van der Meer et al. (14) used 8,000 IU/kg of Darbepoetin alfa, prolong action epoetin, in three different treatment strategies: (1) a single dose of EPO immediately after coronary ligation, (2) repeated doses of EPO every 3 weeks starting the first dose immediately after coronary ligation, and (3) repeated doses of EPO every 3 weeks starting the first dose 3 weeks after coronary ligation. Experiment lasted 9 weeks. At the end of 9 weeks MI size was significantly smaller in groups 1 and 2, i.e., in groups in which treatment was started immediately after coronary ligation. There were no effects on the MI size in group 3 (treatment started 3 weeks after coronary ligation) and there were no differences in the magnitude of effect on the MI size between groups 1 and 2. Capillary density was significantly increased in all three treatment groups and there were no differences in the magnitude of effect.

Study reported by Hale et al. (15) is the only one among studies employing the permanent coronary occlusion model with clearly negative results. 24 h prior coronary ligation rats had been injected subcutaneously with 5,000 IU/kg of Epoetin alfa. The dose had been repeated at the day of ligation (20–25 min before occlusion), and five more times daily. Six weeks after surgery there were no significant effects of treatment either with respect to MI size or with respect to LV volume and function. It was noted, however, that at each given MI size the EF was higher in the EPO treated animals. Notably, after 7 days the hematocrit in the EPO treated rats was elevated to 67%, compared to 45% in untreated animals.

Hirata et al. (16) studied the effect of EPO in the dog experimental model of MI. rhEPO (type of EPO is not reported) had been injected as a single bolus intravenously in three different treatment groups either immediately after permanent coronary occlusion (0 group), or 6 h after occlusion (6 h group), or 1 week after occlusion (1 week group). Six hours after occlusion (0 group) MI was statistically smaller in the EPO treated group vs. control. One week after occlusion the number of progenitor cells was increased in the both EPO treated groups (0 group and 6 h group). Measured 4 weeks after occlusion, the MI was significantly smaller only in the 0 group; EF and capillary density were increased in the 0 group and 6 h group, but not in 1 week group of EPO treated dogs.

Recently (2011) the effect of EPO has been tested in the mini-pig model of MI (17). 7,500 IU/kg of EPO (type is not reported) was injected 30 min following coronary occlusion. The dose had been repeated once, 24 h later. Fourteen days after surgery MI size was smaller and EF was higher in EPO-treated in comparison with placebo-treated animals. In the hearts of the EPO group there were less fibrosis and higher capillary density. Apoptosis in the peri-infarct area was reduced.

Therapeutic window and effective dose of EPO in the rat model of permanent coronary occlusion had been systematically studied by Moon et al. (9). In the first experiment five different doses (3,000, 1,000, 500, 150, and 50 IU/kg) were injected intraperitoneally immediately after coronary ligation. Apoptosis in the AAR 24 h after treatment measured as a number of TUNEL positive nuclei was significantly reduced in all treatment groups except one, treated with the lowest, 50 IU/kg, dose. All doses, but 50 IU/kg were also found effective 4 weeks after coronary ligation with respect to reduction of MI size, LV remodeling and preservation of EF. In the second experiment 3,000 IU/kg had been injected with different delays after coronary ligation (0 min, 4, 8, 12, and 24 h). Four weeks after surgery MI was significantly smaller and EF significantly less reduced in all delayed treatment groups, except when treatment had been delayed by 24 h. In the final experiment the smallest effective dose (150 IU/kg) was injected with the same delays as in the second experiment. Four weeks later it was found that this dose was effective only if injected immediately. Any delay longer than 4 h did not affect the MI size or LV remodeling. In the group where 150 IU/kg treatment was delayed by 4 h, EF was higher and MI was smaller than in untreated group, but with respect to MI size the difference did not reach statistical significance ($p<0.07$).

Finally, using the rat model of permanent coronary occlusion (10), the effect of a single dose of 3,000 IU/kg injected immediately after coronary ligation had been compared with the effect of treatment when immediate dose of 3,000 IU/kg was supplemented with seven more daily injections at the same dose. Comparing with untreated animals 4 weeks after surgery LV remodeling and the reduction of EF were similarly attenuated in single dose and repeated dose groups. However, the reduction of MI size was statistically significant only in the group treated with a single dose.

Therefore, all published studies but one demonstrated that in the experimental model of MI induced by a permanent occlusion of a coronary artery in rats, dogs, and mini-pigs high dose of rhEPO (≥3,000 IU/kg) applied after coronary ligation effectively protects myocardium from ischemic damage by reducing apoptosis in the area at risk resulting in smaller MI size and attenuated LV remodeling. The effect is clearly dose-depended and has a narrow therapeutic window, which is also dose-dependent. If treatment is applied immediately after occlusion, the effective dose can be as low as 150 IU/kg (9, in rats). 1,000 IU/kg was not effective if treatment was delayed by 6 h (16, dogs). A single dose applied immediately after coronary occlusion is as effective as repeated doses (14). Moreover, there are some indications that repeated doses may reduce the effectiveness of treatment (10). In the only study in which EPO treatment did not improve the outcome, the high dose of EPO was applied seven times resulting in the massive elevation of hematocrit.

1.1.2 Ischemia–Reperfusion Model

A summary of published animal experiments in the in vivo model of temporary occlusion of a coronary artery followed by reperfusion are summarized in Table 2.

Based on experimental evidence that EPO attenuates both primary (apoptotic) and secondary (inflammatory) causes of cell death, Calvillo et al. (18) administered 5,000 U/kg i.p. of recombinant human EPO (rhEPO) to rats that underwent 30 min ischemia followed by reperfusion. rhEPO was given 24 and 0.5 h before coronary ligation, then daily for 7 days after coronary ligation rhEPO reduced cardiomyocyte loss by approximately 50%, and accordingly normalized hemodynamic function within 1 week after reperfusion. RhEPO markedly prevented the apoptosis of cultured adult rat cardiac myocytes subjected to 28 h of hypoxia (approximately 3% normal oxygen). This in vitro observation supported the hypothesis that cardiac protective effects of EPO were independent of hematopoiesis.

Parsa et al. (19) evaluated the in vivo protective potential of erythropoietin in the reperfused rabbit heart following 30 min ventricular ischemia followed by 45 min of reperfusion. They showed that 5,000 U/kg i.v. EPO given 12 h before ischemia or at time of coronary ligation limited infarct size and improved global LV function. However, when EPO was administered at the low dose of 1,000 U/kg at any time and/or at the time of reperfusion, it did not improve cardiac function even if significantly reduced infarct size and incidence of apoptosis in cardiac myocytes. EPO activated cell survival pathways in myocardial tissue in vivo and in adult rabbit cardiac fibroblasts in vitro. These pathways, activated by EPO in both whole hearts and cardiac fibroblasts, are also activated acutely by ischemia–reperfusion injury.

As part of a study on the effects of EPO on the tissue/organ injury caused by hemorrhagic shock, endotoxic shock, Abdelrahman et al. (20) showed that 300 U/kg i.v. of EPO after 20 min myocardial ischemia right before reperfusion significantly reduced infarct size in anesthetized rats at 2 h. The administration of the same dose of EPO before resuscitation abolished the renal dysfunction and liver injury in hemorrhagic, but not endotoxic, shock at 4–6 h.

Table 2
Animal experiments with in vivo models of ischemia–reperfusion

First author and references	Date	Species	EPO, type and doses (IU/kg)	Time of treatment: Relative to Isch/Rep	Duration of occlusion (min)	End-points: End-points	End-points: Outcomes
Calvillo (18)	2003	Rat	rhEPO, 5,000 i.p. multiple doses	24 h, 0.5 h before Isch, then daily	30	At 7 days: LVEDP Number of CM	 ↓ $p<0.001$ ↑ $p<0.01$
Parsa (19)	2004	Rabbit	EPO, 1,000 or 5,000 single dose i.v.	12 h before Isch, at Isch or start Rep	45	At 45 min: MI size LV function CM apoptosis	 ↓ $p<0.005$ (*all EPO groups*) ↑ $p<0.05$ (*only EPO 12 h before or at Isch*) ↓ $p<0.05$ (*only EPO 12 h before Isch*)
Abdelrahman (20)	2004	Rat	EPO 300 i.v.	Start of Rep or of resuscitation	20	At 2 h: MI size Renal and liver function	 ↓ $p<0.05$ ↑ $p<0.05$
Lipsic (21)	2004	Rat	EPO 5,000 i.p.	2 h before Isch or at start Isch or right after Rep	45	At 24 h: MI size LV function CM apoptosis	 ↓ $p<0.05$ $p<0.05$ (=NS for 2 h before Isch) ↓ $p<0.05$(=NS for 2 h before Isch)
Rui (22)	2005	Mouse	rhEPO 5,000 i.p.	24 h and 30 min before Isch	30	At 4 h: myocardial MPO	↓ $p<0.05$

(continued)

Table 2 (continued)

First author and references	Date	Species	EPO, type and doses (IU/kg)	Time of treatment: Relative to Isch/Rep	Time of treatment: Duration of occlusion (min)	End-points: End-points	End-points: Outcomes
Hirata (23)	2005	Dog	EPO 100 to 1,000 i.v.	10 min before Rep	90	At 6 h: MI size VF CM apoptosis	 ↓ $p<0.05$ ↓ $p<0.05$ ↓ $p<0.05$
Xu (24)	2005	Rat	rhEPO 3,000 i.p.	24 h before Isch	30	At 2 h: MI size Hsp70 NF-κB	 ↓ $p<0.001$ ↑ $p<0.001$ ↓ $p<0.001$
Bullard (25)	2005	Rat	EPO 5,000 i.v.	Right before Rep	40	At 24 h: MI size	↓ $p<0.05$
Kristensen (26)	2005	Pig	EPO 500 i.v.	1.5 h or 24 h before Isch	45	At 2.5 h: MI size	$p=NS$
Liu (27)	2006	Rat	rhEPO 5,000 i.v.	24 h before Isch	30	At 3 h: MI size MPO NF-κB, Ap-1 TNFα	 ↓ $p<0.001$ ↓ $p<0.01$ ↓ $p<0.05$ ↓ $p<0.01$
Nishihara (28)	2006	Rat	EPO 5,000 i.v.	5 min before Rep	20	At 2 h: MI size Akt phosphorylation	 ↓ $p<0.05$ ↓ $p<0.05$

Olea (29)	2006	Sheep	rhEPO 3,000 i.v.	Right before Rep, then daily for 3 days	90	At 10 weeks: MI size LVEDP LV volume Septum thickening	 $p = NS$ ↑ $p < 0.03$ ↑ $p < 0.05$ ↓ $p < 0.04$
Liu (30)	2006	Rat	rhEPO 5,000 i.p.	24 h before Isch	30	At 24 h: MI size	↓ $p < 0.01$
Gao (31)	2007	Rat	Darbepoetin 2, 7, 11, 30 μg/kg i.v.	Start Isch, 1 or 24 h after Rep	30	At 3 h: MI size Cardiac reserve CM apoptosis	 ↓ $p < 0.05–0.01$ ↑ $p < 0.05–0.001$ ↓ $p < 0.05$
Baker (32)	2007	Rat	Darbepoetin 2.5 μg/kg i.v.	15 min before Isch or start Rep, or 24 h post Rep	30	At 2 h: MI size cTnI	 ↓ $p < 0.05$ ↓ $p < 0.05$
Toma (33)	2007	Pig	Darbepoetin 30 μg/kg i.v.	At Rep	60	At 2 weeks: MI size Fibrosis Capillaries LV wall motion	 $p = NS$ ↓ $p < 0.02$ ↓ $p < 0.003$ ↑ $p < 0.05$
Singh (34)	2007	Rat	rhEPO 5,000 intra-right atrium	15 min before VF or at 10 min VF, start of chest compression	10 min cardiac arrest with VF	At resuscitation: Aortic press Cardiac index	 ↑ $p < 0.05$ ↑ $p < 0.05$ only after 10 min VF
Boucher (35)	2008	Rat	Darbepoetin 30 μg/kg i.v.	Start of Isch	30	At 3 days: MI size CM apoptosis LVEF echo	 ↓ $p < 0.05$ ↓ $p < 0.05$ ↑ $p < 0.05$
Prunier (36)	2009	Rat	Darbepoetin 5,000, or 5,000 and 300 (~1.5 μg/kg) i.p.	Start Rep+-once-a-week	45	At 8 weeks: MI size Thrombosis	 ↓ $p < 0.05$ $p = NS$

(continued)

Table 2
(continued)

First author and references	Date	Species	EPO, type and doses (IU/kg)	Time of treatment: Relative to Isch/Rep	Time of treatment: Duration of occlusion (min)	End-points: End-points	End-points: Outcomes
Shan (37)	2008	Mouse	EPO 1,000	Start Rep	45	At 4 h: MI size LVEF CM apoptosis	 ↓ $p<0.05$ ↑ $p<0.05$ ↓ $p<0.05$
Doue (38)	2008	Rat	EPO 200 i.v.	Start Rep	20	At 30 min: Annexin V area At 2 and 4 weeks: LVEDd LV FS	↓ $p<0.001$ ↓ $p<0.01$ ↑ $p<0.01$
Huang (39)	2008	Mouse	EPO 1,000, 3,000, 5,000 i.v.	3 min after resuscitation from asphyxia-induced cardiac arrest	6.5 or 9.5	At 3 days: Survival dP/dt	 ↑ $p<0.003$ ↑ $p<0.03$ (only at EPO 5,000 U/kg)
Burger (40)	2009	Mouse	rhEPO 2,500 i.v.	24 h before Isch or 5 min after Isch start	45	At 3 or 6 h: MI size CM apoptosis	 ↓ $p<0.05$ ↓ $p<0.05$ (no effect of EPO in HO-1−/− mice)

Burger (41)	2009	Mouse	rhEPO 2,500	24 and 0.5 h before Isch or 5 min after Isch start	45	At 45 min: PVCs and VT At 6 h: MI size	↓ $p<0.05$ ↓ $p<0.05$ (only for EPO pretreatment)
French (42)	2009	Mouse	EPO 10,000 i.p.	Before Isch or right after Rep	3 h	At 72 h: MI size Echo LV FS	 $p=$NS $p=$NS
Schlecht-Bauer (44)	2009	Rat	Darbepoetin 3–30 μg/kg i.v.	Pre Isch	40	At 3 or 28 days: MI size LVEF	 ↓ $p<0.05$ ↑ $p<0.05$
Tamareille (44)	2009	Rat	EPO 1,000 i.v.	Right before Rep	45	At 24 h: MI size CM apoptosis	 ↓ $p<0.05$ ↓ $p<0.05$
Miki (45)	2009	Rat	EPO 5,000 route not reported	15 min before Isch	20	At 2 h: MI size	↓ $p<0.05$, EPO in control rats $p=$NS, EPO in type 2 diabetic rats
Hotta (46)	2010	Rat	EPO 5,000 route not reported	15 min before Isch	20	At 2 h: MI size	↓ $p<0.05$, EPO in control rats $p=$NS, EPO in type 2 diabetic rats

(continued)

Table 2
(continued)

First author and references	Date	Species	EPO, type and doses (IU/kg)	Time of treatment: Relative to Isch/Rep	Time of treatment: Duration of occlusion (min)	End-points: End-points	End-points: Outcomes
Treguer (47)	2010	Rat	Darbepoetin 1,000 i.v.	Right before Rep	45	At 7 days: MI size Transmural MI	 ↓ $p<0.001$ ↓ $p<0.001$ (no benefit on LV function by EPO with conventional echo)
Shen (48)	2010	Rat	rhEPO 100–5,000 i.v.	30 min before Isch	30	At 2 and 4 h: Ventricular arrhythmias TNFα, IL6	 ↓ $p<0.05$ ↓ $p<0.05$
Stein (49)	2010	Nude rat	rhEPO 200 U ± hEPCs intramyocardial	At Rep	30	At 4 weeks: MI size Regional LV wall motion LVEF Vessel density EPC apoptosis	Only EPC + EPO p = NS ↑ $p<0.05$ p = NS ↑ $p<0.05$ ↓ $p<0.05$
Yano (50)	2011	Stroke-prone rats (SHR-SP)	EPO 5,000 i.v.	15 min before Isch	20	At 2 h: MI size	↓ $p<0.05$ in WKY p = NS in SHR-SP

*Only 1.5 dose

Lipsic et al. (21) studied the effect of timing of EPO administration on cardio protection during cardiac ischemia–reperfusion. Male Sprague-Dawley rats were subjected to 45 min of coronary occlusion, followed by 24 h of reperfusion. Animals were randomized to receive saline or single dose of EPO (5,000 IU/kg i.p.) either 2 h before ischemia–reperfusion, at the start of ischemia, or after the onset of reperfusion. Administration of EPO at all three time points resulted in a significant 19–23% reduction in the infarct area/area at risk after24-h reperfusion, which was accompanied by a trend toward better left ventricular hemodynamic parameters only in EPO at start of ischemia and right after reperfusion groups. The same two experimental groups showed a significant decrease in apoptosis, while the decrease was not statistically significant in the group pretreated with EPO. These results suggest that cardiac protection by EPO does not need pretreatment with EPO.

Rui et al. (22) showed that EPO can ameliorate the myocardial inflammatory response in both in vitro and in vivo murine models of ischemia–reperfusion, in agreement with what shown in central nervous system injury. In vivo, in wild type mice, 5,000 U/kg i.p. of rhEPO given 24 h before 30 min ischemia followed by 2 h reperfusion prevented ischemia–reperfusion-induced increase in myocardial myeloperoxidase activity (indicative of polymorphonuclear cells infiltration). With the help of in vitro experiments, the authors attribute this beneficial effect of EPO to the production of NO by eNOS via aPI3-kinase-dependent activation of the nuclear transcription factor AP-1.

Hirata et al. (23) occluded the canine left anterior descending coronary artery for 90 min followed by 6 h of reperfusion. EPO administered (i.v. 100–1,000 IU/kg) just before reperfusion reduced in a dose-dependent way infarct size and the incidence of ventricular fibrillation via the PI3 kinase-dependent pathway in canine hearts. In fact these effects of EPO were abolished by the treatment with wortmannin, an inhibitor of PI3-kinase.

Xu et al. (24) used a rat model of 30 min ischemia followed by 2 h of reperfusion to assess mechanisms of the anti-inflammatory action of recombinant human EPO (rhEPO). A single i.p. injection of 3,000 IU/kg of rhEPO 24 h pre-ligation enhanced the expression of Heat shock protein 70 (Hsp70) and diminished the expression of NF-κB in rat myocardium, and reduced markedly myocardial infarct size, compared to control rats.

Bullard et al. (25) administered 5,000 IU/kg i.v. at the start of reperfusion to rats after 40 min ischemia. Infarct size at 24 h was significantly attenuated in EPO-treated animals. Likewise, wortmannin abrogated EPO-mediated protection, thus suggesting that the protective action of EPO is mediated by ERK and PI3K.

Kristensen et al. (26) tested the effects of rhEPO in a closed chest, catheter-based, porcine coronary occlusion model (45 min of occlusion of the left anterior descending artery). EPO was given i.v. 500 U/kg either 90 min or 24 h and 90 min before coronary occlusion. Infarct size as a proportion of the ischemic area was measured in vivo by myocardial perfusion imaging (MPI) and postmortem by a histochemical procedure (at 150 min of reperfusion). Infarct size relative to the area at risk did not differ significantly between control and EPO groups. These negative results in the pig contrast with all positive results obtained in small rodents, even if the authors cannot exclude possible beneficial effects at later times on apoptosis and/or on left ventricular remodeling.

Liu et al. (27) further investigated mechanisms of the anti-inflammatory action of EPO in a rat model of cardiac 30 min ischemia and 3 h reperfusion. Injection of 5,000 U/kg of rhEPO 24 h before insult remarkably reduced infarct size and neutrophil infiltration. EPO greatly attenuated ischemia–reperfusion-induced NF-κB and AP-1 activation with decreased TNF-alpha, IL-6, andI-CAM-1 production and enhanced the production of the anti-inflammatory cytokine IL-10.

Nishihara et al. (28) showed that both preconditioning (PC) with 5-min ischemia–5-min reperfusion and EPO (5,000 U/kg i.v.), given 5 min before reperfusion reduced infarct size after 20-min ischemia in rat hearts in situ followed 2 h of reperfusion. The results suggest that EPO and PC afford additive infarct size-limiting effects by additive phosphorylation of GSK-3beta at the time of reperfusion by Akt-dependent and -independent mechanisms.

While high-dose EPO has been found to be cardioprotective in experimental acute MI in small rodents, in large mammals, however, results are controversial and long-term follow-up data are lacking. Olea et al. (29) assessed the long-term effects of high-dose EPO on left ventricular infarct size and function in an ovine model of reperfused MI. After 90 min of coronary occlusion, at the onset of reperfusion, sheep received recombinant human erythropoietin (rhEPO) 3,000 U/kg on 3 consecutive days (rhEPO group) or vehicle (placebo group). Infarct size, evaluated as percent fibrotic myocardium (morphometry) and by hydroxyproline quantification, was similar in both groups. Left ventricular contractile function was decreased in rhEPO group, compared to controls, left ventricle was more dilated, and left ventricular end-diastolic pressure was increased 10 weeks after coronary occlusion.

Liu et al. (30) studied the possible role of cyclooxygenase (COX)-2 in the delayed phase of the cardioprotection induced by recombinant human erythropoietin (rhEPO). Rats were given 5,000 U/kg i.p. of rhEPO 24 h before 30 min of ischemia followed by 3 h of reperfusion. Infarct size reduction in rhEPO group was abolished by the COX-2 inhibitor celecoxib given i.p. at the dose of 5 mg/kg 24 h before ischemia. It appears that COX-2 plays an essential part in cardioprotection of the delayed phase of EPO preconditioning in rats, which might be related to actions of PGE(2) or PGI(2), or both.

In Gao et al. (31) dose–response study, darbepoetin (DA) 2, 7, 11, and 30 μg/kg or vehicle was administered as a single i.v. bolus at the start of 30-min ischemia. To determine the time window of potential cardioprotection, a single high dose of DA (30 μg/kg) was given at either the initiation or the end of ischemia or at 1 or 24 h after reperfusion. After 3 days, cardiac function and infarct size were assessed. DA significantly reduced infarct size in a dose-dependent manner. Treatment with DA up to 24 h after the beginning of reperfusion still demonstrated a significant reduction in infarct size. Consistent with infarction data, DA improved in vivo cardiac reserve compared with vehicle, as assessed by left ventricular dP/dt at increasing doses of isoproterenol. Finally, DA significantly decreased myocyte apoptosis and caspase-3 activity after ischemia–reperfusion.

Baker et al. (32), found that a single intravenous darbepoetin treatment immediately before 30 min of regional ischemia in the rat reduced myocardial necrosis following 120 min of reperfusion in a dose-dependent manner. Optimal protection with darbepoetin-alfa against MI was manifest at a dose of 2.5 μg/kg among doses ranging from 0.25 to 30 μg/kg. Darbepoetin was cardioprotective when administered after the onset of ischemia and at the start of reperfusion. Darbepoetin-alfa (2.5 μg/kg) also reduced infarct size and Troponin I leakage 24 h after reperfusion. Inhibition of p42/44 MAPK (PD98059), p38 MAPK (SB203580), mitochondrial ATP-dependent potassium (KATP) channels (5-HD), sarcolemmal KATP channels (HMR 1098), but not phosphatidylinositol-3 (PI3) kinase/Akt (Wortmannin and LY 294002) abolished darbepoetin-alfa-induced cardioprotection.

Toma et al. (33) randomized to darbepoetin 30 μg/kg i.v. or saline at the time of reperfusion after 60 min ischemia 16 domestic pigs. Ischemia was induced by inflating an angioplasty balloon in the proximal left circumflex artery. Darbepoetin did not reduce infarct size, but a limited decrease in interstitial fibrosis, increased capillary area and regional functional improvement in darbepoetin-treated animals was observed. However, this did not translate to improved wall thickening of the left ventricle.

Singh et al. (34) induced ventricular fibrillation electrically and maintained it, untreated, for 10 min. Chest compression and ventilation were then started and electrical defibrillation was attempted 8 min later. Rats were randomized to receive rhEPO (5,000 U/kg) in the right atrium at baseline, 15 min before induction of VF (rhEPOBL –15-min), or at 10 min of VF, immediately before the start of chest compression (rhEPOVF 10-min), or to receive saline (control). Post-resuscitation, rats in the rhEPOVF 10-min group displayed higher mean aortic pressure associated with numerically higher cardiac index, stroke work index, and systemic vascular resistance index. In this model of short lasting global in vivo cardiac ischemia followed by reperfusion, rhEPO may rapidly induce myocardial protection.

Boucher et al. (35) subjected rats to ischemia–reperfusion and treated them with IGF-1, DA, and a combination of IGF-1 and DA 30 μg/kg i.v., or vehicle at the start of 30 min ischemia. Both IGF-1 and DA reduced infarct size and improved cardiac function 3 days after ischemia–reperfusion compared to vehicle. In the reperfused heart, apoptosis was reduced with either or both IGF-1 and DA treatments as measured by reduced TUNEL staining and caspase-3 activity.

Prunier et al. (36) examined the thrombogenic effects of a chronic EPO therapy after MI. Rats underwent coronary occlusion followed by reperfusion. They were assigned to one of the following groups: EPO-A, single i.p. injection of EPO 5,000 U/kg at the time of reperfusion; EPO-C, injection of EPO 5,000 U/kg at the time of reperfusion followed by 300 U/kg/week; PBS-C, injection of vehicle only. After 8 weeks of treatment they were exposed to a prethrombotic test based on partial stenosis of the inferior vena cava. As compared to the rats receiving vehicle only, the rats treated with EPO exhibited a significant reduction in MI size, the hematocrit was significantly increased in EPO-C, but the proportion of rats in which a thrombus occurred was similar in all groups.

Shan et al. (37) subjected mice to 45 min ischemia followed by 4 h reperfusion; EPO 1,000 U/kg, administered right before reperfusion, reduced infarct size assessed by TTC staining. Echocardiography examination suggested that EPO administration significantly improved cardiac function following ischemia–reperfusion. TUNEL assay indicated that EPO treatment decreased apoptosis. EPO administration also significantly increased the level of nuclear GATA-4 phosphorylation in the myocardium which was positively correlated with the reduction of MI. Activation of GATA-4 may be one of the mechanisms by which EPO induced protection against myocardial ischemia–reperfusion injury.

Doue et al. (38) studied rats with left coronary artery occlusion randomized to receive either an intravenous injection of 200 U/kg of EPO or saline immediately after release of the occlusion. This study demonstrated that a single treatment with EPO immediately after release of coronary ligation suppressed cardiac remodeling and functional deterioration. (99m)Tc-annexin V autoradiographs and TUNEL staining confirmed that this change was due to a decrease in the extent of myocardial apoptosis in the ischemic–reperfused region.

A model of murine cardiac arrest was employed by Huang et al. (39). Cardiopulmonary resuscitation was started after 6.5 or 9.5 min of asphyxia-induced cardiac arrest. The resuscitated animals received either erythropoietin (1,000, 3,000, or 5,000 U/kg) or placebo intravenously 3 min after return of spontaneous circulation. Erythropoietin 5,000 U/kg improved post-resuscitation myocardial dysfunction and short-term survival only when cardiac arrest lasted 6.5 min.

Burger et al. (40) investigated the role of heme-oxygenase (HO)-1 in the anti-apoptotic effects of EPO in isolated fetal murine cardiac myocytes and in vivo. Pretreatment with 2,500 U/kg i.v. of EPO 24 h and 30 min before coronary ligation decreased infarct size as well as ischemia–reperfusion-induced apoptosis in wild-type mice. However, these effects were significantly diminished in HO-1(−/−) mice. Furthermore, EPO given during ischemia reduced infarct size in mice subjected to I/R, and this effect was blocked by chromium mesoporphyrin, a selective inhibitor of HO-1, in wild-type mice. Moreover, inhibition of p38 diminished the cardioprotective effects of EPO. The authors conclude that upregulation of HO-1 expression via p38 signaling contributes to EPO-mediated cardioprotection during myocardial ischemia–reperfusion.

Burger et al. (41) studied in vivo the antiarrhythmic effects of EPO: wild-type (WT) and nNOS null mice were anesthetized and, after a baseline measurement, subjected to myocardial I/R to provoke ventricular arrhythmias. Pretreatment with 2,500 U/kg of i.v. EPO 24 h and 30 min before ischemia increased nNOS expression and significantly reduced the number of premature ventricular contractions and the incidence of ventricular tachycardia in WT mice. In contrast, treatment with EPO had no effect on premature ventricular contractions or the incidence of ventricular tachycardia in nNOS null mice.

French et al. (42) induced myocardial ischemia for 3 h in 10-week-old C57BL6 mice followed by 72 h of reperfusion. EPO (10,000 U/kg) or an equivalent amount of saline vehicle alone was injected i.p. before ligation or immediately after the onset of reperfusion. Assays of residual left ventricular creatine kinase (CK) activity and calculation of left ventricular CK depletion were performed on left ventricular homogenates harvested 72 h after onset of reperfusion for measurement of infarct size, and echocardiography was performed immediately before harvest of tissue for measurement of function. Both mice administered EPO before ligation or after reperfusion had similar infarct size and echo scores compared with those in control mice administered saline.

Schlecht-Bauer et al. (43) subjected rats to 40 min of coronary artery ligation followed by 72 h or 4 weeks reperfusion and received either darbepoetin (DA) (3–30 μg/kg) or vehicle prior to ischemia. In the DA groups reperfused for 72 h, left ventricular shortening fraction and left ventricular ejection fraction were higher than that in the control rats, in agreement with a smaller left ventricular infarct size. The cardioprotective effect of DA was dose-dependent. DA treatment activated the JAK2/Akt signaling pathway, lowered cleaved caspase-3, and increased both phosphorylated-Bad and phosphorylated-GSK-3beta proteins. This was consistent with the decrease of reactive oxygen species production and the lowered binding of Bad to Bcl-xL and Bcl-2 in a DA30 group of rats. Accordingly, at 4 weeks, in the DA group, left ventricular function was better compared to controls.

Tamareille et al. (44) studied rats after 45 min ischemia and 24 h of reperfusion. The control group was compared with ischemic post-conditioning (three cycles of 10 s reperfusion/10 s ischemia) and EPO (1,000 IU/kg i.v.) at the onset of reperfusion. EPO decreased significantly more infarct size and transmurality than ischemic post-conditioning.

Miki et al. (45) tested the hypothesis that endoplasmic reticulum stress in diabetic hearts impairs phospho-glycogen synthase kinase (GSK)-3beta-mediated suppression of mitochondrial permeability transition pore (mPTP) opening, compromising myocardial response to cytoprotective signaling. Infarction was induced by 20-min coronary occlusion and 2-h reperfusion. Administration of 5,000 U/kg of EPO 15 min before ischemia induced phosphorylation of Akt and GSK-3beta and reduced infarct size (percent risk area) from in control hearts. However, neither phosphorylation of Akt and GSK-3beta nor infarct size limitation was induced by EPO in rats with type 2 diabetes. The authors suggest that disruption of protective signals leading to GSK-3beta phosphorylation and increase in mitochondrial GSK-3beta are dual mechanisms by which increased ER stress inhibits EPO-induced suppression of mPTP opening and cardioprotection in diabetic hearts.

Infarct size after 20-min ischemia followed by 2-h reperfusion was larger in a rat model of type 2 diabetes (Otsuka-Long-Evans-Tokushima fatty (OLETF) rat) than in its control (46). Activation of Jak2-mediated signaling by erythropoietin, given 15 min before ischemia (dose and route of administration not reported), limited infarct size in control rats but not in OLETF rats. The results suggest that the diabetic heart is refractory to protection by Jak2-activating ligands such as EPO because of AT1 receptor-mediated upregulation of calcineurin activity.

Treguer et al. (47) used speckle tracking imaging in a rat model of ischemia–reperfusion to accurately detect the reduction of infarct size and transmural extension of MI induced by 1,000 U/kg of EPO i.v. as early as 24 h after reperfusion. A single bolus of EPO was administered at the onset of reperfusion after 45 min ischemia. EPO significantly decreased infarct size and transmural extension of MI. None of the conventional echocardiography parameters was significantly different between the MI-EPO and MI-control groups. These results support the usefulness of speckle tracking imaging in the early evaluation of a cardioprotective strategy in a rat model of ischemia–reperfusion.

Shen et al. (48) studied the cardioprotective and anti-inflammatory effects of EPO in a rat model of 30-min ischemia followed by 3 h reperfusion. Rats receiving single i.v. boluses of rhEPO of 100–5,000 IU/kg 30 min before ischemia exhibited dose-dependent significant reduction in the incidence of ventricular arrhythmias caused by reperfusion, and markedly decreased serum concentrations of IL-6, IL-8, and tumor necrosis factor-alpha compared with the untreated group.

Stein et al. (49) comparatively assessed the beneficial effects of intra-myocardial application of umbilical cord blood expanded endothelial progenitor cells (eEPCs) and EPO as compared to eEPCs or EPO alone in experimental MI. Nude rats underwent ligation of the left anterior descending coronary artery for 30 min followed by reperfusion. After intra-myocardial transplantation of eEPC and 200 U of EPO, CD68+ leukocyte count and vessel density were enhanced in the border zone of the infarct area. Moreover, apoptosis of transplanted CD31 + TUNEL + eEPC was decreased as compared to transplantation of eEPCs alone. Regional wall motion of the left ventricle was measured using Magnetic Resonance Imaging 4 weeks after infarction. After injection of eEPC in the presence of EPO regional wall motion significantly improved as compared to injection of eEPCs or EPO alone, while global ejection fraction was not affected. In conclusion, intra-myocardial transplantation of eEPC in the presence of EPO during experimental MI improves regional wall motion.

Yano et al. (50) induced MI by 20-min coronary occlusion followed by 2 h reperfusion in spontaneously hypertensive stroke-prone rats (SHR-SPs) and their controls (Wistar-Kyoto rats (WKYs)). Infarct size expressed as a percentage of area-at-risk was larger by 29% in SHR-SPs than in WKYs. Pretreatment with 5,000 U/kg of EPO i.v. 15 min before ischemia significantly limited infarct size in WKYs but not in SHR-SPs. The results suggest that enhanced opening of mitochondrial permeability transition pores by reactive oxygen species and modification of the signal downstream of phospho-glycogen synthase kinase-3β in the mitochondria underlie the increased vulnerability to infarction and the lack of anti-infarct tolerance by EPO, respectively, in hypertensive hypertrophied hearts.

The straightforward clinical transferability of results obtained in models of cardiac ischemia–reperfusion requires some final comments.

In vivo animal studies with EPO in cardiac ischemia–reperfusion model suggest the following:

1. Practically all studies on small rodents show that doses of EPO as low as 100 U/kg reduce infarct size and improve left ventricular function
2. Most dose regimens included a dose given before ischemia (clinically unfeasible), other studies showed that EPO was protective even when administered right before or even after the onset of reperfusion (21, 38, 42); in one study the darbepoetin was effective even administered 24 h after the onset of reperfusion (31).

3. Two out of four studies on larger animals (dog, sheep and two studies with pig) reported the lack of effect of EPO, even if EPO was administered before ischemia and/or right before reperfusion (23, 26, 29, 33); one study in the pig showed better left ventricular function than controls, in the presence of unchanged size of infarct.

1.2 Clinical trials

Pharmacokinetics and pharmacodynamics of erythropoietin in healthy volunteers have been well established (51). A single subcutaneous dose of erythropoietin as high as 2,400 IU/kg is well tolerated by healthy people (higher dose was not tested), and while with respect to erythropoiesis the effect of EPO is dose-dependent, the dose higher than 1,800 IU/kg not followed by further increase of reticulocytes in blood. However, the safety and tolerance of erythropoietin in patients, especially patients with developing MI needed to be tested. 33,000 IU of rhEPO (about 450 IU/kg) had been well tolerated by patients when injected during the first 3 days after stroke (52). 200 U/kg had been also injected intravenously to 29 patients during the first 3 days of acute MI with ST-segment elevation additionally to PCI and standard therapy (53). EPO treatment did not result in any complications. The bleeding time, platelet function assay closure time and plasma levels of von Willebrand factor were not altered by the EPO treatment. Safety of EPO in patients in acute stage of MI was also evaluated in the safety arms of clinical trials described below and it was found mostly safe.

Published to date results of completed phase I and II clinical trials are summarized in the Table 3. The first small study (20 patients with first acute ST-segment elevation MI and indications for PCI) was concluded in 2006 in the Netherlands (54). 60,000 IU of Darbepoetin-alfa (long-acting EPO with half-life of 21 h), which is approximately equivalent to 800 IU/kg in the average patient, or placebo were injected intravenously immediately before PCI. Time between PCI and symptom onset was not reported. The serum EPO level was significantly increased 24 h and the progenitor endothelial cells were significantly increased 72 h after treatment in EPO treated vs. placebo cohorts. EPO treatment was well tolerated and no adverse effect of EPO treatment had been noted during 30 days after treatment. However, the LV function measured 4 months after intervention by radionuclide ventriculography showed no differences in LVEF between two groups.

A pilot study involving 51 patients with non-STEMI was also conducted in the Netherlands (55). Single intravenous dose of 40,000 IU of Epoetin-alfa was given within 8 h from elevated troponin level to 26 patients. Enzymatic MI size measured at non-STEMI was not different between experimental and placebo groups. Elevation of BP was noted among patients treated with EPO.

Table 3
Clinical trials

				Average time to drug		Primary end-point		Secondary end-point	
Name/year/ references	Country	Number of pt	EPO, type and dose	From onset	From PCI	Measurements	Outcome	Measurements	Outcome
Lipsic 2006 (21)	The Netherlands	20	Darbepoetin alfa 60,000 IU	NA	0 min	LVEF at 4 m	NS	NA	NA
Liem, 2009 (22)	The Netherlands	51	Epoetin alfa 40,000 IU	8 h from elevated tro-ponin	NA	MI size (enzymes)	NS	NA	NA
Binbrec, 2009 (23, 24)	Dubai	236	Epoetin beta 30,000 IU	6 h	At the time of thrombolysis	MI size (enzymes)	NS	Echo at discharge	NS
EPO/AMI-1, 2010 (25, 26)	Japan	36	Epoetin beta 12,000 IU	Within 48 h	Within 24 h	LVEF 4 day and 6 month	NS	NA	NA
HEBE III 2010 (27, 28)	The Netherlands, Germany	529	Epoetin alfa 60,000 IU	12–24 h	Within 3 h	LVEF at 6 weeks	NS	MI size (enzymes)	NS
REVIVAL-3, 2010(29)	Germany	138	Epoetin beta 33,300 IU	Within 24 h	0 min (add. doses at 24 and 48 h)	LVEF at 6 month	NS	ΔLVEF and MI	NS
EPOC-AMI, 2010(30)	Japan	35	Epoetin beta 6,000 IU	Within 24 h	0 min (add. doses at 24 and 48 h)	MI size 4 days and 6 month	NS	NA	NA

(continued)

Table 3
(continued)

				Average time to drug		Primary end-point		Secondary end-point	
Name/year/ references	Country	Number of pt	EPO, type and dose	From onset	From PCI	Measurements	Outcome	Measurements	Outcome
REVEAL, 2011(31, 32)	USA	131	Epoetin alfa 60,000 IU	Within 8 h	Within 4 h	MI size 2 to 6 days and 12 weeks	NS	LV remodeling	NS
Ferrario et al., 2011 (33)	Italy	30	Epo 33,000 IU	Within 6 h	0 min (add. doses at 24 and 48 h)	MI size (enzymes) LVEF at admission and at 6 months	MI ↓ $p<0.02$ EF ↑ $p<0.05$	Progen. cells at 72 h	↑ $p<0.002$

In the study conducted in Dubai (56, 57) 236 patients with STEMI, admitted less than 6 h after onset of chest pain were subjected to thrombolytic treatment. 115 of them were additionally treated with intravenous injection of 30,000 of epoetin-beta prior to thrombolytic therapy. The primary end-point was an enzymatically estimated MI size, secondary end-point—echocardiographic measurements at the time of discharge. There were no differences between groups at any of end-point parameters.

Results of a small (16 control, 20 EPO) prospective randomized, single-blind, multicentered trial (EPO/AMI-1) conducted in Japan were reported in 2010 (58, 59). AMI patients with ST-segment elevation were admitted within 24 h after onset of symptoms. They undergo successful PCI and thrombolysis. EPO, 12,000 IU (epoetin-beta) was given within 24 h after PCI. LVEF measured via photon emission CT 4 days after PCI was not different between groups, but at 6-month follow-up LVEF was significantly higher in EPO treated group. Nevertheless, the LVEF increments between 4 days and 6 months measurements were not different between groups.

Combined efforts of clinical groups in the Netherlands and Germany resulted in prospective, randomized, blinded end-points study, HEBE III (60, 61). Patients with first ST-segment elevation MI were hospitalized within 12–24 h of symptom onset and randomized after successful PCI. A single bolus of 60,000 IU of epoetin-alfa or placebo was injected intravenously within 3 h after PCI. MI size measured enzymatically was slightly (but statistically not significant) reduced. However, LVEF 6 weeks after MI was not different between groups.

In the German clinical trial REVIVAL-3 (62) patients diagnosed with ST-segment elevation MI were undergoing primary PCI within 24 h of symptom onset. Epoetin-beta (33,300 IU/kg) was given immediately after PCI. Additional doses were given 24 and 48 h later. Cardiac MRI was performed 5 days and 6 months after MI. There were no differences between EPO treated and control groups in any measured parameter at 5 days or 6 months. The changes of MI size and EF between two groups were also similar.

Japanese clinical trial, EPOC-AMI (63) was similar in design with REVIVAL-3, but EPO dose was lower. Patients with SR-segment elevated MI were subjected to a primary PCI within 12 h of symptom onset and treated with 6,000 IU of epoetin-beta immediately after that. This dose was repeated on day 2 and 4. MI size was measured via Tc^{99m}-tetrofosmin SPECT 4 days and 6 months after PCI. There were no differences between treated and untreated groups at either measurement. However, comparison of these two measurements showed the reduction of MI and some increase of EF with time in EPO-treated group, while changes in untreated group were not statistically significant.

The only clinical trial in the USA was recently completed (64, 65). It was randomized, double-blind, placebo-controlled, multicenter phase I–II trial evaluating the effect of epoietin-alpha in the patients with large MI. In the dose escalating safety phase the dose of EPO was tested in the cohorts receiving 15,000, 30,000, and 60,000 IU. In the phase II, 65 patients received EPO within 4 h of PCI (within 8 h of symptom onset). Equal number of patients received placebo at that time. Main outcome was an MI size estimated as a percent of LV mass measured by cardiac magnetic resonance imaging 2–6 days after treatment and 12 weeks later. MI size was not reduced by EPO measured either during the first CMR, or 12 weeks later. There was a tendency of increased MI size and the number of adverse effect in older patients.

A small clinical trial, which authors considered a pilot, completed recently in Italy (66) was the only one so far with encouraging outcome. It was a single-center, randomized, placebo controlled, double-blind study. Thirty patients with a first ST-segment elevation AMI underwent PCI within 6 h of symptoms onset. EPO, 33,000 IU (type is not reported) or placebo were delivered intravenously in 50 mL of saline over 30 min, starting before and continued during PCI. This dose was repeated 24 and 48 h after PCI. MI size was estimated enzymatically. Echocardiography and cardiac magnetic resonance studies were performed shortly after admission and repeated after 6 months. Seventy-two hours after initiation of treatment the level of CD34+ cells was reported four times higher in the EPO group compared with placebo group ($p=0.002$). Enzymatically measured MI was smaller ($p<0.03$) in the EPO group. LVEF measured by CMR improved after 6 months in the EPO group vs. placebo. EF measured by echocardiography was not different at 6 months between groups, but in the EPO group EDV did not change and ESV was smaller, while in placebo group these changes were reversed.

Finally, a clinical trial, EPAMINONDAS (Exogenous erythropoietin in Acute Myocardial Infarction New Outlook aNd Dose), is underway in Italy (67). The results are not available at this time. It is a multicenter, randomized controlled trial that enrolls >100 patients with ST-segment elevation MI. Two doses of epoetin-alfa will be tested (100 and 200 IU/kg). EPO treatment (or placebo) will be repeated in the first 3 days after primary PCI with the first dose given within 12 h of procedure. Primary end-points will be an enzymatically measured MI size and results of cardiac MRI. Secondary end-points will be 12-month cardiac remodeling and safety events.

Therefore, all completed clinical trials except one (66) failed to demonstrate any or almost any beneficial effects of EPO treatment in the setting of acute stage of MI. All these trials have one common trait—timing of application of experimental drug is different from that used in the experiment on animals. Experimental data obtained in animal research clearly indicated that effectiveness of EPO treatment reversely proportional to a time elapsed since a coronary occlusion. In experiments involving the model of permanent coronary occlusion the best result were achieved when EPO was applied at the time of occlusion. In ischemia–reperfusion model the most effective results were observed when treatment was applied no later than at the time of reperfusion, i.e., 30–90 min from coronary occlusion. In reviewed clinical trials, even if EPO was injected at the time of PCI, the time from the symptom onset was sometime as long as 24 h. In the only trial with promising outcome (66) the EPO was injected at the time of PCI and the PCI was done within 6 h of the beginning of symptoms.

We propose that critical therapeutic window for using EPO in MI patients should be counted from the time of symptom onset, not from the time of reperfusion. With this proposal in mind we conducted an experiment to test a hypothesis that design of above clinical trials made them impossible to succeed (68). Experiment was designed to closely replicate one of the clinical trials, REVEAL (64, 65). In five groups of 2-month old male Wistar rats, the left descending coronary artery was occluded either by a permanent (two groups) or a temporary ligature (three groups). In the three temporary ligature groups the occluding ligature was released 2 h after occlusion. All animals were sacrificed 24 h after beginning of occlusion and MI size was measured histologically. Animals from one of the permanent occlusion groups and one of the temporary occlusion groups remained untreated. Animals from one of the temporary occlusion groups received intraperitoneal injection of 3,000 IU/kg of epoetin-alfa immediately after reperfusion. Another group received the same dose of EPO 4 h after reperfusion, i.e., 6 h after coronary occlusion. The last group (permanent occlusion) also received the same dose of EPO 6 h after occlusion. Results of MI size 24 h after occlusion are presented in Fig. 1. Our hypothesis had been confirmed: MI size 24 h after coronary occlusion was 42 ± 2% of the area at risk in untreated animals; it was reduced by ~50% (19 ± 2% of the AAR, $p < 0.0001$) only in the group that was treated at the time of reperfusion; MI size in groups in which treatment was delayed by 4 h after reperfusion or by 6 h after permanent occlusion was not different from untreated animals.

1.3 Conclusion

Comparison of experimental data with design of several completed to date clinical trials clearly suggests that failure of these trials to improve clinical outcome by using erythropoietin in the acute phase of MI might be related to design of clinical trials. Experimental animal data definitely indicated that therapeutic window for erythropoietin is narrow and dose-dependent. In the doses that substantially exceed customary therapeutic doses for erythropoietin, the window between coronary occlusion and EPO injection could not

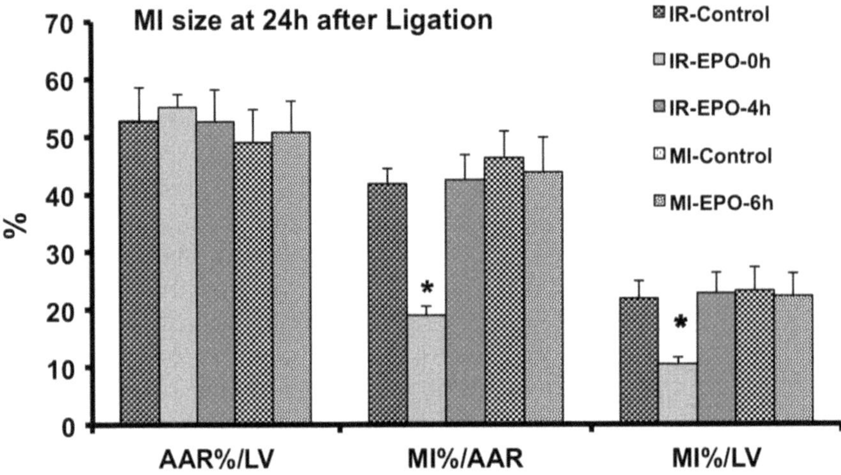

Fig. 1 Area of myocardium at risk (AAR) and size of myocardial infarction (MI) 24 h after permanent occlusion of the left descending coronary artery or after 2 h of occlusion followed by 22 h of reperfusion. rhEPO is administered at the time of reperfusion (IR-EPO-0 h) or 4 h after beginning of reperfusion (IR-EPO-4 h) or 6 h after permanent coronary occlusion (MI-EPO-6 h). Data presented as means ± SEM. AAR is presented as percent of left ventricle (LV). MI is presented as percent of AAR or percent of LV. $^{*}p<0.05$ vs. all other group (Bonferroni post hoc comparison)

exceed 12 h. In doses approaching acceptable therapeutic doses the efficacy of EPO was lost even with small delay from the time of coronary occlusion. These considerations drawn on the basis of animal experiments were disregarded in the clinical trials design. The issue whether this chasm stemmed from regulatory difficulties in conducting clinical trial involving a patient in clinical emergency that creates a pressure on clinicians to shorten the door-to-balloon time and makes it difficult to obtain informed consent early enough, or the problem is much deeper and have its roots in a growing culture among clinical scientist not to accept data obtained from animal research for anything more than proof of concept that can serve only as a trigger to start independent clinical evaluation deserves special discussion elsewhere.

1.4 Experimental Models of Myocardial Infarction

Two, mostly used experimental MI models are permanent occlusion of a coronary artery or temporary coronary occlusion followed by a reperfusion (I/R model). Permanent occlusion is the most "pure" experimental model, which allows studying and intervening into the effects of ischemic damage of the myocardium per se. Nevertheless, reperfusion following acute MI is the most effective intervention to reduce the extent of injury and to improve clinical outcomes, and contemporary treatment of MI always includes reperfusion either by thrombolysis or by percutaneous coronary intervention (PCI). In the light of this clinical mandate I/R experimental model is more clinically relevant model with respect to translation of experimental finding into clinical practice, However, in the past 30 years, it has become apparent that reperfusion may bring additional damage to a part of the viable myocardium via ROS; inflammatory response to reperfusion following ischemia has been identified as a major player in the so called "reperfusion injury". In large animals (dogs, minipigs) MI can be induced through balloon occlusion of a coronary artery; this technique is similar to clinical percutaneous angioplasty (PCI) and is not discussed here.

The most common small rodents used for this kind of experiments are the mouse and the rat.

The instruments, reagents. and apparatuses used for the two most widely used models of MI in small rodents, permanent ischemia and ischemia–reperfusion, are the same. Therefore, materials are listed once with specifications when requested by differences between the two models.

2 Materials

1. Operating table with safe aspiration of anesthetic gas.
2. Perspex chamber for induction of anesthesia, before tracheal intubation.
3. Inhalation anesthesia: isoflurane, oxygen.
4. Vaporizer.
5. Ventilator for rodents.
6. 23-gauge intravenous cannula (for mouse intubation).
7. Drugs: ampicillin; polyvinylpyrrolidone; Evans blue 5% w/v; tetrazolium chloride (TTC).

3 Methods

3.1 Surgery

1. All procedures are conducted in conformity with the institutional guidelines that are in compliance with national and international laws and policies.
2. Anesthesia is induced in a small chamber at 5% isoflurane in oxygen and animals are quickly moved from the chamber to the operating table for tracheal intubation.
3. Anesthesia is maintained with isoflurane 1.5% in oxygen under mechanical ventilation with a ventilator (tidal volume 8–10 mL/kg; 120 strokes/min for the mouse and 70 strokes/min for the rat) through the endotracheal cannula.
4. The left anterior descending coronary artery is ligated with a 7-O silk (Ethicon) suture after exteriorization of the heart through a small opening at the fourth-intercostal space.

5. An overhand knot is tied with two pieces of suture to arrest blood flow for 20–90 min before release of the knot (reperfusion); a permanent knot is tied in case of permanent ischemia. Body temperature of the mouse/rat must be strictly controlled and maintained at 37°C to obtain reproducible infarcts with ischemia–reperfusion. In case of permanent ischemia, temperature is not so critical since the procedure is much shorter: chest is closed under negative pressure right after coronary ligation.
6. Proper placement of the ligature is confirmed by the appearance of ventricular ectopy and blanching of the myocardium. The chest is the closed under negative pressure and mice are weaned from mechanical ventilation. Postsurgical analgesia is achieved by buprenorphine (0.1 mg/kg s.c. q12h for 1 day).

3.2 In vivo Imaging for serial assessment of progression of cardiac remodeling, function and MI size

The two most widely used noninvasive imaging techniques in studies in small rodents are *echocardiography and magnetic resonance imaging*.

Echocardiography, which can easily be performed on awake or lightly sedated mice, has become the procedure of choice for most studies of the mouse heart for its flexibility. While echocardiography in small rodents, first in the rat then in the mouse, has been performed with apparatuses used for clinical activity, choosing mostly vascular probes of 7–15 MHz, in the last decade, new instruments designed for small animal exams have been made commercially available. The high-frequency probes (30 MHz) allow a higher spatial resolution (30 μm) and, consequently, a more accurate assessment of heart dimensions and function. However, the major difficulties of this technique are the geometric assumptions, image positioning errors, and use of subjective visual methods thus making necessary a skilled sonographer. Other limitations are LV wall motion assessment because of postsurgical adhesions and chest deformations, and a not well visualization of the right ventricle.

Magnetic resonance imaging (MRI) represents the gold standard for measurement of cardiac morphology and function either in humans or in small rodents. It is noninvasive, accurate, free from geometric assumptions compared to echocardiography. The contiguous stack of short-axis images of the heart covers the entire length of the LV and allows for an accurate estimation of both left and right ventricular mass, volumes, and ejection fraction. Even though MRI is a more reliable method than echocardiography for the characterization of the pathophysiological consequences of experimental MI and of treatments response in rodents, MRI exams usually last more than 1 h, require deep anesthesia, and is much more expensive. For these reasons it is not routinely used in large series of animals.

Transthoracic echocardiography can be performed with an instrument for clinical use on conscious, previously trained mice with the use of a 12–15 MHz linear transducer at high frame rate imaging (102 Hz) and a 7.5 MHz phased array probe for pulsed-wave Doppler measurements. An echocardiograph, especially design for mice, is used to examine anesthetized (0.5–1.5% isoflurane) animals. High frequency of its probes, 30 MHz, allows higher spatial resolution. The rat cannot be easily trained, therefore, needs to be evaluated under sedation, with ketamine and imidazolam for example, or under anesthesia with 0.5–1.5% isoflurane.

Short and long-axis two dimensional (2D) views and M-mode at the level of infarction is analyzed in real time and recorded on magneto-optic disk for off-line analysis by a sonographer blind to study groups. Left atrial diameter, end-diastolic and end-systolic wall thicknesses and left ventricular internal dimensions are measured, as recommended by the American Society of Echocardiography (see Note 1).

As for echocardiography, more than one model and set up for MRI exist on the market. Since the detailed description of the instrumental settings goes far beyond the aims of the present chapter, the setting used in the laboratory of one of the authors (RL) is described (see Note 2).

1. All animals are anesthetized by facemask with isoflurane (induction in chamber at 3–5%, maintenance 0.5–1.5%) and 0.3 L/min O_2.
2. Animals are positioned in a purpose-built cradle and ECG electrodes are attached to the front paws. NOTE: ad hoc properly shielded ECG leads and cables must be purchased in order to avoid the instrument-derived noise that would completely mask the ECG signal.
3. A pressure-transducer for respiratory gating is positioned above the abdomen.
4. A fiber optic probe is used for monitoring the rectal temperature.
5. ECG and respiratory signals are processed and displayed using a gating device and temperature is maintained at 37 ± 0.5°C during the whole exam (60–90 min).
6. After the image plane orientation from coronal, axial and oblique LV long-axis, a 2 chamber long-axis view is obtained and orthogonal to it a 4 chamber long-axis is acquired followed by 1 mm serial short-axis slices covering the entire LV length. Seven to ten short-axis slices are acquired from base to apex. Sixteen frames per slice for one cine sequence are saved. Images are exported in DICOM format and analyzed off-line.
7. End-diastolic and end-systolic variables are measured in selected frames according to the visual estimation of the maximal and minimal ventricular cavity respectively.

8. LV anterior and posterior diameters and wall thicknesses are measured at the mid-papillary level. End-diastolic and systolic endocardial areas of each slice are traced manually. LV volumes are calculated by the modified formula for Simpson's rule.
9. As for echocardiography, it is recommended that two investigators blind to the experimental conditions read MRI recordings even though the better definition of the images reduce the inter and intra-reader variability.

3.3 Hemodynamics

Prior to euthanasia LV pressure or pressure–volume loop analyses can be performed. Anesthesia can be provided by different routes and with different agents. The authors routinely use (a) isoflurane (2% in oxygen), rats are intubated and mechanically ventilated, or (b) pentobarbital 40–60 mg/kg i.p.

Two approaches are possible:

1. The less invasive approach is by pushing a Millar pressure catheter (indicatively SPC-320 for the rat, SPR-671 for the mouse) or a Millar pressure–conductance catheter through the right carotid artery to the left ventricle.
2. The open chest approach requires a bilateral thoracotomy in fourth and fifth intercostal. After opening the pericardium, a catheter (Millar Instruments Inc., Houston, TX) is inserted into the LV from the apex.

Naturally, pressure catheters measures are limited to a LV pressures and their derivatives. Pressure-conductance catheter allow for a full array of pressure–volume loop analyses: LV end-diastolic pressure (EDP), end-diastolic (EDV) and end-systolic(ESV) volumes, stroke volume (SV), $+\mathrm{d}P/\mathrm{d}t$, $-\mathrm{d}P/\mathrm{d}t$, isovolumic relaxation time (tau), and arterial elastance (Ea) are determined in 10–20 digitally averaged cardiac cycles while the ventilator is stopped. LV end-systolic elastance (Ees), preload recruitable stroke work (PRSW) and end-diastolic stiffness (Eed) are measured using a graded preload reduction technique. Arterio-ventricular (AV) coupling is calculated as Ea/Ees. The test is concluded by advancing the catheter into thoracic aorta to measure arterial blood pressure.

3.4 Gross Pathology and Histological Assessment

The hearts and lungs are removed and weighed (wet weight). Hearts are processed and the histological analyses are performed at different time depending on experiment design. Three approaches are used by the authors:

1. Hearts are cut into two pieces through the short axis. The basal half is fast frozen and stored for different assays, and the apical half is used for histological analysis.
2. Hearts are slightly frozen to reach a consistency which favors thin and precise cutting, are cut with a chopper into 1 mm-thick

sections which are then incubated in TTC for determination of infarct size and the area at risk (AAR is stained in the red color).

3. Hearts are cut into two to three transverse sections (perpendicular to the long axis) starting from slightly above the coronary ligature (in case of permanent ischemia the thread is in place and visible), or above the visible area of injury (in case of ischemia–reperfusion the thread is preferably not left in place to decrease the occurrence of adhesions between epicardium and chest wall). The sections are put in 10% buffered formalin to be processed within 1 month for paraffin embedding.

Methods 1 and 2 are usually employed in case of long-term experiments (>7 days follow-up), while method 2 is for short-term experiments (2–24 h).

Myocardial tissue sections are subjected to Masson's trichrome and hematoxylin–eosin staining. Myocyte cell size and density are measured in H&E-stained sections (see Note 3).

To avoid likely biases, it is crucial that the person assessing all histological slides is blinded to a grouping.

4 Notes

1. Fractional shortening (FS) is calculated from the composite LV internal, diastolic (LVIDd) and systolic (LVIDs) dimensions as: $FS = ((LVIDd - LVIDs)/LVIDd) \times 100$ from M-mode short axis view. Left ventricular (LV) volumes and LVEF are calculated by the modified simple plane Simpson's rule from the parasternal long-axis view. The MI size at the mid-papillary muscle level is estimated from 2D short axis LV images and expressed as a percent of the LV endocardial circumference. Infarct area is identified as a sharply demarcated section of the LV free wall that fails to thicken during systole. The length of the akinetic part of the LV endocardial circumference is measured from freeze-frame images at end-diastole. Aortic outflow and diastolic transmitral LV inflow velocities are measured from 4 to 5 apical chamber views respectively by pulsed-wave Doppler with a sample volume length of 3.5–7.5 mm; the ultrasound beam is aligned as parallel as possible to Color Doppler flow and to record the highest velocities.

2. Usually the anesthetized rat/mouse is placed in a horizontal bore 7 T Biospec preclinical scanner with a shielded gradient insert (BGA 12, 40 mT/m; rise time, 110 ms). A quadrature or linear volume coil and rat/mouse surface coil are used to transmit/receive the magnetic resonance signals (Quadrature or Linear volume coil inner diameter, 72 mm; rat/mouse surface coil inner diameter, 20 mm/10 mm). After localization

of the heart, scans (segmented double-gated FLASH imaging) are run to ensure correct positioning; slice-selective shimming and flip angle calibration are performed manually before each experiment. The parameters of Cine-MRI pulse-sequence are: matrix, 256×128; FOV, 40.0×20.0 mm; echo time, 2.98 ms; flip angle, 15°; slice thickness, 1 mm; number of averages, 6; repetition time, 12 ms.

3. Only myocytes which nuclei are clearly identified are counted. Myocyte diameter is measured as the shortest distance across the nucleus in transverse cell sections. Diameters of 100 myocytes from five randomly selected microscope fields (×200 magnification) from the LV posterior wall are averaged to represent the myocyte diameter of a given specimen. Myocyte density is calculated from the same area in the same fashion. Myocardial tissue fibrosis is measured in Masson's trichrome-stained sections and is expressed as a fraction of a microscopic field (×100 magnification) of the LV posterior wall. An average of five randomly selected fields represented results of a given specimen. Collagen content in the thoracic aorta wall is measured on 7 μm fresh frozen sections stained with Masson's trichrome. Digital images of stained sections are obtained from light microscopy and analyzed using a digital imaging analysis system (MCID, InterFocus Imaging Ltd, Cambridge, UK). The collagen content in aortic wall is calculated as a percentage of the total wall thickness or tunica media.

References

1. Bogoyevitch MA (2004) An update on the cardiac effects of erythropoietin cardioprotection by erythropoietin and the lessons learnt from studies in neuroprotection. Cardiovasc Res 63(2):208–216, Review
2. Koul D, Dhar S, Chen-Scarabelli C, Guglin M, Scarabelli TM (2007) Erythropoietin: new horizon in cardiovascular medicine. Recent Pat Cardiovasc Drug Discov 2(1):5–12, Review
3. Riksen NP, Hausenloy DJ, Yellon DM (2008) Erythropoietin: ready for prime-time cardioprotection. Trends Pharmacol Sci 29(5):258–267, Review
4. Latini R, Brines M, Fiordaliso F (2008) Do non-hemopoietic effects of erythropoietin play a beneficial role in heart failure? Heart Fail Rev 13(4):415–423, Review
5. Vogiatzi G, Briasoulis A, Tousoulis D, Papageorgiou N, Stefanadis C (2010) Is there a role for erythropoietin in cardiovascular disease? Expert Opin Biol Ther 10(2):251–264, Review
6. Lipsic E, Schoemaker RG, van der Meer P, Voors AA, van Veldhuisen DJ, van Gilst WH (2006) Protective effects of erythropoietin in cardiac ischemia: from bench to bedside. J Am Coll Cardiol 48(11):2161–2167, Epub 2006 Nov 9. Review
7. Tramontano AF, Muniyappa R, Black AD, Blendea MC, Cohen I, Deng L, Sowers JR, Cutaia MV, El-Sherif N (2003) Erythropoietin protects cardiac myocytes from hypoxia-induced apoptosis through an Akt-dependent pathway. Biochem Biophys Res Commun 308(4):990–994
8. Moon C, Krawczyk M, Ahn D, Ahmet I, Paik D, Lakatta EG, Talan MI (2003) Erythropoietin reduces myocardial infarction and left ventricular functional decline after coronary artery ligation in rats. Proc Natl Acad Sci USA 100(20):11612–11617
9. Moon C, Krawczyk M, Paik D, Lakatta EG, Talan MI (2005) Cardioprotection by recombinant human erythropoietin following acute experimental myocardial infarction: dose response and therapeutic window. Cardiovasc Drugs Ther 19(4):243–250
10. Moon C, Krawczyk M, Lakatta EG, Talan MI (2006) Therapeutic effectiveness of a single vs. multiple doses of erythropoietin after

experimental myocardial infarction in rats. Cardiovasc Drugs Ther 20(4):245–251

11. Moon C, Krawczyk M, Paik D, Coleman T, Brines M, Juhaszova M, Sollott SJ, Lakatta EG, Talan MI (2006) Erythropoietin, modified to not stimulate red blood cell production, retains its cardioprotective properties. J Pharmacol Exp Ther 316(3):999–1005
12. Ahmet I, Tae HJ, Juhaszova M, Riordon DR, Boheler KR, Sollott SJ, Brines M, Cerami A, Lakatta EG, Talan MI (2011) A small non-erythropoietic helix B surface peptide based upon erythropoietin structure is cardioprotective against ischemic myocardial damage. Mol Med 17(3–4):194–200. doi:10.2119/molmed.2010.00235
13. Parsa CJ, Matsumoto A, Kim J, Riel RU, Pascal LS, Walton GB, Thompson RB, Petrofski JA, Annex BH, Stamler JS, Koch WJ (2003) A novel protective effect of erythropoietin in the infarcted heart. J Clin Invest 112(7):999–1007
14. van der Meer P, Lipsic E, Henning RH, Boddeus K, van der Velden J, Voors AA, van Veldhuisen DJ, van Gilst WH, Schoemaker RG (2005) Erythropoietin induces neovascularization and improves cardiac function in rats with heart failure after myocardial infarction. J Am Coll Cardiol 46(1):125–133
15. Hale SL, Sesti C, Kloner RA (2005) Administration of erythropoietin fails to improve long-term healing or cardiac function after myocardial infarction in the rat. J Cardiovasc Pharmacol 46(2):211–215
16. Hirata A, Minamino T, Asanuma H, Fujita M, Wakeno M, Myoishi M, Tsukamoto O, Okada K, Koyama H, Komamura K, Takashima S, Shinozaki Y, Mori H, Shiraga M, Kitakaze M, Hori M (2006) Erythropoietin enhances neovascularization of ischemic myocardium and improves left ventricular dysfunction after myocardial infarction in dogs. J Am Coll Cardiol 48(1):176–184, Epub 2006 Jun 2
17. Chua S, Leu S, Lin YC, Sheu JJ, Sun CK, Chung SY, Chai HT, Lee FY, Kao YH, Wu CJ, Chang LT, Ko SF, Yip HK (2011) Early erythropoietin therapy attenuates remodeling and preserves function of left ventricle in porcine myocardial infarction. J Investig Med 59(3):574–586
18. Calvillo L, Latini R, Kajstura J, Leri A, Anversa P, Ghezzi P, Salio M, Cerami A, Brines M (2003) Recombinant human erythropoietin protects the myocardium from ischemia-reperfusion injury and promotes beneficial remodeling. Proc Natl Acad Sci USA 100:4802–4806
19. Parsa CJ, Kim J, Riel RU, Pascal LS, Thompson RB, Petrofski JA, Matsumoto A, Stamler JS, Koch WJ (2004) Cardioprotective effects of erythropoietin in the reperfused ischemic heart: a potential role for cardiac fibroblasts. J Biol Chem 279:20655–20662
20. Abdelrahman M, Sharples EJ, McDonald MC, Collin M, Patel NS, Yaqoob MM, Thiemermann C (2004) Erythropoietin attenuates the tissue injury associated with hemorrhagic shock and myocardial ischemia. Shock 22:63–69
21. Lipsic E, van der Meer P, Henning RH, Suurmeijer AJ, Boddeus KM, van Veldhuisen DJ, van Gilst WH, Schoemaker RG (2004) Timing of erythropoietin treatment for cardioprotection in ischemia/reperfusion. J Cardiovasc Pharmacol 44:473–479
22. Rui T, Feng Q, Lei M, Peng T, Zhang J, Xu M, Abel ED, Xenocostas A, Kvietys PR (2005) Erythropoietin prevents the acute myocardial inflammatory response induced by ischemia/reperfusion via induction of AP-1. Cardiovasc Res 65:719–727
23. Hirata A, Minamino T, Asanuma H, Sanada S, Fujita M, Tsukamoto O, Wakeno M, Myoishi M, Okada K, Koyama H, Komamura K, Takashima S, Shinozaki Y, Mori H, Tomoike H, Hori M, Kitakaze M (2005) Erythropoietin just before reperfusion reduces both lethal arrhythmias and infarct size via the phosphatidylinositol-3 kinase-dependent pathway in canine hearts. Cardiovasc Drugs Ther 19:33–40
24. Xu B, Dong GH, Liu H, Wang YQ, Wu HW, Jing H (2005) Recombinant human erythropoietin pretreatment attenuates myocardial infarct size: a possible mechanism involves heat shock Protein 70 and attenuation of nuclear factor-kappa B. Ann Clin Lab Sci 35:161–168
25. Bullard AJ, Govewalla P, Yellon DM (2005) Erythropoietin protects the myocardium against reperfusion injury in vitro and in vivo. Basic Res Cardiol 100:397–403
26. Kristensen J, Maeng M, Rehling M, Berg JS, Mortensen UM, Nielsen SS, Nielsen TT (2005) Lack of acute cardioprotective effect from pre-ischaemic erythropoietin administration in a porcine coronary occlusion model. Clin Physiol Funct Imaging 25:305–310
27. Liu X, Xie W, Liu P, Duan M, Jia Z, Li W, Xu J (2006) Mechanism of the cardioprotection of rhEPO pretreatment on suppressing the inflammatory response in ischemia-reperfusion. Life Sci 78:2255–2264
28. Nishihara M, Miura T, Miki T, Sakamoto J, Tanno M, Kobayashi H, Ikeda Y, Ohori K, Takahashi A, Shimamoto K (2006) Erythropoietin affords additional cardioprotection to preconditioned hearts by enhanced phosphorylation of glycogen synthase kinase-3 beta. Am J Physiol Heart Circ Physiol 291:H748–H755

29. Olea FD, Vera Janavel G, De Lorenzi A, Cuniberti L, Yannarelli G, Cabeza Meckert P, Cearras M, Laguens R, Crottogini A (2006) High-dose erythropoietin has no long-term protective effects in sheep with reperfused myocardial infarction. J Cardiovasc Pharmacol 47:736–741
30. Liu X, Zhou Z, Feng X, Jia Z, Jin Y, Xu J (2006) Cyclooxygenase-2 plays an essential part in cardioprotection of delayed phase of recombinant human erythropoietin preconditioning in rats. Postgrad Med J 82:588–593
31. Gao E, Boucher M, Chuprun JK, Zhou RH, Eckhart AD, Koch WJ (2007) Darbepoetin alfa, a long-acting erythropoietin analog, offers novel and delayed cardioprotection for the ischemic heart. Am J Physiol Heart Circ Physiol 293:H60–H68
32. Baker JE, Kozik D, Hsu AK, Fu X, Tweddell JS, Gross GJ (2007) Darbepoetin-alfa protects the rat heart against infarction: dose-response, phase of action, and mechanisms. J Cardiovasc Pharmacol 49:337–345
33. Toma C, Letts DP, Tanabe M, Gorcsan J III, Counihan PJ (2007) Positive effect of darbepoetin on peri-infarction remodeling in a porcine model of myocardial ischemia-reperfusion. J Mol Cell Cardiol 43:130–136
34. Singh D, Kolarova JD, Wang S, Ayoub IM, Gazmuri RJ (2007) Myocardial protection by erythropoietin during resuscitation from ventricular fibrillation. Am J Ther 14:361–368
35. Boucher M, Pesant S, Lei YH, Nanton N, Most P, Eckhart AD, Koch WJ, Gao E (2008) Simultaneous administration of insulin-like growth factor-1 and darbepoetin-alfa protects the rat myocardium against myocardial infarction and enhances angiogenesis. Clin Transl Sci 1:13–20
36. Prunier F, Pottier P, Clairand R, Mercier A, Hajjar RJ, Planchon B, Furber A (2009) Chronic erythropoietin treatment decreases post-infarct myocardial damage in rats without venous thrombogenic effect. Cardiology 112:129–134
37. Shan X, Xu X, Cao B, Wang Y, Guo L, Zhu Q, Li J, Que L, Chen Q, Ha T, Li C, Li Y (2009) Transcription factor GATA-4 is involved in erythropoietin-induced cardioprotection against myocardial ischemia/reperfusion injury. Int J Cardiol 134:384–392
38. Doue T, Ohtsuki K, Ogawa K, Ueda M, Azuma A, Saji H, Strauss HW, Matsubara H (2008) Cardioprotective effects of erythropoietin in rats subjected to ischemia-reperfusion injury: assessment of infarct size with 99mTc-annexin V. J Nucl Med 49:1694–1700
39. Huang CH, Hsu CY, Tsai MS, Wang TD, Chang WT, Chen WJ (2008) Cardioprotective effects of erythropoietin on post-resuscitation myocardial dysfunction in appropriate therapeutic windows. Crit Care Med 36: S467–S473
40. Burger D, Xiang F, Hammoud L, Lu X, Feng Q (2009) Role of heme oxygenase-1 in the cardioprotective effects of erythropoietin during myocardial ischemia and reperfusion. Am J Physiol Heart Circ Physiol 296:H84–H93
41. Burger DE, Xiang FL, Hammoud L, Jones DL, Feng Q (2009) Erythropoietin protects the heart from ventricular arrhythmia during ischemia and reperfusion via neuronal nitric-oxide synthase. J Pharmacol Exp Ther 329:900–907
42. French CJ, Zaman AK, Sobel BE (2009) Failure of erythropoietin to render jeopardized ischemic myocardium amenable to incremental salvage by early reperfusion. Coron Artery Dis 20:295–299
43. Schlecht-Bauer D, Antier D, Machet MC, Hyvelin JM (2009) Short- and long-term cardioprotective effect of darbepoetin-alpha: role of Bcl-2 family proteins. J Cardiovasc Pharmacol 54:223–231
44. Tamareille S, Ghaboura N, Treguer F, Khachman D, Croué A, Henrion D, Furber A, Prunier F (2009) Myocardial reperfusion injury management: erythropoietin compared with postconditioning. Am J Physiol Heart Circ Physiol 297:H2035–H2043
45. Miki T, Miura T, Hotta H, Tanno M, Yano T, Sato T, Terashima Y, Takada A, Ishikawa S, Shimamoto K (2009) Endoplasmic reticulum stress in diabetic hearts abolishes erythropoietin-induced myocardial protection by impairment of phospho-glycogen synthase kinase-3beta-mediated suppression of mitochondrial permeability transition. Diabetes 58:2863–2872
46. Hotta H, Miura T, Miki T, Togashi N, Maeda T, Kim SJ, Tanno M, Yano T, Kuno A, Itoh T, Satoh T, Terashima Y, Ishikawa S, Shimamoto K (2010) Angiotensin II type 1 receptor-mediated upregulation of calcineurin activity underlies impairment of cardioprotective signaling in diabetic hearts. Circ Res 106:129–132
47. Treguer F, Donal E, Tamareille S, Ghaboura N, Derumeaux G, Furber A, Prunier F (2010) Speckle tracking imaging improves in vivo assessment of EPO-induced myocardial salvage early after ischemia-reperfusion in rats. Am J Physiol Heart Circ Physiol 298:H1679–H1686
48. Shen Y, Wang Y, Li D, Wang C, Xu B, Dong G, Huang H, Jing H (2010) Recombinant human erythropoietin pretreatment attenuates heart ischemia-reperfusion injury in rats by suppressing the systemic inflammatory response. Transplant Proc 42:1595–1597
49. Stein A, Knödler M, Makowski M, Kühnel S, Nekolla S, Keithahn A, Weidl E, Groha P, Schürmann M, Saraste A, Botnar R,

Oostendorp RA, Ott I (2010) Local erythropoietin and endothelial progenitor cells improve regional cardiac function in acute myocardial infarction. BMC Cardiovasc Disord 10:43

50. Yano T, Miki T, Tanno M, Kuno A, Itoh T, Takada A, Sato T, Kouzu H, Shimamoto K, Miura T (2011) Hypertensive hypertrophied myocardium is vulnerable to infarction and refractory to erythropoietin-induced protection. Hypertension 57:110–115
51. Cheung WK, Goon BL, Guilfoyle MC, Wacholtz MC (1998) Pharmacokinetics and pharmacodynamics of recombinant human erythropoietin after single and multiple subcutaneous doses to healthy subjects. Clin Pharmacol Ther 64(4):412–423
52. Ehrenreich H, Hasselblatt M, Dembowski C, Cepek L, Lewczuk P, Stiefel M, Rustenbeck HH, Breiter N, Jacob S, Knerlich F, Bohn M, Poser W, Rüther E, Kochen M, Gefeller O, Gleiter C, Wessel TC, De Ryck M, Itri L, Prange H, Cerami A, Brines M, Sirén AL (2002) Erythropoietin therapy for acute stroke is both safe and beneficial. Mol Med 8(8):495–505
53. Tang YD, Hasan F, Giordano FJ, Pfau S, Rinder HM, Katz SD (2009) Effects of recombinant human erythropoietin on platelet activation in acute myocardial infarction: results of a double-blind, placebo-controlled, randomized trial. Am Heart J 158(6):941–947
54. Lipsic E, van der Meer P, Voors AA, Westenbrink BD, van den Heuvel AF, de Boer HC, van Zonneveld AJ, Schoemaker RG, van Gilst WH, Zijlstra F, van Veldhuisen DJ (2006) A single bolus of a long-acting erythropoietin analogue darbepoetin-alfa in patients with acute myocardial infarction: a randomized feasibility and safety study. Cardiovasc Drugs Ther 20(2):135–141
55. Liem A, van de Woestijne AP, Bruijns E, Roeters van Lennep HW, de Boo JA, van Halteren HK, van Es TP, Jukema JW, van der Laarse A, Zwinderman AH, van Veldhuisen DJ (2009) Effect of EPO administration on myocardial infarct size in patients with non-STE acute coronary syndromes; results from a pilot study. Int J Cardiol 131(2):285–287, Epub 2007 Nov 1
56. Binbrek AS, Mittal B, Rao KN, Sobel BE (2007) The potential of erythropoietin for conferring cardioprotection complementing reperfusion. Coron Artery Dis 18(7):583–585, Review
57. Binbrek AS, Rao NS, Al Khaja N, Assaqqaf J, Sobel BE (2009) Erythropoietin to augment myocardial salvage induced by coronary thrombolysis in patients with ST segment elevation acute myocardial infarction. Am J Cardiol 104(8):1035–1040
58. Ozawa T, Toba K, Suzuki H, Kato K, Iso Y, Akutsu Y, Kobayashi Y, Takeyama Y, Kobayashi N, Yoshimura N, Akazawa K, Aizawa Y, EPO/AMI-I Pilot Study Researchers (2010) Single-dose intravenous administration of recombinant human erythropoietin is a promising treatment for patients with acute myocardial infarction—randomized controlled pilot trial of EPO/AMI-1 study. Circ J 74(7):1415–1423
59. Yoshimura N, Toba K, Ozawa T, Aizawa Y, Hosoya T (2010) A novel program to accurately quantify infarction volume by (99 m)Tc MIBI SPECT, and its application for re-analyzing the effect of erythropoietin administration in patients with acute myocardial infarction. Circ J 74(12):2741–2743
60. Belonje AM, Voors AA, van Gilst WH, Anker SD, Slart RH, Tio RA, Zijlstra F, van Veldhuisen DJ, HEBE III investigators (2008) Effects of erythropoietin after an acute myocardial infarction: rationale and study design of a prospective, randomized, clinical trial (HEBE III). Am Heart J 155(5):817–822
61. Voors AA, Belonje AM, Zijlstra F, Hillege HL, Anker SD, Slart RH, Tio RA, van't Hof A, Jukema JW, Peels HO, Henriques JP, Ten Berg JM, Vos J, van Gilst WH, van Veldhuisen DJ, HEBE III Investigators (2010) A single dose of erythropoietin in ST-elevation myocardial infarction. Eur Heart J 31(21):2593–2600
62. Ott I, Schulz S, Mehilli J, Fichtner S, Hadamitzky M, Hoppe K, Ibrahim T, Martinoff S, Massberg S, Laugwitz KL, Dirschinger J, Schwaiger M, Kastrati A, Schmig A, REVIVAL-3 Study Investigators (2010) Erythropoietin in patients with acute ST-segment elevation myocardial infarction undergoing primary percutaneous coronary intervention: a randomized, double-blind trial. Circ Cardiovasc Interv 3(5):408–413
63. Taniguchi N, Nakamura T, Sawada T, Matsubara K, Furukawa K, Hadase M, Nakahara Y, Nakamura T, Matsubara H (2010) Erythropoietin prevention trial of coronary restenosis and cardiac remodeling after ST-elevated acute myocardial infarction (EPOC-AMI): a pilot, randomized, placebo-controlled study. Circ J 74(11):2365–2371
64. Melloni C, Rao SV, Povsic TJ, Melton L, Kim RJ, Kilaru R, Patel MR, Talan M, Ferrucci L, Longo DL, Lakatta EG, Najjar SS, Harrington RA (2010) Design and rationale of the reduction of infarct expansion and ventricular remodeling with erythropoietin after large myocardial infarction (REVEAL) trial. Am Heart J 160(5):795–803, e2
65. Najjar SS, Rao SV, Melloni C, Raman SV, Povsic TJ, Melton L, Barsness GW, Prather K, Heitner JF, Kilaru R, Gruberg L, Hasselblad V, Greenbaum AB, Patel M, Kim RJ, Talan M, Ferrucci L, Longo DL, Lakatta EG, Harrington RA, REVEAL Investigators (2011) Intravenous erythropoietin in patients with ST-segment

elevation myocardial infarction: REVEAL: a randomized controlled trial. JAMA 305(18): 1863–1872

66. Ferrario M, Arbustini E, Massa M, Rosti V, Marziliano N, Raineri C, Campanelli R, Bertoletti A, De Ferrari GM, Klersy C, Angoli L, Bramucci E, Marinoni B, Ferlini M, Moretti E, Raisaro A, Repetto A, Schwartz PJ, Tavazzi L (2011) High-dose erythropoietin in patients with acute myocardial infarction: a pilot, randomised, placebo-controlled study. Int J Cardiol 147(1):124–131

67. Andreotti F, Agati L, Conti E, Santucci E, Rio T, Tarantino F, Natale L, Berardi D, Mattatelli A, Musumeci B, Bonomo L, Volpe M, Crea F, Autore C (2009) Update on phase II studies of erythropoietin in acute myocardial infarction. Rationale and design of Exogenous erythroPoietin in Acute Myocardial Infarction: New Outlook aNd Dose Association Study (EPAMINONDAS). J Thromb Thrombolysis 28(4):489–495

68. Ahmet I, Lakatta EG, Talan MI (2011) Therapeutic window for cardioprotection by recombinant human erythropoietin (rhEPO) in the rat model of ischemia-reperfusion: implications for clinical trial design. AHA 2011, Abstract

Chapter 18

Using Plethysmography to Determine Erythropoietin's Impact on Neural Control of Ventilation

Tommy Seaborn, Max Gassmann, and Jorge Soliz

Abstract

The evaluation of respiratory parameters often requires the use of anesthetics (that depress the neural network controlling respiration), and/or ways to restrain the animal's mobility (that produces a stress-dependent increase of respiration). Consequently, the establishment of plethysmography represented an invaluable technique in respiratory physiology. Plethysmography, indeed, allows the assessment of ventilatory parameters on living, unanesthetized, and unrestrained animals. The conception of the barometric plethysmography relies on the fact that an animal placed inside a hermetically closed chamber generates through its breathing a fluctuation of pressure in the chamber than can be recorded. Thus, the respiratory frequency and the tidal volume can be directly measured, while the animal's ventilation is calculated indirectly by the multiplication of these two parameters. In our hands, plethysmography was a key tool to investigate the impact of erythropoietin (Epo) on the neural control of hypoxic ventilation in mice.

Key words Lung capacity, Mouse, Respiration, Respiratory frequency, Tidal volume, Ventilation, Hypoxia

1 Introduction

Plethysmography is a reliable method to measure ventilation in several animal species, including mouse (1, 2), rat (3), shrew (4), guinea pig (5), cat (6), dog (7), pig (8), sheep (9), horse (10), and primates (11). In humans, this technique is also commonly used as a clinical tool to determine the functional residual capacity as well as the lung's total capacity. Concerning small mammals in which the evaluation of gases in blood is challenging, plethysmography allows to obtain a subsampling of air that in turn allows the evaluation of metabolic parameters (O_2 consumption and CO_2 production). The metabolic assessment is essential in order to determine whether changes in ventilation and/or the ventilatory pattern are due to regulations on the neuronal network control system, or reflect only the alteration of metabolic parameters.

Pietro Ghezzi and Anthony Cerami (eds.), *Tissue-Protective Cytokines: Methods and Protocols*, Methods in Molecular Biology, vol. 982, DOI 10.1007/978-1-62703-308-4_18,

The parameters directly provided by plethysmography are respiratory frequency (fR) and tidal volume (V_T), while ventilation must be calculated by multiplying fR with V_T. Plethysmography is technically demanding and requires a relatively complex set-up to accurately generate the desired air regimen and adequately acquire, register, and analyze data. Accordingly, the influence of several parameters, such as O_2 exposure regimen, air flow, ambient temperature within the chamber, evaporating water, and animal body temperature, must be taken into account. Plethysmography is an invaluable technique in respiratory physiology since it allows determination of the ventilatory parameters of an animal kept under physiological conditions (unanesthetized and unrestrained) during a relatively long period of time and inside a controlled environment. Once again, the concept underlying the barometric plethysmography system is to measure, in a hermetically closed chamber, the fluctuation of pressure produced by the respiration of the animal.

In our laboratory, plethysmography was a key method to determine that erythropoietin (Epo), is implicated in the modulation of the neural control system both at central (brainstem) as peripheral (carotid bodies) levels. We specifically demonstrated that Epo in adult mice improves the acute hypoxic ventilatory response, as well as the ventilatory acclimatization to hypoxia (2). Indeed, we have shown for the first time that neural respiratory and erythropoietic systems are tightly interconnected, thus playing a complementary role improving the tissue oxygenation upon hypoxia (2, 12). Moreover, we have further expanded these findings showing that chronic hypoxic exposure produces a drastic down-regulation of the soluble Epo receptor (sEpoR—a truncated form of the Epo receptor, that binds and inactivates endogenous Epo) in the central nervous system in mice. Furthermore, when sEpoR was chronically infused in the mouse's central nervous system, the process of ventilatory acclimatization to chronic hypoxia was abolished. These results showed that neural regulation of Epo and its antagonist sEpoR play a contra-balancing role in the central nervous system in establishing ventilatory activity and ensuring systemic oxygen delivery under low O_2 conditions (1). Finally, still by using plethysmography, we demonstrated that the impact of Epo on ventilation occurs in a sex-dependent manner. Keeping in mind that women are less susceptible to several respiratory diseases than men, these findings suggest that Epo plays a key role in sexually dimorphic hypoxic ventilation (1). All together, these results foresee that Epo has a potential therapeutic use as treatment for hypoxia-associated ventilatory diseases. This review describes how to measure basal ventilation and hypoxia ventilatory response in adult mice (see Note 1) under unrestrained conditions (see Note 2).

2 Materials

2.1 Plethysmography System Components

An overview of the plethysmography system as well as the connections between its components is schematized in Fig. 1.

1. Pump for regulated air supply (EMKA Technologies) (see Note 3).
2. Flowmeter (E & E Process Instrumentation) (see Note 4).
3. Plethysmography chamber for whole-body plethysmography on unrestrained mice (EMKA Technologies).
4. Thermocouple monitoring thermometer with probe (see Note 5).
5. Differential pressure transducer (EMKA Technologies) (see Note 6).
6. Amplifier for differential pressure transducer's signal (EMKA Technologies).
7. Humidity meter (see Note 7).
8. Subsampling pump (Sable Systems International) (see Note 8).
9. CO_2 analyzer (Sable Systems International).
10. O_2 analyzer (Sable Systems International).

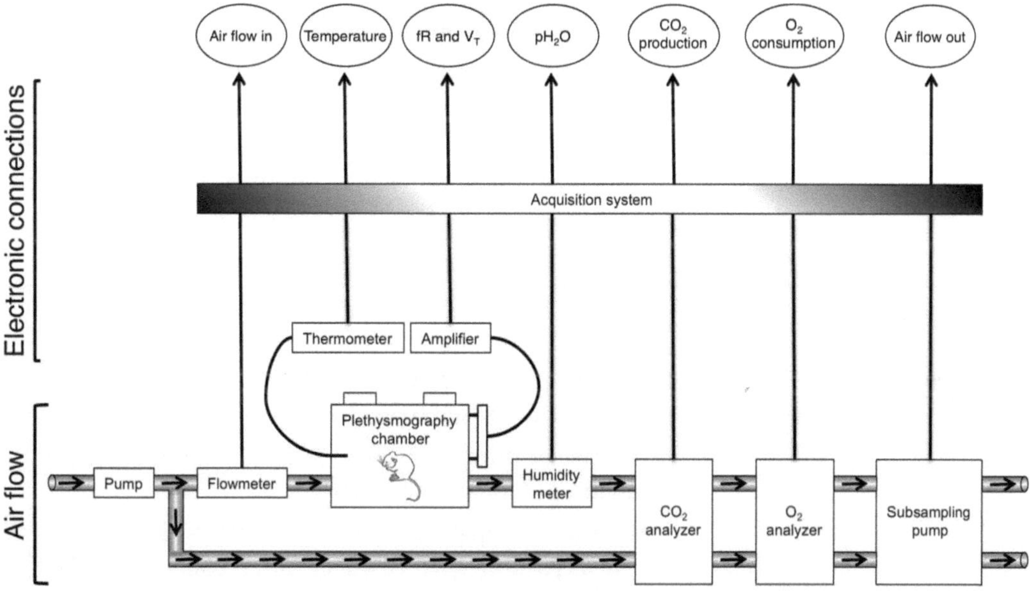

Fig. 1 Schematic overview of the system used for whole-body unrestrained plethysmography in adult mice. The tubings containing *arrows* represent the air path, while *dark lines* symbolize the electronic connections transporting analog data which are converted by the acquisition system into digital data to provide the experimental parameters included in the *ovals*

11. An acquisition system made up of an interface box (EMKA Technologies, Cat.# itfl6) connected to an acquisition card (National Instruments, Cat.# PCI-6023E) inserted to a computer (see Note 9).
12. A computer with an installed acquisition, registering and processing software (iox version 1.8.9, EMKA Technologies) (see Note 10).

2.2 Experimental Gases

1. Calibrating cylinders for gas analyzers zero: 100% N_2, certified (see Note 11).
2. Calibrating cylinders for gas analyzers span: 21% O_2, 5% CO_2, and balance N_2, certified.
3. Cylinders containing 15, 10, and 6% O_2, respectively (each with 5% CO_2 and balance N_2).
4. Each gas cylinder should be equipped with a pressure regulator and a manometer corresponding to the gas contained in the cylinder.
5. Flexible PVC tubing nontoxic (conform to FDA standard) and nonporous (see Note 12).
6. Nylon connector kit (Harvard Apparatus) (see Note 13).
7. Desiccant membrane air dryer (Perma Pure) (see Note 14).

2.3 Mice and Epo

1. Sufficient number of adult mice (wild-type or transgenic) of a comparable age and of determined sex to generate reliable and reproducible mean group data (see Note 15).
2. Thermocouple monitoring thermometer and stainless steel rectal probe for mice (see Note 16).
3. In case of experiments involving injection of exogenous Epo (to wild type mice, typically), human recombinant Epo (see Note 17).

3 Methods

3.1 Calibrations

1. Calibrate the air flow in to 700–800 ml/min by adjusting the air supplying pump (see Note 18).
2. Calibrate the air flow out to 600–700 ml/min by adjusting the subsampling pump (see Note 19).
3. Calibrate the CO_2 analyzer by using a 100% N_2 cylinder to calibrate the zero and a certified cylinder containing 5% CO_2 (with 21% O_2 and balance N_2) to calibrate the span.
4. Calibrate the O_2 analyzer by using a 100% N_2 cylinder to calibrate the zero and a certified cylinder containing a percentage

of O_2 in the range of the experimental design (e.g., 21% O_2, 5% CO_2, balance N_2) to calibrate the span.

5. Calibrate the ventilation volume by injection of 1 ml of air in about 2–3 s into the animal chamber.

3.2 Effect of Epo on Basal Ventilatory State

1. Weight the mouse (see Note 20).
2. Put the mouse into the plethysmography chamber, close the chamber, and let the animal acclimatized until it reach a calm and quiet state reflected by a regular respiratory signal (typically between 30 and 60 min).
3. Open the plethysmography chamber and gently manipulate the animal to take body temperature with rectal probe (this is the initial body temperature) (see Note 16).
4. Put the mouse back to the plethysmography chamber and let it acclimatized again.
5. Once the mouse is calm, start to register the acquired data during 10 min with the animal exposed to room air (or to a cylinder containing 21% O_2, 5% CO_2, balance N_2 for more accuracy). These data provide the baseline state to which the experimental state (severe hypoxia in this example) will be compared with.

3.3 Effect of Epo on Ventilatory Response to Severe Hypoxia (See Note 21)

1. In continuity with the basal state registering, turn open the gas cylinder containing 15% O_2, 5% CO_2, and balance N_2.
2. At this moment, a hand-made tag could be added on the recording stream to help to localize the O_2 regimen modification at the time of analysis.
3. Thereafter, a gradual decrease in O_2 exposure is generated by sequentially changing at each 5 min for a cylinder containing 10 and 6% O_2, respectively (with 5% CO_2 and balance N_2 in each case) (see Note 22). Tag each modification of O_2 regimen on the recording stream.
4. Continue to register the acquired data during 10 min with the animal exposed to hypoxia at 6% O_2. When compared with basal state, hypoxia should increase both frequency and amplitude of respiratory signal.
5. When hypoxia exposure is ended, turn off the gas cylinder generating the hypoxic air. This moment could be tagged on the stream as the end of hypoxia.
6. Continue to register the signal during 10 min with the animal exposed to room air (or to a cylinder containing 21% O_2, 5% CO_2, and balance N_2). These data characterized which is called the recovery period.

3.4 End of the Ventilation Measurements

1. Via the software control, stop to register the acquired data.
2. Rapidly open the plethysmography chamber and take body temperature with the rectal probe (this is the final body temperature) (see Note 16).
3. Return the mouse in its original cage or anesthetize the animal for euthanasy before harvesting tissues, depending on the experimental protocol.
4. Turn off the gas cylinder(s) if applicable, the air supply pump, the flowmeter, the thermometer, the humidity meter, and the subsampling pump (see Note 23).
5. Be sure all the acquired data have been saved on the computer for further analysis.
6. Turn off the acquisition system and the computer.
7. Carefully clean the plethysmography chamber (see Note 24).

3.5 Analyzing Ventilation and Metabolism Data (See Note 25)

For visual example purpose, typical results obtained by analysis of data from whole-body unrestrained plethysmography before and after hypoxia with mice injected or not with Epo are presented on Fig. 2.

1. Respiratory frequency (fR) is directly evaluated as the number of respiratory cycles per minute (resp. per min).
2. Tidal volume (V_T) is obtained with previously described calculations (13) based on an equation firstly described by Drorbaugh and Fenn (14) in which V_T is expressed in ml per 100 g of body weight (ml per 100 g) in BTPS (body temperature and pressure, saturated) conditions.
3. Minute ventilation ($\dot{V}E$) is calculated as the product of fR and V_T and normalized to 100 g of body weight (ml/min/100 g).
4. O_2 consumption ($\dot{V}O_2$) is calculated as the difference between the fractions of O_2 before and after the plethysmography chamber corrected for STPD (standard temperature and pressure in air-dry) conditions and normalized for body weight (ml_{STPD}/min/100 g).
5. CO_2 production ($\dot{V}CO_2$) is calculated as the difference between the fractions of CO_2 after and before the plethysmography chamber corrected for STPD (standard temperature and pressure in air-dry) conditions and normalized for body weight (ml_{STPD}/min/100 g).

4 Notes

1. It is well possible to perform similar experiments with younger mice, even from neonates, by using a specifically designed insert in the plethysmography chamber. Such an insert increases the relative impact of the small pressure fluctuations of the

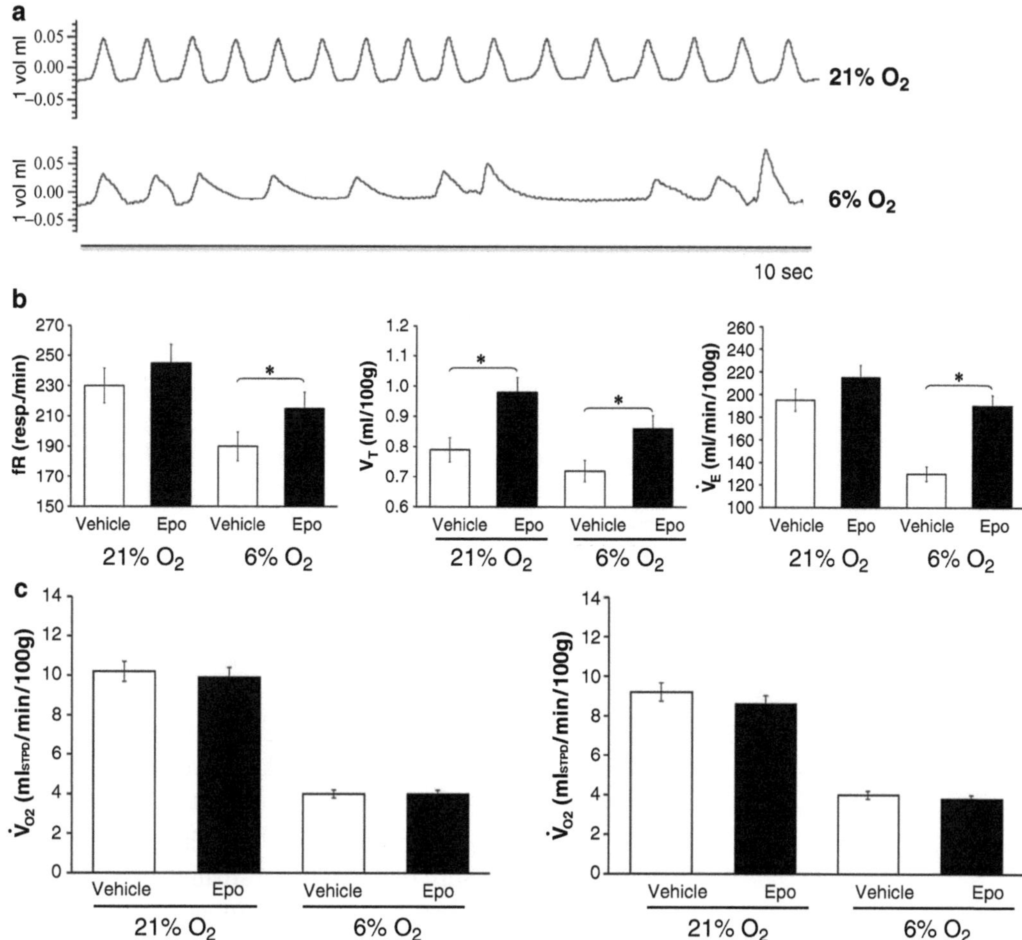

Fig. 2 Typical results obtained by analysis of data from whole-body unrestrained plethysmography achieved on female adult mice. (**a**) Representative examples of respiratory traces registered at 21% O_2 (normoxia) and at 6% O_2 (severe hypoxia). Example of results obtained for ventilatory (**b**) and metabolic (**c**) parameters in mice injected with Epo or vehicle (i.p.). Significant difference ($p < 0.05$) between vehicle and Epo groups are delineated by an *asterisk*

smaller animal by reducing the total air volume in the animal chamber. Moreover, controlling the temperature within the plethysmography chamber is necessary when experiments were carried out on younger animals that do not have yet the physiological capacity to thermoregulate adequately (<15 postnatal days in mice and rats).

2. Plethysmography can be achieved in a whole-body or double-chamber/head-out set-up. The whole-body system allows to make measurements on unrestrained and conscious animal during a long period of time without anesthetic bias. The relatively reasonable size of mice (and rats!) allows the use of this approach.

3. The pump is responsible to provide, at a constant and precise flow rate, the air mixture from the ambient air or gas cylinder(s) to the plethysmography chamber. There is a wide range of quality in the commercially available pumps, the main differences between them being the flow regulation. These parameters are very important to prevent any accumulation of CO_2 and evaporated water inside the chamber and to minimize the interaction of the provided air with the animal's breathing pattern. The importance of the latter is increasing while the size of the tested animal is decreasing, becoming crucial in plethysmography experiments carried out on neonatal mice.
4. The flow meter is a device that directly and precisely measures, in a real-time manner, the flow of the air entering in the plethysmography chamber. It should be installed just before the arrival of the air to the chamber, in order to take into account all eventual air resistance upstream to the chamber. A precise flow meter is essential to a rigorous plethysmography system since it should provide an exact evaluation of the air flow the mice is exposed to, as well as a way to verify a constant flow in order to minimize the interaction with the animal's breathing pattern registered.
5. The ambient temperature in the plethysmography chamber should be monitored in order to make corrections for STPD conditions in the calculation of O_2 consumption and CO_2 production by the mice (cf. Subheading 3.5, steps 4 and 5).
6. The differential pressure transducer is a small device connected directly to the plethysmography chamber with one port exposed to the animal chamber and another one exposed to the reference chamber which is empty and subjected to the atmospheric pressure. Thanks to this configuration, the transducer can measure the pressure fluctuations within the plethysmography chamber which results from mice breathing.
7. The pressure odf the evaporated water in the plethysmography chamber should be monitored in order to make corrections for BTPS conditions in the calculation of V_T (cf. Subheading 3.5, step 2).
8. The subsampling pump helps to avoid the generation of a positive pressure within the plethysmography chamber and, most importantly, it allows obtaining the air composition before being modified by the animal breathing inside the chamber (Fig. 1). This subsampled air is essential to determine $\dot{V}O_2$ (cf. Subheading 3.5, step 4) and $\dot{V}CO_2$ (cf. Subheading 3.5, step 5).
9. A data acquisition system is necessary to receive analog data (electrical signals) from the measurements devices (flow meter, thermometer, amplifier of pressure transducer, humidity meter,

CO_2 and O_2 analyzers) and to convert them into digital numeric values that can be processed by a computer. There are several acquisition systems commercially available and each of them has its pros and cons. The main concerns for the experimenter in a choice of acquisition system is the compatibility with analyzing software (especially in term of acquisition frequency) and measurement devices connections (usually BNC connectors).

10. Once converted by the acquisition system, the collected data should be registered and then analyzed on a computer. Several comprehensive software are available for data acquisition and real-time analysis. Some software present two distinct modules, a first one which register the data inputs and a second one devoted to data analysis. Usually, the software used for plethysmography analysis generate both a stream file of the registered traces and data files containing numeric values obtained from the traces or calculated from them.
11. Although this is not necessary the gas content of the experimental cylinders to be certified, it would increase the homogeneity in the results (especially if one would compare experiments done by using more than one gas cylinder). If mixing the gases in the lab, we strongly recommend the use of an appropriate gas analyzer(s) located within the system to specifically assess the gas content of the air entering in the plethysmography chamber.
12. The caliber (internal and external diameter) of the tubing should be optimized as much as possible in order to avoid over-pressure and to limit the air dead volume within the air path. Also in order to limit the air dead volume, the length of the tubing should be kept at the necessary minimum.
13. A wide variety of connectors is necessary to properly and conveniently interconnect tubing together or with devices.
14. It is essential to dry the air before its arrival in the gas analyzers because evaporated water causes interference with CO_2 measurement which is carried out by infrared optical detection. For this reason, an air dryer should be inserted in the air path just before entering in the CO_2 analyzer.
15. Our experience teaches us that between 6 and 10 mice per group is usually sufficient to obtain satisfying and representative plethysmography results. Nevertheless, the sample size has to be determined empirically for each experimental design and increased until the standard errors are reasonable in comparison to means. If wild-type mice are used, the extrapolation and comparison with previous or published results should be carefully leaded since some differences in the breathing phenotype of different mice strains has been reported (15).

Alternatively, transgenic mice over-expressing Epo only in the brain (Tg21 mice) or in both brain and plasma (Tg6 mice) have been generated (16, 17) and they represent an interesting tool to study the effect of Epo on the respiratory control. In addition, since a sex difference exists in the ventilatory pattern, in the ventilatory response to hypoxia, and in Epo impact on this response, sex of the mice is an important variable and it should be determined and annotated. Indeed, the experimental groups should be designed in order to equally represent males and females or, ideally, to consider each gender separately, in both sampling and analysis.

16. The rectal temperature of mice should be measured before and after the experimental protocol in order to make corrections for BTPS conditions in the calculation of V_T (cf. Subheading 3.5, step 2). The initial body temperature is considered to correct all the data, except the 5 last minutes (recovery period), which are corrected by using the mice temperature at the end of the protocol (the final body temperature).
17. Indeed, the dose- and time-effect of Epo should be firstly inferred from data available in the literature and/or determined specifically for the considered experimental context. The time of injection and the doses regimen of Epo are other parameters depending on the specific experimental design. Alternatively, other Epo derivatives are commercially available, some of them presenting only the non-erythropoietic characteristics of Epo and being a potential useful tool to study the impact of Epo on the control of respiration. Obviously, control animals should be injected with the vehicle used to resuspend the recombinant Epo.
18. The air flow in depends on the size of the animal and the value proposed here for adult mice has been determined empirically from our previous experiments.
19. The flow of the subsampling pump should always be a little bit lower than the air flow in to avoid the creation of a negative pressure within the plethysmography chamber.
20. In order to facilitate comparisons between mice, minute ventilation ($\dot{V}E$), O_2 consumption ($\dot{V}O_2$), and CO_2 production ($\dot{V}CO_2$) are normalized to 100 g body weight. For this reason, mice should be weighted before the beginning of the experiment.
21. In addition to severe hypoxia as tested in this example, it is also possible to test by using a similar procedure the impact of other gas regimen on ventilation and/or of Epo treatment on ventilation (e.g., hyperoxia, normoxia, mild hypoxia, anoxia, hypercapnia, normocapnia, hypocapnia).

22. The fraction of O_2 in the plethysmography chamber should be gradually decreased from 21 to 6% O_2 during 15 min in order to avoid mice death due to a too rapid O_2 deprivation. However, if mice are exposed to 10% O_2 or higher during the hypoxia regimen, the decrease in the fraction of O_2 do not necessitate intermediate step(s) and could be done directly from 21 to 10% O_2 without risk of asphyxia for mice.
23. It is highly suggested to not turn off the O_2 and CO_2 analyzers and the amplifier after each utilization in order to maintain their functioning temperature as stable as possible.
24. Proceed as recommended by the supplier and, importantly, do not use any chemicals that may damage the plethysmograph chamber's integrity (such as ethanol) or influence respiration (or stress) of the next experimental mice.
25. There is several ways to analyze plethysmography data, depending on the experimental context or the plethysmography set-up used. We present and propose here one of these possibilities, which is appropriate and relevant for the whole-body unrestrained plethysmography carried out on adult mice with the set-up we presented. However, this is not the only appropriate method to analyze this type of results.

Acknowledgments

The authors would like to thank Drs Vincent Joseph and Cécile Julien as well as Mr Raphaël Lavoie and Miss Hanan Khemiri for helpful discussions and valuable advices. J.S. is supported by the Respiratory Health Network of the FRSQ (Fonds de la Recherche en Santé du Québec), the Foundation of Stars for the Children's health research, the Molly Towell Perinatal Research Foundation (MTPRF). M.G. is supported by the Zurich Center for Integrative Human Physiology (ZIHP).

References

1. Soliz J, Gassmann M, Joseph V (2007) J Physiol 583:329–336
2. Soliz J, Joseph V, Soulage C, Becskei C, Vogel J, Pequignot JM, Ogunshola O, Gassmann M (2005) J Physiol 568:559–571
3. Joseph V, Soliz J, Pequignot J, Sempore B, Cottet-Emard JM, Dalmaz Y, Favier R, Spielvogel H, Pequignot JM (2000) Am J Physiol Regul Integr Comp Physiol 278:R806–R816
4. Tashiro N, Kataoka M, Ozawa K, Ikeda T (2007) J Am Assoc Lab Anim Sci 46:81–85
5. Vargas MH, Sommer B, Bazan-Perkins B, Montano LM (2010) Vet Res Commun 34:589–596
6. Kirschvink N, Leemans J, Delvaux F, Snaps F, Clercx C, Gustin P (2007) Vet J 173:343–352
7. Murphy DJ, Renninger JP, Schramek D (2010) J Pharmacol Toxicol Methods 62:47–53
8. Halloy DJ, Kirschvink NA, Vincke GL, Hamoir JN, Delvaux FH, Gustin PG (2004) Vet J 168:276–284

9. Bedenice D, Bar-Yishay E, Ingenito EP, Tsai L, Mazan MR, Hoffman AM (2004) Am J Vet Res 65:1259–1264
10. Nolen-Walston RD, Kuehn H, Boston RC, Mazan MR, Wilkins PA, Bruns S, Hoffman AM (2009) J Vet Intern Med 23:631–635
11. Iizuka H, Sasaki K, Odagiri N, Obo M, Imaizumi M, Atai H (2010) J Toxicol Sci 35:863–870
12. Soliz J, Soulage C, Hermann DM, Gassmann M (2007) Am J Physiol Regul Integr Comp Physiol 293:R1702–R1710
13. Joseph V, Soliz J, Soria R, Pequignot J, Favier R, Spielvogel H, Pequignot JM (2002) Am J Physiol Regul Integr Comp Physiol 282:R765–R773
14. Drorbaugh JE, Fenn WO (1955) Pediatrics 16:81–87
15. Menuet C, Kourdougli N, Hilaire G, Voituron N (2011) J Appl Physiol 110:1572–1581
16. Wiessner C, Allegrini PR, Ekatodramis D, Jewell UR, Stallmach T, Gassmann M (2001) J Cereb Blood Flow Metab 21:857–864
17. Ruschitzka FT, Wenger RH, Stallmach T, Quaschning T, de Wit C, Wagner K, Labugger R, Kelm M, Noll G, Rulicke T, Shaw S, Lindberg RL, Rodenwaldt B, Lutz H, Bauer C, Luscher TF, Gassmann M (2000) Proc Natl Acad Sci U S A 97: 11609–11613

Chapter 19

Cerebral Malaria: Protection by Erythropoietin

Anne-Lise Bienvenu and Stephane Picot

Abstract

Cerebral malaria (CM) is still responsible for unacceptable death rate, while new antimalarial drugs were recently developed. CM pathophysiology shares essential biological features with cerebral ischemia. Because erythropoietin (Epo) was demonstrated to reduce mortality rate during experimental cerebral ischemia (1), in the early 2000, we wondered whether Epo could help to reduce the burden of CM. There is now evidence that Epo high doses could prevent early mortality during cerebral malaria. This evidence was obtained first using mice model of cerebral malaria, and later confirmed by prospective clinical trial in endemic area. High doses of Epo are needed to cross the blood–brain barrier (see Note 1) and to favor the cytoprotective versus hematopoietic effect of this pleiotropic cytokine (see Note 2).

Key words Erythropoietin, Cerebral malaria, Neuronal apoptosis, Endothelial apoptosis, Murine model, Neuroprotection, CEPO, Epo-biosimilars

1 Introduction

Intravenous artesunate is now the first line treatment for severe *P. falciparum* malaria due to high antimalarial activity and low toxicity (2). Despite the use of these highly effective drugs, 10–20% mortality rate still occurs mostly within the first 24 h of the disease (3) and more in remote endemic areas.

Adjunctive therapies during cerebral malaria (pentoxifyllin, anti-TNF antibody, and deferoxamine) were previously tested, but no clear evidence for increase in survival rate was provided (4–6).

High doses recombinant human erythropoietin are increasingly used for cytoprotective properties (1, 7) in order to prevent cell and tissue injury during acute ischemic diseases (e.g., stroke), neurological chronic diseases (e.g., Alzheimer) or acute neurological infections (e.g., meningitis). For most of the cases, first evidence of Epo cytoprotective effects was demonstrated using murine models of the disease, and later confirmed in humans.

For cerebral malaria, the accuracy of the murine model has been subjected to debate for a while. It is clear that difference

Pietro Ghezzi and Anthony Cerami (eds.), *Tissue-Protective Cytokines: Methods and Protocols*, Methods in Molecular Biology, vol. 982, DOI 10.1007/978-1-62703-308-4_19, © Springer Science+Business Media, LLC 2013

between mice and humans do exist, mostly based on the type of sequestered cells in brain microvasculature. However, since there is no alternative to study severe malaria and to develop new concepts of treatment, the murine model of cerebral malaria provides experimental evidences that need to be confirmed in humans (see Note 3).

2 Materials

Animal experiments are conducted according to the Institutional Animal Care and Use Committee guidelines and in agreement with international rules. The model used in the lab is CBA/J mice 6 weeks old inoculated with 10^6 *Plasmodium berghei ANKA* (see Note 4).

2.1 Inoculum: Plasmodium berghei ANKA Parasitized Red Blood Cells

1. *P. berghei* ANKA aliquots of 1 mL (concentration = 10^7 GRP/mL) are stored in liquid nitrogen. Frozen aliquots must be discarded after 1 year.
2. Preheat sterile phosphate buffered saline (PBS) at 37°C.
3. Remove aliquots of *P. berghei* ANKA from liquid nitrogen and proceed to a quick defrost of the aliquot at 37°C.
4. Slowly add 1 mL of sterile PBS to 1 mL aliquot of *P. berghei* ANKA and homogenize the suspension carefully.
5. Start intraperitoneal (IP) infection less than 5 min after inoculum preparation.

2.2 Mice: CBA/J (See Note 5)

Six-week-old female CBA/J mice weighting 18–20 g are used. Allow mice to adapt to their environment for 7 days before experiments. Choose the number of mice needed per group according to the expected effect and statistical analysis (see Note 6).

2.3 Drugs

1. Erythropoietin beta (see Notes 7 and 8): Dilute erythropoietin beta in sterile saline to obtain a final concentration of 100 IU/mL. After dilution, Epo is stable for 2 h. Calculate the volume of Epo according to each mouse weight. For a mouse weighting 20 g and an expected posology of 1,000 IU/kg, inject 200 μL of EPO 100 IU/mL using intraperitoneal route.
2. Artesunate (Molecular weight: 384.42; Molecular formula: $C_{19}H_{28}O_8$): Dilute 8 mg of artesunate powder in 133 μL of sodium bicarbonate 4.2%. Then, add 877 μL of sterile saline to obtain 8.0 g/L concentration. Calculate the injected volume of artesunate solution according to each mouse weight. For a mouse weighting 20 g and an expected posology of 40 mg/kg, inject 100 μL of artesunate 8.0 g/L using intraperitoneal route.

3 Methods

The flow chart describing the experimental procedures is shown in Fig. 1.

3.1 Infection

1. During the injection procedure, keep *P. berghei* ANKA inoculum at 37°C using a mobile water bath.
2. Take a 1 mL sterile syringe with 0.33 × 12 mm needle.
3. Fill in the syringe with 200 μL *P. berghei* ANKA suspension.
4. Hold the mice by the neck using forefinger and the skin using other fingers.
5. Bulge the abdomen of the mice.
6. Proceed to IP injections of 200 μL *P. berghei* ANKA suspension per mouse.

3.2 Treatment

1. At day 6, randomized mice in four groups of 15 mice.
2. At day 6, day 7, and day 8, for a mouse weighing 20 g, proceed to intraperitoneal injection of:
   - 200 μL of saline to control group.
   - 200 μL of Epo 100 IU/mL per day to Epo group.
   - 100 μL of artesunate 8.0 g/L per day to AS group.
   - 200 μL of Epo 100 IU/mL and 100 μL of artesunate 8.0 g/L per day to AS–Epo group.

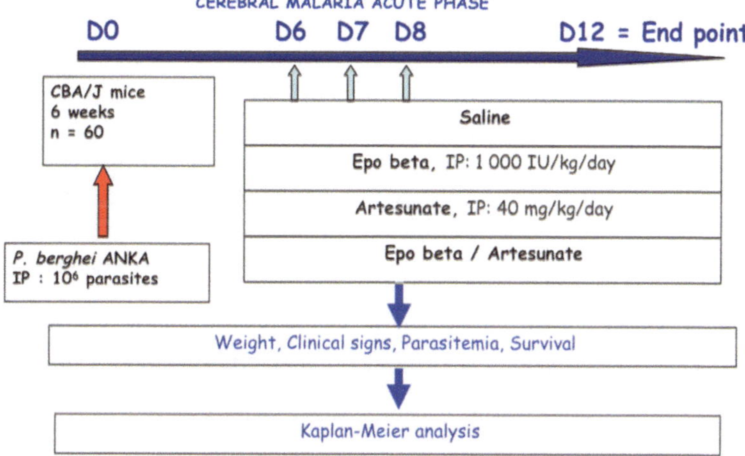

Fig. 1 Flow chart of murine cerebral malaria infection and treatment with Epo and artesunate

3.3 Follow-Up

1. From day 0 to day 12, record body weight and clinical presentation every day before and after treatment.
2. In order to evaluate the clinical presentation of cerebral malaria, use the classification in clinical stages (from 0 to 4) named BIENVENU score described below (8):

   Stage 0: healthy mice

   Stage 1: symptoms not associated with cerebral malaria (ruffled fur)

   Stage 2: ruffled fur and motor impairment

   Stage 3: respiratory distress

   Stage 4: convulsions and/or coma

3.4 Parasitemia

From day 4 to day 12, measure parasitemia of each mouse as follow.

1. Collect blood in the using tail vein a 1 mL sterile syringe with 0.33 × 12 mm needle.
2. Take a drop of blood and place it on a slide.
3. Spread the drop of blood to obtain a thin film.
4. Dry the slide.
5. Fix in Diff Quick Fixative (or methanol) for 20 s.
6. Stain with Diff Quick solution I for 20 s.
7. Counterstain with Diff Quick solution II for 30 s.
8. Rinse in tap water to remove excess stain.
9. Use one drop of oil to watch the slide at immersion ×1000.
10. Calculate the number of parasitized erythrocytes per 100 erythrocytes.
11. Experiment end-point is day 12 post-infection and is defined as the day that at least 90% of the control group mice are dead.

3.5 Statistical Analysis

Calculate cumulative long-term survival according to the Kaplan–Meier method. Compare groups with the log rank test. Survival time is the dependent variable. p-value of <0.05 is considered significant. From the last day of treatment (day 8 post-infection), cumulative long-term survival analyses are performed every day.

3.6 Expected Biological and Clinical Presentation

Biological and clinical presentation of mice post-infection without treatment is expected as follow (Table 1, Fig. 2):

- Day 3: positive parasitemia.
- Day 4: increased parasitemia not yet associated with clinical signs.
- Day 5: most mice with ruffle fur (first stage) related to fever.

Table 1
Typical evolution of biological and clinical signs during the follow-up of untreated CM mice

Day post-infection	Clinical stage	Parasitemia	Death
0–3	0	<0.01	0
4	0	1–3%	0
5	1	3–5%	0
6	2	5–8%	0
7	3	8–12%	20%
8–10	3–4	>12%	80%

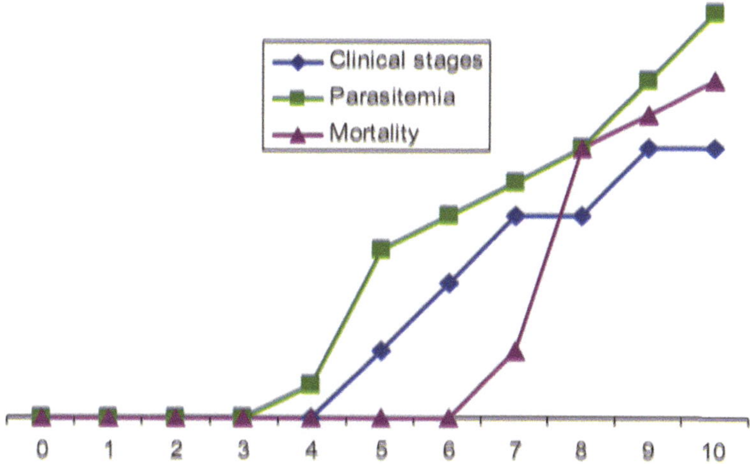

Fig. 2 Clinical presentation, parasitemia, and death rate of CM-untreated mice

- Day 6: most mice at the second stage of cerebral malaria.
- Day 7: severe clinical signs or death for a minority of mice.
- Days 8–10: severe clinical signs or death for a majority of mice.
- Day 10: more than 80% of mice are dead.

Clinical presentations of untreated mice, mice treated with Epo, Artesunate, and Artesunate Epo according to BIENVENU score (8) are presented below (Fig. 3):

Survival of mice treated with Epo, Artesunate, and Artesunate Epo compared to Control is presented below (Fig. 4):

The analysis of variance (ANOVA, $p<0.001$) and the chi(2) square test ($p<0.01$) will be used to compare the effects of the different therapeutic options.

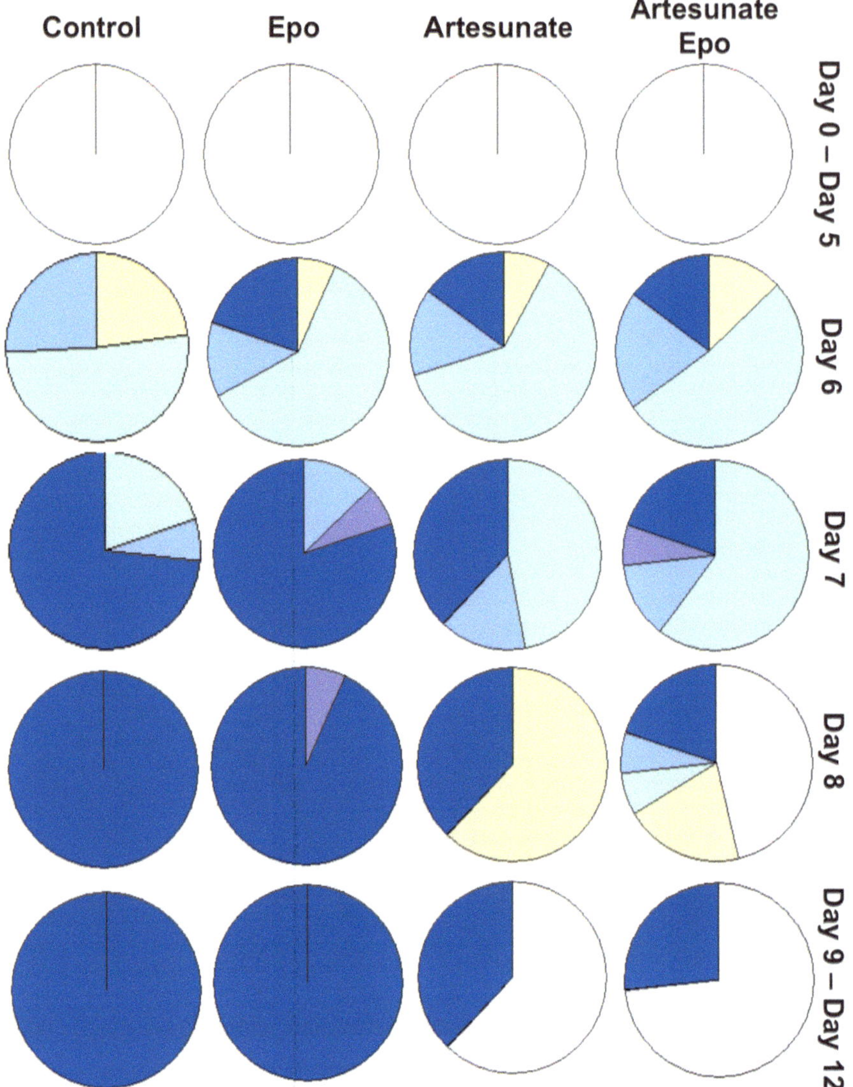

Fig. 3 BIENVENU score showing clinical ranking of infected mice without treatment (Control), infected mice treated with Epo (Epo), infected mice treated with artesunate (Artesunate), and infected mice treated with artesunate and Epo (Artesunate Epo). Stage 0 = white; stage 1 = yellow; stage 2 = green; stage 4 = light blue; death = dark blue

4 Notes

1. Epo has a poor bioavailability as demonstrated in recent studies: when rhuEpo is administrated at low dose used to promote erythropoiesis (400 IU/kg), no increase of cerebrospinal fluid (CSF) Epo concentration is observed. At doses ranging

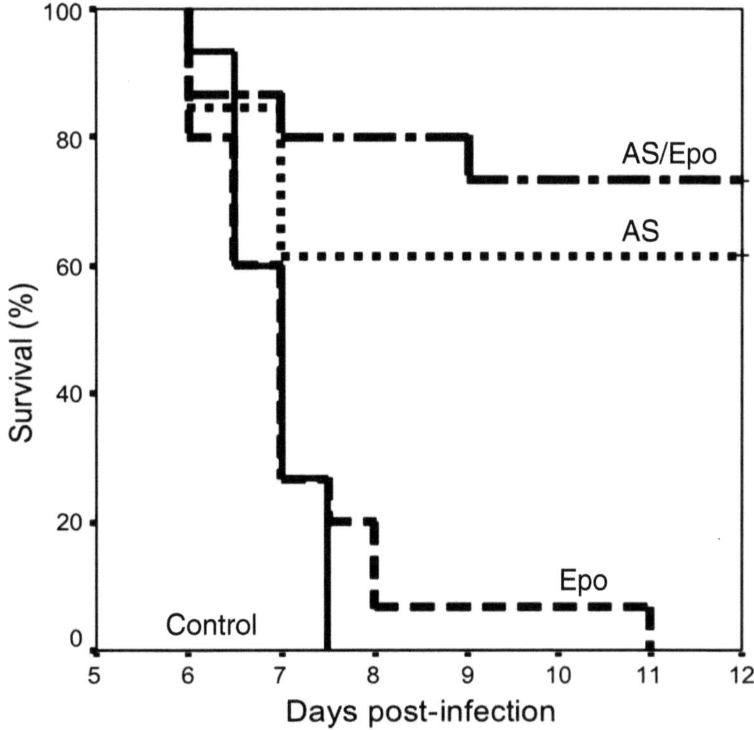

Fig. 4 Univariate analysis of survival (spss 11.5) showing benefit of the drug combination artesunate–erythropoietin (AS/Epo) on survival compared to artesunate treatment alone (AS)

500–5,000 IU/kg, a small proportion of circulating rhuEpo does cross the BBB (9). rhuEpo large molecular size (37 kDa), high glycosylation and negative charge could explain the small proportion of BBB passage (10).

2. It has been recently described that the hematopoietic and non-hematopoietic functions of Epo are mediated through different receptors (Fig. 2). Whereas Epo receptor homodimers mediate the hematopoietic response, the tissue protective activities of Epo are mediated by a heterodimeric receptor composed of Epo receptor associated with the common beta receptor subunit, also known as CD131 (11). Binding of a tissue-protective molecule initiates signaling via phosphorylation of JAK-2 activating multiple downstream pathways that promote tissue protection (11).

3. There is at least one major difference between the mouse model of cerebral malaria and human cerebral malaria that needs to be considered. During human cerebral malaria, the predominant cell type sequestered on the endothelium of the cerebral microvasculature is PRBCs. Whereas mice develop

distinctive cerebral lesions closed to humans, cerebral vessels obstruction is monocytes predominant by adhesion to vessels walls (12). Moreover, PbA cerebral malaria lacks the systemic involvement such as hypoglycaemia and the liver damage observed in *falciparum* malaria (13). Another difference commonly found in the literature is that over 80% of susceptible mice infected with PbA died with cerebral symptoms and BBB breakdown (14), whereas very few humans develop cerebral complications leading to death. This statement could be explained by the experiment homogeneity: animals (sex, age, and genetic background), clone of *Plasmodium*, and treatment. Such conditions are not applicable to humans. Experimental models cannot reproduce exactly the brain pathology observed in humans, but there is a growing number of similarities between models and the human disease. In particular, monocytes are also seen in the cerebral vessels of African children (15). The accumulation of platelets in the brain vessel of these children is also significantly more important than in non cerebral malarial patients (16). While only slight differences in the sequestration process were described (17), this model has contributed greatly to general understanding of cerebral malaria.

4. Four species and 13 subspecies of rodent *Plasmodium* exist (18), but *P. berghei* and *P. yoelii* are commonly used for CM studies. *P. berghei* was first isolated from an infected *Anopheles dureni* in Congo in 1948 (19). Different strains were isolated in the same region of Congo and four of them, *P. berghei* SP11, ANKA (Anvers-Kasapa) (PbA), NK65, and Kyberg 173 (K173), were used to study CM The extent of experimental cerebral malaria (ECM) induction is also dependent of the amount of inoculum (20). In (20), a 2×10^6 inoculum led to the maximum incidence of ECM in C57BL/6 mice (80–100%). Changes in the amount of the inoculum could lead to the release of a different array of lymphokines by activation of CD4 Th2 cells and could explain this observation. The K173 was also used to study CM because histopathological analysis of the brains of infected mice showed cerebral involvement (21). Ball et al. (22) demonstrated that low inoculum of K173 induces CM-neurological signs. Despite this demonstration, the PbA strain is the most common strain for cerebral malaria studies: from 2000 to 2008, only five CM studies used Pb K173, comparing to 97 for PbA and 10 for *P. yoelii* (Pubmed "cerebral malaria and *P. berghei* K173 or *P. berghei* ANKA or *P. yoelii*").
5. PbA in CBA/Ca and C57BL/6 mice and PbK173 in C57BL/6 have been the most studied mouse/parasite combination. Alternative CM models using *P. yoelii yoelii* 17XL or YM in CF1, Swiss and BALB/c mice have also been proposed (23),

but cerebral involvement is variable. The genetic background of mice can affect the course of parasitemia and disease. The selection of various virulent clones is regulated by the mouse genetic background and the dynamic interplay between the clones and their hosts. Marked differences were observed in Amani's study (20) in susceptibilities to ECM between the mouse strains, but the levels of parasitemia were similar in all the strains. Among the strains, C57BL/6, 129Sv/eV, and CBA/Ca were the most susceptible. It is also reported that the age of the mouse inversed correlated with CM susceptibility (24).

6. To obtain significant data, 15 mice are included in each group (60% of increase in recovery, type I error of 0.05, power of 80%).
7. Epoietin alpha differ from epoietin beta in *in vivo:in vitro* bioactivity and in content in basic isoform. Epoietin beta bioactivity is 20% higher than epoietin alpha. This may be linked to epoietin beta isoform content more basic than epoietin alpha, responsible for a longer half-life after intravenous and subcutaneous administration (25).
8. Nonhematopoietic Epo derivatives are available (26) such as asialoEpo and carbamylated Epo (CEPO) that maintain their tissue-protective activities, but do not increase the concentration of erythrocytes (27). AsialoEpo was demonstrated to provide significant and equal neuroprotection as Epo when administered 4 h prior to hypoxia-ischemia in 7-day-old rats (28). CEPO is obtained by incubation of Epo with cyanate at 37°C. The injection of CEPO demonstrated no change from baseline in erythropoietic parameters (29).

References

1. Sakanaka M, Wen TC, Matsuda S et al (1998) *In vivo* evidence that erythropoietin protects neurons from ischemic damage. Proc Natl Acad Sci U S A 95:4635–4640
2. World Health Organization (2010) Guidelines for the treatment of malaria. Second edition, Rev 1
3. Newton CR, Krishna S (1998) Severe *falciparum* malaria in children: current understanding of pathophysiology and supportive treatment. Pharmacol Ther 79:1–53
4. Di Perri G, Di Perri IG, Monteiro GB et al (1995) Pentoxifylline as a supportive agent in the treatment of cerebral malaria in children. J Infect Dis 171:1317–1322
5. van Hensbroek MB, Palmer A, Onyiorah E et al (1996) The effect of a monoclonal antibody to tumor necrosis factor on survival from childhood cerebral malaria. J Infect Dis 174:1091–1097
6. Thuma PE, Mabeza GF, Biemba G et al (1998) Effect of iron chelation therapy on mortality in Zambian children with cerebral malaria. Trans R Soc Trop Med Hyg 92:214–218
7. Maiese K, Li F, Chong ZZ (2005) New avenues of exploration for erythropoietin. JAMA 293:90–95
8. Bienvenu AL, Ferrandiz J, Kaiser K et al (2008) Artesunate-erythropoietin combination for murine cerebral malaria treatment. Acta Trop 106:104–108
9. Juul SE, McPherson RJ, Farrell FX et al (2004) Erythropoietin concentrations in cerebrospinal fluid of nonhuman primates and fetal sheep following high-dose recombinant erythropoietin. Biol Neonate 85:138–144

10. Statler PA, McPherson RJ, Bauer LA et al (2007) Pharmacokinetics of high-dose recombinant erythropoietin in plasma and brain of neonatal rats. Pediatr Res 61:671–675
11. Brines M, Cerami A (2008) Erythropoietin-mediated tissue protection: reducing collateral damage from the primary injury response. J Intern Med 264:405–432
12. Grau GE, Del Giudice G, Lambert PH (1987) Host immune response and pathological expression in malaria: possible implications for malaria vaccines. Parasitology 94: S123–S137
13. Clark IA (1987) Cell-mediated immunity in protection and pathology of malaria. Parasitol Today 3:300–305
14. Thumwood CM, Hunt NH, Clark IA et al (1988) Breakdown of the blood–brain barrier in murine cerebral malaria. Parasitology 96:579–589
15. Taylor TE, Fu WJ, Carr RA et al (2004) Differentiating the pathologies of cerebral malaria by postmortem parasite counts. Nat Med 10:143–145
16. Grau GE, Mackenzie CD, Carr RA et al (2003) Platelet accumulation in brain microvessels in fatal pediatric cerebral malaria. J Infect Dis 187:461–466
17. de Souza JB, Riley EM (2002) Cerebral malaria: the contribution of studies in animal models to our understanding of immunopathogenesis. Microbes Infect 4:291–300
18. Landau I, Boulard Y (1978) Life cycles and morphology. In: Killick-Kendrick R, Peters W (eds) Rodent malaria. Academic, London, pp 53–84
19. Vincke IH, Lips MAH (1948) Un nouveau *Plasmodium* d'un rongeur sauvage du Congo *Plasmodium berghei* n. sp. Ann Soc Belg Med Trop 28:97–104
20. Amani V, Boubou MI, Pied S et al (1998) Cloned lines of *Plasmodium berghei* ANKA differ in their abilities to induce experimental cerebral malaria. Infect Immun 66:4093–4099
21. Polder T, Jerusalem C, Eling W (1983) Topographical distribution of the cerebral lesions in mice infected with *Plasmodium berghei*. Tropenmed Parasitol 34:235–243
22. Ball HJ, MacDougall HG, McGregor IS et al (2004) Cyclooxygenase-2 in the pathogenesis of murine cerebral malaria. J Infect Dis 189:751–758
23. Yoeli M, Hargreaves BJ (1974) Brain capillary blockage produced by a virulent strain of rodent malaria. Science 184:572–573
24. Hearn J, Rayment N, Landon DN et al (2000) Immunopathology of cerebral malaria: morphological evidence of parasite sequestration in murine brain microvasculature. Infect Immun 68:5364–5376
25. Storring PL, Tiplady RJ, Gaines Das RE et al (1998) Epoetin alfa and beta differ in their erythropoietin isoform compositions and biological properties. Br J Haematol 100:79–89
26. Brines M, Patel NS, Villa P et al (2008) Nonerythropoietic, tissue-protective peptides derived from the tertiary structure of erythropoietin. Proc Natl Acad Sci U S A 105: 10925–10930
27. Leist M, Ghezzi P, Grasso G et al (2004) Derivatives of erythropoietin that are tissue protective but not erythropoietic. Science 305: 239–242
28. Wang X, Zhu C, Wang X et al (2004) The non-erythropoietic asialoerythropoietin protects against neonatal hypoxia-ischemia as potently as erythropoietin. J Neurochem 91:900–910
29. Mun KC, Golper TA (2000) Impaired biological activity of erythropoietin by cyanate carbamylation. Blood Purif 18:13–17

Index

Pietro Ghezzi and Anthony Cerami (eds.), *Tissue-Protective Cytokines: Methods and Protocols*, Methods in Molecular Biology, vol. 982, DOI 10.1007/978-1-62703-308-4, © Springer Science+Business Media, LLC 2013

MIX
Papier aus verantwortungsvollen Quellen
Paper from responsible sources
FSC® C105338

If you have any concerns about our products,
you can contact us on
ProductSafety@springernature.com

In case Publisher is established outside the EU,
the EU authorized representative is:
Springer Nature Customer Service Center GmbH
Europaplatz 3, 69115 Heidelberg, Germany

Printed by Libri Plureos GmbH
in Hamburg, Germany